INVENTING THE WORLD

INVENTING THE WORLD

VENICE AND THE TRANSFORMATION
OF WESTERN CIVILIZATION

MEREDITH F. SMALL

PEGASUS BOOKS
NEW YORK LONDON

INVENTING THE WORLD

Pegasus Books, Ltd.
148 West 37th Street, 13th Floor
New York, NY 10018

First Pegasus Books cloth edition December 2020

Interior design by Christopher Brian King

Library of Congress Cataloging-in-Publication Data is available.

ISBN: 978-1-64313-538-0

10 9 8 7 6 5 4 3 2 1

Printed in the United States of America
Distributed by Simon & Schuster
www.pegasusbooks.com

For Francesca

Mi so tua e ti, ti xe mia.

CONTENTS

An invention is not a revolution, it is only the beginning of a journey.
Printing Evolution Exhibition,
Correr Museum Venice,
September, 2018

There are men in this city and around it attracted by its excellence and magnificence, many men of different origins, with ingenious minds and the capacity to imagine and discover different artificial things. And if it were possible to ensure that others could not reproduce the works and the artifices that they have invented, and guarantee them their honor, then these men would use their minds to discover things of great utility to our republic.
The First Patent Legislation
The Republic of Venice,
March 19, 1474

Preface

IT ALL STARTED with a comic book and a cocktail.

I had come to Venice, Italy, for a month long Italian language course. Monday through Friday I sat in a classroom with a bunch of other foreigners from all over the world. The only thing we shared was a desire to communicate in Italian and the willingness to study. For four hours every day we had grammar points drilled into our heads while trying to navigate the confusion of elementary conversation in a new language. It was brain numbing. The only way to deal with it was to adhere to the venerable Venetian tradition of stopping at a local bar for a spritz of Aperol and prosecco on the way home. On this particular day, I had wandered into a bar near the Rialto Market, a bar frequented mostly by locals. As per usual, there were no tables, just a counter where people stood drinking and a narrow strip of wood placed in front of the open window. I asked for a spritz and parked myself on a stool looking outward, comic book in hand. As I bent my head and tried to translate the simple language of a thief and his girlfriend from Italian to English, an authoritative voice floated through the window from the little street outside. Not a surprise. Because this city is so densely populated and has no car or traffic noises, the sound of people talking is ubiquitous. Voices and conversations float down the calli over the bridges and fill the public spaces. It's a great place for eavesdropping. It was a few English-speaking tourists and a guide who was obviously explaining the long financial history of the Rialto district. "We know Venetians were great shipbuilders," said the guide. "But they also invented double-entry bookkeeping, bank-to-bank transfers, and government bonds." I stopped reading, fished a pencil out of my purse, and quickly wrote down those statements in the margins of my comic book.

It was the word "invented" that grabbed me. As an anthropologist, I am always intrigued by any mention of cultures changing and developing and always interested in what exactly pushes that change forward. I am also interested in the apparently universal human urge to bring something new to the table. Novel ideas occur to us all the time, and as a species we also seem to appreciate new things; the very roots of our evolutionary success surely must underlie both those processes. And here a complete stranger was telling me that Venetians had invented some very important economic concepts that have lasted into modern-day culture. What struck me was not just the particular inventions but the fact that those inventions had lasted for centuries and had, in fact, formed our present-day ideas about banking, accounting, and investments. Like most tourists, I had just thought of the city as a pretty face in an unusual setting. But comic book in hand, I suddenly felt like I was being initiated into a whole new perspective on Venice as a place of action and innovation, that there might be an undiscovered current running through the lagoon that the guidebooks and history books never talk about. Before that moment, I had never thought of Venice as an instigator of human cultural change or considered, even for a moment, that this city continued to be a great influence on how we operate today. Now I had a new perspective on Venice, a possible path to follow that no one had noted before.

I was right to pay attention to that pivotal moment. After the revelation at the bar, I was forever attuned to any mention of inventions by Venetians. I spent my free time the rest of that month visiting museums and confraternities where I quietly came across other things that Venetians had invented or discovered. I quickly moved from writing those "firsts" above the speech bubbles of a comic book to keeping a list of Venetian inventions in a special notebook. Once back home, I began to check that list, item by item, to confirm if what I had overheard, seen, or read was true. I took that list online and into the university library and checked and rechecked every claim of invention from every possible angle. My most trusted resource was academic research by historians of Venice because they, too, are honor bound to get their facts correct. With my growing skills in Italian, I was also able to peruse academic research written in Italian, and that opened up the world of Venetian archaeology and Venetian Renaissance history. The list of inventions and new ideas grew so fast and so broadly over so many disciplines that it became obvious I had compiled the framework for a book, and a new perspective on Venice.

I soon realized that it wasn't just the fact of those inventions but the context in which they appeared that made for a good story, one that includes scholars, mercenaries, peasants, noblemen, priests, scallywags, businessmen, explorers, and inventors. It also has to do with boats, money, plague, face cream, opera, semicolons, tiramisu, and many other familiar objects, events, and laws. But these Venetian inventions and new ideas were not just disjointed quirky events that rated only superficial acknowledgement. As I organized my list into topics (rather than chronologically), it was obvious that many of these Venetian firsts were still with us, part of the fundamentals of Western culture, how we think and how we operate. I was stunned to realize that many of these ideas and inventions that I was bringing to the forefront had significantly shaped our contemporary notions of institutions and conventions that we in Western culture think of as "normal." The thread of Venetian thought is woven into how we now envision community, health care, economics, consumerism, and globalization, all the "big picture" subjects that are the scaffolding of our lives. Venetians, I soon realized, invented our world. I found this revelation especially interesting because Venice has none of the characteristics that we usually associate with sophisticated innovation. There had been no royal, wealthy, or powerful class throwing patronage at anyone in Venice, no intellectual body that revered smart people, and no real pressing need for inventing anything. Even more interesting, most of those original Venetian ideas came before or after that grand period of cultural newness, the Renaissance.

I made many trips to Venice post–comic book, usually staying for months at a time, and embedding myself further by joining the Bucintoro Rowing Society to learn traditional Venetian rowing, taking lace making lessons on Burano, participating in protest marches and local festivals, and going to just about any event that showed up on a poster. I also walked for hours every day into the nooks and crannies of Venice hunting down the sites of various inventions and placing myself in the very atmosphere that had produced these firsts. Venice is unique in that it maintains its medieval geography that easily catapults a visitor into the past. Others have said that if Casanova were dropped into Venice today he could still easily find his way around because the geography of the city has barely changed in centuries. If Marco Polo arrived in Venice on a cruise ship today he'd probably be bringing bits of some foreign culture to add to the mix of Venice, just as he did long ago. And both men would recognize the Venetian health care and banking systems; the kiosks full of newspapers and paperback books would be familiar, a continuity of culture and society that is also the mark of Venice and its twelve hundred years of continuous history. Like a detective looking for admissible evidence, I combed the city for inventions and went farther afield on the mainland to cities that were once part of the Venetian Republic, such as Padua.

After two years of building that list of inventions and ideas, I knew that it was time to step back and make sense of what I had come across. I realized that these stories could provide the framework for a very different way of looking at this city both historically and culturally. Instead of a standard history book, *Inventing the World* is the tale of how one small place had an outsized influence on the development of Western culture. The book begins with the simple but profound question, Why does anyone invent anything? Chapter 1 addresses this most basic of human urges. Cultures and societies, of course, do not construct themselves. Instead, they are built by people solving puzzles, inventing things, and figuring out how to survive in a particular ecology. That meant I needed to dig deeply into Venice's archaeology to understand the reality of its founding environment and the pressures that must have produced the Venetian urge to create. Chapter 2, therefore, outlines how these land-based people made a spectacular leap from life in the countryside into a vast lagoon. The first Venetians disconnected themselves from Roman rule and built a city surrounded by water, and in doing so, they invented themselves. From that initial creation they also produced a unique identity that adapted to unusual economic, cultural, and historical circumstances. For the next seven chapters, the core of the book, I present the list of inventions and discoveries by subject and not lumped together and described chronologically as a history book would do. These new ideas cross many disciplines, including science, medicine, art, writing and publishing, politics, the art of war, and commerce that also confirms the need to group them by type rather than by year. Gathering them by subject also makes for interwoven stories over time, stories that often end with how we view that subject today. Chapter 3 focuses on the Venetian connection with the sea, and the many firsts that came from that unique birthright. That chapter ranges from sea exploration to map making and how Venetians early on changed the way we saw the rest of the world. In chapter 4, I delve into the broad organization of the city as a community. Disparate lagoon islands banded together and instituted a community of rules—a government, that is—in 697 with the election of a *dux* (or doge). That spirit of connection, combined with the idea of social equality and personal freedom, made Venice a radically different place from other Italian city-states of the time and initiated so many firsts that we consider normal today. Chapter 5 focuses on how Venice began the idea of public health. As I write this today, March 2020, the world is facing the threat of a rapidly spreading pandemic. Countries and communities are initiating standard quarantine practices that were first developed and put into place by the Venetian government in 1348. Venice also supported medical and scientific breakthroughs at the University of Padua, including the first evidence-based medicine, the first understanding of contagion, and the first notion of human metabolism among so many new ideas that are the fundamentals of modern medical practice. Economically, Venice thrived as the first consumer culture, and chapter 6 explains how its merchant traders fostered the beginnings of capitalism and the art of selling things globally. The beating heart of Venice has always been money, so it's no surprise that Venetians instituted many important monetary changes, which chapter 7 is all about. Venetians invented double-entry bookkeeping, the accounting structure used by large and small business all over the world today, bank-to-bank transfers, letters of credit, and government bonds. Chapter 8 brings to light all the many inventions that happened during the heyday of the Venetian printing and publishing boom in the late 1400s and early 1500s. This small city once boasted one of the largest publishing industries in Europe. But more interesting, Venetians pushed that industry forward with the invention of the paperback book that gave common people, not just the rich, access to reading matter. Venetians were the first to recognize the moneymaking potential of mass printing and selling books and they established the first publishing houses. Printers and publishers there also invented italics, the comma, the semicolon, various typefaces, the first book on punctuation, and the first thesaurus. At the same time, the written word was first protected in Venice with the invention of the copyright. Venetians have also affected how we now spend our free time, which is the focus of chapter 9. The first public gambling casino is in Venice, and it is now tucked into a renovated modern hotel. Venetians were mad for opera and theater, and the city opened the first ever public theaters where common folk could be entertained.

As I wrote the final chapters, I was also slogging through bouts of high water (*acqua alta*) in Venice and trying to skirt the mobs of tourists so I could go about my research. It became clear that in good conscience I would have to address those two issues that currently threaten the very existence of the city. I knew that a book on Venice from an anthropological point of view, or any point of view, would not be complete, or even valid, without exploring the mass tourism that has overrun the city in the past decade and the increasing destruction of the high water brought on by climate change. Even though there are no specific inventions in these subjects to highlight, *Inventing the World* ends with an afterword about tourism and the destruction of the infrastructure of the city due to flooding. This crisis also has the potential to inspire some creative solutions and technological inventions that fend off the major destruction of the city and those new ideas could also be applied elsewhere today.

We are at a point in human history where our species is capable of traveling the globe, when capitalism is the desired economic policy, and where money is king. Western culture appears to be the dominant culture on earth, yet we have little understanding of why this particular formation is considered "successful" or "desirable." There are, in fact, downsides to Western culture. Even with economic wealth, we Westerners struggle with the concept of community, or the idea that our kind also need each other for a good and interesting life. By turning to Venice, the birthplace of so many enlightened ideas, a place with a bustling history where so many occupations and endeavors moved forward, we might discover ourselves. Today, no one thinks of Venice as a mover and shaker of human culture. It's perceived as an open-air museum, preserved in amber. We do not consider, even for a moment, how this city continues to be a great influence on how we operate today.

Western culture didn't appear out of nowhere. It came from Venice, where they invented the world as we know it and live it.

INVENTING
THE WORLD

CHAPTER ONE

The Spark: Why Humans Invent, Create, and Explore

We're the species that invents. Our brains have evolved for problem solving, planning for future disasters, and subsisting in environments like the Artic, where by all rights we should not be able to dwell.

—Pagan Kennedy, *Inventology*
2016

Genius is one percent inspiration and ninety-nine percent perspiration. Accordingly, a "genius" is often merely a talented person who has done all of his or her homework.

—Thomas Edison, Inventor
1903

THE STORY OF invention changing history begins with a young monkey on a beach halfway around the world from Venice. Her name was Imo and she was a macaque, a member of a very successful species of monkey that has survived in various habitats around the globe. If macaques are very adaptable and very clever monkeys, Imo must have been the cleverest monkey of all.

She lived with her troop on Koshima Island, a thirty-hectare forested rock with high-sided cliffs, set three hundred meters offshore in the Sea of Hyuga at the southernmost tip of Japan. This island is ringed by pebbly shores and pockets of sandy beaches, and the monkeys have been observed on Koshima since 1947. The typical macaque social system is matrilineal, so Imo would have been in a troop with her mother and sisters, along with other female lineages. Males leave at maturity and look for another troop, so the males in Imo's troop were most likely arrivals from some other troop on the island. Macaques are also highly social animals. There is lots of inter-monkey interaction all day long, from grooming to threatening, and their world is full of the kind of social hum that comes when primates—humans and monkeys alike—go about their daily business. Starting in the 1950s, Japanese primatologists began coaxing Imo's troop out of the hills and onto the beach with food. In particular, they dumped a pile of sweet potatoes on the beach every few days, and the monkeys were happy to come out and eat them, giving the researchers the first close-up look at how macaques behave in the wild. Then in 1952, a singular act by Imo changed everything everyone had ever thought about the process of innovation. In a eureka moment, this young female monkey picked a sweet potato off the sand, took it to a freshwater stream that emptied onto the beach, and washed the grit off her meal. Even more telling, that move spread bit by bit to other members of her troop, disseminating to other females first, then males started to pick up the trick. Imo soon had another eureka moment—one day she took her potato to the sea and dipped it in salt water, thereby washing off the sand and getting a little seasoning at the same time.[1] Once again, Imo's singular behavior spread to almost all the monkeys in her troop, with adult males the last to change their food prep strategy. What can only be called a cultural change among these monkeys took at least a decade, but Imo had started something—she had enacted a dietary revolution.

This series of behavioral innovations was possible, of course, because the potatoes had been provided to the monkeys by the researchers.[2] That was the opportunity. Add to that all the water, fresh and salted, at Imo's disposal. Imo also understood her environment. She had already experienced wetness, the flavor of salt, and what it was like to eat natural vegetation without a layer of sand. What makes Imo a genius monkey is that she was surrounded by opportunity, connected the dots, and made dinner better for everyone. She could have simply grabbed her spuds and eaten them as is, as she had always done, or hidden her behavior from her troop mates. Instead, her monkey mind came up with something entirely new, including the idea that potatoes could be "processed" into something easier to eat and better tasting, and the innovation caught on. As a recent exhibit in Venice on the effect of the printed book in the 1500s explained, "Inventions introduce innovation into our society, but revolutions happen when innovation spreads and dramatically changes our everyday life."[3] Imo's potato washing was a quiet cultural revolution that slowly but surely revolutionized her troop's way of life. Many human inventions also happen quietly, and some don't even make it to the revolution stage because they are unknown to society. Other inventions never come off the mental drawing board. But our kind has a long history of making and doing, and as our history shows, these innovations can change the world. And like in Imo's world, that's what happened in Venice.

THE EVOLUTION OF INVENTION

Imagine a life where nothing is new. Such an idea is anathema today because every minute of every day we are touched by all sorts of new ideas and things. Or at least it feels that way. But we aren't living in some Jetsons world. Instead, we are living in a world that not only has new stuff but also has lots of old stuff. In fact, we are surrounded by things and ideas that were invented in the past, and because of that florid past we wake to an alarm clock on a cell phone, put on clothes that someone else designed and sewed, and get into a car, train, or bus that someone else built to work at an office where what we do is taken to unknown places and people. We also tote a lunch full of goods grown by pretty recent agricultural advances or grab a sandwich made who knows where by who knows whom and often made with ingredients that are manufactured rather than grown. As Pagan Kennedy, the author of *Inventology* puts it, "If you scan the environment where you're sitting right now, you will discover that most of the technologies that you depend on were developed centuries ago—from the laces in your shoes to the ceramic mug that holds your coffee."[4] In other words, our day relies on the good ideas of others, and all day long we might even have some good ideas of our own, some realized, some not. This mishmash of inventions speaks to the fact that human history is one long path chockful of change. First, some words: *invention, innovation, creativity, novel, new, unexpected, clever, imaginative, life-changing.* These words inhabit any new idea or invention that is successful. Sure, there are lots of bad inventions, ones that cause harm and ones that don't work or are meaningless, but overall, we tend to meet every new idea with a smile of expectation. And in that cheery expectation, that welcoming nod to a new idea, is a universal feature of the human mind and something that pretty much defines who we are. As such, we are born change makers.

Embedded in our daily experience of living through the inventions of others also lies the deepest of questions—why do people invent? Certainly, inventions spring from a creative spirit as well as the impulse to fix something, solve a problem, or try something new. Why do they come up with new things or new ideas, and what elements must be present to foster such ideas? Exploring that question is essential to this book because the answer provides the foundation upon which the next chapters rest.

There are several places to look for answers. First of all, all that human change, Venetian or not, is an outgrowth of the natural course of life, from the beginning of life itself. The very process of evolution is one of change and taking advantage of something that comes up. Geography, or environment, has a lot to do with all that natural change. All animals living today had ancestors who crawled or scampered or flew about the earth. Some tried out a new food source, a new type of shelter, or a new way to escape from predators. If that act helped in their survival or helped them pass on genes to the next generation, that seemingly innocent reach for the "new" became part of their offspring and the generations that followed. Invention, defined as taking advantage of an opportunity and solving a puzzle, is hardwired into all organisms because they all are subject to natural selection and evolution. Some might suggest, however, that taking advantage of something new isn't exactly invention as we think it, but surely recognizing something different and then using it is a kind of innovation. This means that nature itself is the engine of invention as it selects for those minds that move forward, those that look for something new and exploit it. And it's the primates, ourselves and our close genetic relatives, that seem to excel in both exploitation and out-and-out invention that often goes beyond simple survival.

We didn't hear about the genius monkey Imo for decades because the observation was published in Japanese. But in the 1960s, animal behaviorist Jane Goodall watched as an adult male chimpanzee in the wilds of Tanzania picked up a long blade of grass and jammed it into a small hole in a very tall mud nest full of African termites. As the chimp pulled out the blade, Goodall saw that it was covered in termites. The chimp then passed the grass through his lips and ate every single insect. Goodall realized she was witnessing a moment that would topple the long-held notion that only humans are smart enough to fashion and use tools. Goodall and others went on to witness chimps wield inventions—leaves for sponges to soak up drinking water and stones to smash nuts and release their protein. Then there were sightings of capuchin monkeys in the wild carefully choosing their stones before using them to crack nuts.[5] And orangutans using twigs to open a fruit and squeegee out the seeds.[6] More interesting, comparisons across chimpanzee sites in Africa show that chimps are also multicultural.[7] For example, groups of chimpanzees in West Africa routinely smash palm nuts as a dietary resource but chimps in Central Africa do not, even though they have the nuts and big stones to carry out the operation. No one knows which chimp or chimps first figured out nut cracking, but the West African chimps know from each other how to do it. Researchers in the Taï forest of Ivory Coast have watched little chimps spend hours, days, months, and years trying to place a palm nut on the right stone anvil and hit it smack in the middle to break open the tough shell. They have to learn how to crack nuts, and they do that by closely watching their elders.

Primatologists have found that chimps across their range in Africa exhibit thirty-nine variable behaviors that cannot be explained by where they live, their genes, or anything else. In other words, these are learned behaviors and, as such, are "cultural."[8] Keep in mind that chimpanzees have brains one third the size of modern humans. And yet, what you see is chimps and other primates inventing tools and new behaviors and improving their ability to survive. These observations also suggest that inventing and culture are very deeply engrained parts of human nature. As Goodall famously said, "If you see something in chimpanzees and you see it in humans it means it was in the common ancestor." What do chimp tools, and the capacity to invent them, have to do with human invention? It has to do with brain size and intelligence.

This fable begins with a fact: the human fossil record shows a sharp increase in brain size (half again as much) about 1.5 million years ago, and those early hominids were found alongside crude tools. Surely, the rumination went, tool use must be linked to human intelligence. Although paleontologists and others sincerely believed that making and using tools were instrumental in evolving complex human mentality, the first evidence that it might be so was discovered in the 1960s. After digging for decades in East Africa, paleontologists Louis and Mary Leakey found the very first stone tools associated with a hominid. Those tools, chunks of rock with razor-sharp flakes peeled off, were found in association with hominid bones christened *Homo habilis*, or "handy man."

The anthropological community and the public were so invested in finding a link between tools and human brain size that Goodall's tool-using chimps in no way shook that hypothetical connection. Tools became the benchmark of humanity, a benchmark that keeps moving back in time, ultimately eliminating the tool use/brain size idea.

In 2010 a team of paleontologists from the California Academy of Sciences reported on a series of bones found in a remote area of Ethiopia called Dikika that dates back 3.4 million years.[9] The team found two non-hominid bones of note: first, a rib from what they call a "cow-sized" animal; and second, a piece of the lower leg bone of a "goat-sized" animal that was clearly food for some other creature. Paleontologists know this because the bones are covered with all sorts of cuts, pits, and scrape marks, the kind of marks that can only get on a bone as meat is purposefully hacked off. And the only possible butcher was *Australopithecus afarensis*, little, upright-walking, small-brained (their brains were the size of chimpanzees) prehumans who apparently had a taste for meat and the innovative idea to pick up a sharp stone flake and carve dinner off the bone. Marks on these specimens also show that clever hominids (the name of all human and human-like creatures) were taking bigger stones and smashing the bones to remove the inner marrow. That move was evolutionarily important because marrow is full of protein and omega-3 fatty acids that are necessary for brain growth. As the lead researcher of the group that works on this site, Zeresenay Alemseged said, "Tool use fundamentally altered the way our early ancestors interacted with nature, allowing them to eat new types of food and exploit new territories. It also led to tool making—a critical step in our evolutionary path that eventually enabled such advanced technologies as airplanes, MRI machines, and iPhones."

Alemseged also calls this earliest example of human tool use "game changing."[10] Metaphorically, we stand on the shoulders of giants. In fact, we stand on the very short and small shoulders of Australopithecines, relatives with very small brains.

The connection between tools and brain size seemed obvious before Dikika, one of those "missing link" moments that paleontologists often declare and then retract, but the focus on tools as a defining feature of humanity was a mistake. And a mistake on many levels. Focusing on stone tools as the great leap forward gave a skewed picture of life back then. First off, what if those ancestors, or ones before them, fashioned things out of sticks or grass, or crumpled up leaves for sponges, as chimps do? All those natural materials would have quickly composted away. And for many, tools meant early humans had a particular social life, lived in groups, and practiced cooperative hunting. This line of conjecture led to thinking that tool use was instrumental in selecting for big brains and making humans into groups of socially interwoven individuals who not only used those tools but appreciated them. In other words, tools turned into the touchstone of humanity and maybe they aren't. What if those early stone tools were only used by a few in the group, by only their maker, by only men? Why then would non-users have big brains? The calculus of evolution just doesn't work out well with a singular focus on tools as the innovative spark of all humanity or as the basis for our evolutionary success as we spread across the globe. Believing tools to be the singular sign of culture was also not an accurate leap of faith—since chimps make tools, we eventually had to expand the definition of cultural animals to include them. Finding those tools imposed on early humans a kind of intellectual stature that maybe they didn't deserve. In other words, we used the tools to define "ourselves" because tools are familiar to us now and we often revere them as the key to our very important modern lives. It is true that if we include all the gadgets and things that pass through our hands in one day, or even one hour, we see ourselves as master tool users, but maybe using tools once they have been invented by someone else or a team of someone elses is not such a big deal. In fact, tools per se are probably a distraction. A tool is a thing that can be used to help open, close, reach, manipulate, or connect with something else. And the list of Venetian inventions includes many of them, including the fork, the telescope, and the body scale. But are these helpful objects more important to human history than the creation of other items such as, say, boats or face cream?

Tool use is not the only thing that makes us smart creatures, but the invention of tools fits right into our human playbook and is the first thing that someone expects in a book about inventions. But mostly the focus on tools and using them has been used to shore up the idea that humans, more than any other animals, are smart. By "smart" we tend to mean that humans clearly know when they are faced with a puzzle and the smartest people are those who figure out how to solve that puzzle. At a very basic evolutionary level we dream up new things because we have a very busy brain. Along with these big and complex brains that we humans evolved for reasons that are still not clear (funny how we laud big brains but seem unable to harness all that brain tissue to figure out how we got them), our self-knowledge, our sense of collective, and our ability to see the past and think about the future are key to explaining the general human penchant for invention.

Invention is simply an offshoot of thinking way too much. And we can't seem to stop all that thinking and the inventing.

THEORIES OF INVENTION

Written history is a cluttered trail of inventions, innovations, and new ideas. Many of these inventions became the exclamation marks of human history, but it's important to recognize that they didn't appear out of nowhere. The seeds of those ideas come from various directions. Many newfangled ideas grow out of an irritating problem, a repeated frustration that one person or a team wants to fix, or a wish to go further and accomplish more. When Venetian publisher and printer Aldus Manutius first used italic type in 1501 (chapter 8), he was looking for a way to present the printed word in a way that didn't irk people who were used to reading books in handwritten script. We are also the species that has the brain power, the insight, to look into the past or toward the future, and new ideas come from that gaze as well. Human consciousness means having an innate curiosity about who we are and where we come from and what makes us tick. As such, self-reflection is a singular human trait, even when it results in self-absorption. Girolamo Fracastoro, a Venetian physician, spent his life curious about himself (chapter 5). He invented the body scale so that he could weigh himself every day and chart the input and exit of food and water. He just wanted to know how the human body worked, and in the process, he invented the concept of metabolism. Some inventions have also been spurred by economic possibilities. This book is full of them. In our consumer culture, making money out of an invention seems like the great instigator for innovations, and in fact, chapters 7 and 9 are all about ideas and inventions that created something new and earned someone money. But many inventions make no money at all, and some inventors didn't have money in mind when they thought up something new. When Venetians invented the sailing galleon in the 1500s it wasn't about making money but stopping pirates. Venetian priest Paolo Sarpi wrote in 1606 that church and state should be separate, but he didn't make a dime, or a ducat, off this profound idea that became the very basis of American democracy. Even Galileo wasn't in it for the money. In fact, he wasn't rewarded but severely punished for his novel ideas.

But how do invention and innovation actually happen? Does anyone know?

Using our very big brains, we often light upon new ideas, or they bubble up from some odd internal space, or we force them to do so. Our brains can be focused or not, but the fact that there are lots of possible neuronal connections helps trigger newness. If we were lizards, for example, our pea-sized brains probably wouldn't come up with a way to get a man or a woman on the moon, or even develop a heat source to keep our home rock warm. A group of lizards might stare at their reflections for years in a tiny pool of water and still not come up with the idea for a mirror, as Venetian glassmakers did in the 1300s. The answer to how we invent, it seems, is a very messy, very twisted, and often contradictory and overlapping map of human action. And it has birthed the convoluted discipline called the "science of invention."

We owe this science to an Uzbekistani science fiction writer named Genrich Altshuller. According to journalist Pagan Kennedy, the twenty-year-old Altshuller joined the Soviet Navy in the 1950s. He worked on keeping ships ship shape but eventually landed a job in the local military patent office in Baku, Azerbaijan.[11] As a connoisseur of science fiction books, Altshuller was already interested in why some things get invented and who thinks them up. He started to look through the patent applications in his military office for patterns that might explain the process of invention. That search led Altshuller to his theory of invention, which, he claimed, took only two things—creative insight and how that insight is manifested. From there Altshuller pushed for everyone to pay attention to science fiction, because in those books and stories, he felt, lay the future of invention. Altshuller also believed that it is possible to learn to be an inventor and that rummaging through the stockpile of past inventions and understanding them was essential.[12] His work was the first time anyone had actually laid out a conscious method for spurring invention, and he assumed that being inventive was a good and noble calling. In other words, Altshuller brought credibility not to inventions but to inventors. In doing so, Pagan Kennedy says, he invented "Inventology," or the Science of Invention. Put simply, Altshuller made thinking about the process of invention credible.

Altshuller also brought inventors out of the closet, or rather out of the basement, and redefined what it means to invent. Previously, these tinkerers (and yes, that's how inventors are usually portrayed) were considered either rare geniuses (when an invention worked and proved important over time, like Steve Wozniak fiddling around in Steve Jobs's garage) or obsessed nutcases when inventions didn't work or didn't sell (the military Vespa or the outdoor hanging cage for baby care come to mind). The common notion of an inventor, in fact, looks something like Christopher Lloyd in the movie *Back to the Future*—working in obscurity, making something unreasonable or impossible, and putting it together with common household objects. And these inventors are usually men, although an analysis by the website InnoCentive.com shows that winners of anonymous invention challenges are, in fact, more often women than men.[13] What all these crazy inventors wait for, so we all think, is the "eureka moment," the act of invention that has been sadly likened to a lightbulb. As the saying goes, "Then a lightbulb when on." Surely Galileo, all alone there in his academic office at the University of Padua in the Republic of Venice, must have had a sudden switch go off and his brain light up when he invented the telescope. Or the Trivisian cook in the province of Venice who first put together tiramisu must have thought, "Mascarpone, whipped cream, eggs, and (lightbulb) ladyfingers soaked in coffee!" In fact, almost all inventions are not eureka moments per se. They might look that way to the uninformed, or those late to the show, but germs of that particular innovation might have been in the works for decades. In that sense, new ideas also come with a lot of practice and hit-or-miss.

How the mind pops out something new is still a mystery, but there are theories. Tech writer Stephen Johnson says that invention actually breeds on chaos—mental chaos, that is.[14] He says that the brain is full of whizzing bits of whatever, and sometimes two bits hook up and an idea flashes. He calls these millisecond brain crashes "accidental connections." Perhaps that's the reason we all sometimes say, "This just popped into my mind" when the "this" is some new idea or thought or strategy. It seems like it came out of nowhere, but apparently it came from the neuronal mess of our minds. Sometimes we call such thoughts serendipity. Or a bit of nonsense, especially since many of these "accidental connections" are stupid. But sometimes they are also brilliant and that's when the thought needs to be written down, because these brain flashes are usually ephemeral.

A new idea might also start quite pointedly with a problem or puzzle, something that needs to be conquered or overcome for things to move along.

Sometimes what needs to be solved is a life-threatening issue, say a new kind of medical device or an agricultural tool that increases available food. In both cases, the puzzle is there for just about anyone to ponder and anyone to solve. And the invention probably includes a history of frustration. Various inventologists have suggested that necessity is indeed often the mother of invention in most cases, because not having a proper tool is certainly frustrating.[15] That frustration is more likely to motivate invention when someone has faced that same impasse over and over. That's why new tools are often invented by a person faced with a repetitive and boring task, or by a group of people who experience the same frustration. And that's why glassblowers in Venice invented reading lenses in the 1200s and put them in frames so that anyone over fifty years old, including themselves, could see close up. Today we think this means reading, and it might have back then as well, but up-close vision is also critical for running a business or buttoning up your shoes, or being a glassblower and working with hot molten glass without burning your fingers.

New ideas or ideologies can also come from individual or collective concern. Fear, in other words. Venetians came up with child labor laws in 1284 because they were scared for their kids. They likewise thought up the idea of quarantine because they were afraid of disease-bearing ships. We don't know the actual authors of these ideas, but as they were enacted by the state, it's safe to assume someone offered up the bright idea and then everyone voted "yes" because the idea made sense and was, in hindsight, obvious. The group, the collective, society let's say, is also deeply involved in the process of invention because we are the ones who determine the success of an invention. Our accepting something new, even paying for it, and certainly using it is a vote of confidence that whatever it is makes life better.

Sometimes new inventions are simply a reworking of other ideas and inventions. As Stephen Johnson puts it, "We take the ideas we've inherited or that we've stumbled across, and we jigger them together into some new shape."[16] The reading glasses example above is actually an invention made following the invention of eyeglasses, lenses that were first invented for nearsighted people. No one is really sure who came up with this product, but the first image of someone wearing glasses was frescoed in 1352 in the cathedral in Treviso, a city that used to belong to Venice. In other words, glasses had been in the air for a century and were so normal that they rated a place in art. In 1378 the Venetian navy was the first to mount guns on ships, but they had the guns, they had the ships, it didn't take a genius to see the advantage, and so they cobbled it together. Johnson also feels that human history is full of what he calls "adjacent possibilities," which also explains why some inventions occur simultaneously by different people in different locations. Everyone at the time also shares the same knowledge base up to that moment, so it's no surprise that different groups come up with the same next idea.

Ideas can also pop up in a collective, a group of people working toward a common goal or just sitting around having a glass of wine. That kind of collective can be made, and often works well, according to biologist E. O. Wilson and anthropologist Agustín Fuentes, because we humans are on a common mental plane, and one that connects us to each other.[17] Wilson sees this connection as hope for humanity, while Fuentes believes that working collectively is what makes humans who we are.

Speaking of "adjacency possibilities," both of these approaches take advantage of an idea that several other biologists independently came up with decades earlier. Called the social intelligence hypothesis, this theory suggests that nonhuman primates have big brains compared with other mammals to keep track of our very complex social relationships. The hypothesis suggests that those who are socially the most intelligent know their group members well and can make alliances or manipulate others to their survival and reproductive advantage.[18] Some think that humans are not just intelligently social but that we are also naturally compelled to share our ideas.[19] Although this idea is easily applied to humans, and might in fact explain why we have such big brains, using social intelligence to paint a picture of humans as naturally cooperative people, as Wilson and Fuentes have done, is a bit of a stretch. Both academics are trying way too hard to think of people as naturally nice and cooperative. Wouldn't we all like to believe that humans have worked peacefully together for hundreds of thousands of years, sharing our minds and cooperating? And that in doing so we have pushed our culture forward with our collectively shared ideas and inventions? Contrary to that approach, Venetians invented patents in 1474 and copyright in 1486 to stop people from reading the minds of others and stealing their ideas, clearly more self-serving than a collective act.

There are times, however, when thinking together or borrowing from other places makes very good sense. Sometimes inventors borrow ideas or systems from another discipline and put it to novel use. In recent years, businesses have come to understand that people working together, even just sitting where they can see each other, promotes ideas. The "open office" with adjoining desks and no walls is the offshoot of a new wave of thinking about invention. Instead of waiting for one guy or gal to come up with an idea (and for businesses this idea must be a money maker), managers hope that simply being part of what is physically a collective will turn the office into a creative team. This approach works well with those who are gregarious but not for the inventor in the basement who thinks best alone.

Paying attention to an "accidental connection" and thinking about "adjacent possibilities" for the social intelligence hypothesis also produces another hypothesis to explain human intelligence and our big brains. Perhaps social intelligence is not about tracking social relationships per se, and not about forming a collective, but actually about the very act of invention. When the first hominid took the long view of an antelope leg and picked up a stone and flaked off a sharp edge, getting ready to fillet that leg, she might have been observed by another member of her group who then did the same thing and off they went butchering together. Maybe the real answer to why humans have big brains is not so that we can use tools, not so that we can share our minds and form a cooperative collective, but down to the nitty-gritty of inventions and innovations. As Richard Ford puts it, "Make no mistake, creativity is as biologically and intellectually innate a characteristic to human beings as thought itself."[20] In this scenario, being social means an arena of collective or shared innovation. The early hominid inventor had many advantages, such as butchered meat and crafted digging sticks to dig up tubers. Both inventions would have probably translated into better survival and maybe even better reproductive success when the inventor could better support a pregnacy by supplying food to a mate. Or maybe that early inventor had a better chance of passing on genes when the opposite sex asked for some of those newly processed meals. In that sense, invention and innovation would be selecting for bigger and more complicated brains. Combined with evolutionary theory, this hypothesis uses tools-meets-social-intelligence combined with evolutionary theory as an explanation for human smarts. In other words, we don't invent because we have big brains, but we have big brains because our ancestors were master inventors. And so, our so-called social intelligence is not about people identifying, aligning with, or manipulating each other but about identifying, gathering, using, and modifying inventions, or even stealing inventions from others. Evidence to support this idea comes from the now. We might not invent a steam engine or an electric toothbrush as we skate through the day, but we are able to figure out the dashboard of a rental car or unhinge a new kind of coat clasp because we've been listening and looking, gathering information like a sponge and staying alive because of this ability. Babies and children are often referred to as "sponges" because they seem to be always looking, listening, and trying things out. They are actually always inventing and problem solving in their own little world. But that doesn't stop in adulthood. We continue to be sponges because it's to our advantage to absorb everything around us, and in particular we have been selected to notice and be enamored of the "new."

Humans are not a species that simply solves puzzles or makes new things; we are the species of ideas and innovations. And we actually think of ourselves as such. Pagan Kennedy cites a *Time* magazine poll from 2013 in which one third of the people in seventeen countries said they were inventors.[21] The only response to that number is, "Really?" Yes, really, we are the species of inventors, and our self-identity as such underscores how important new ideas are to our species, and how dependent we have always been on this ability. Now seems like the right time to move beyond any idea of "social intelligence" and into the realm of "invention intelligence" or "idea intelligence" as the force that shaped the evolution of our species.

THE CITY AS AN INVENTIVE PLACE

Taken together, these various philosophies of invention suggest that ideas pop up all the time, but that they also have a particular time and place. Invention often grows from knowledge and opportunity, and both those parameters are fostered when people are together. Therefore, there might be some reason that Venice, as a locality, was the right place for some of that time. And the most obvious characteristic of Venice in the social sense is that it was, and still is, a city full of people living, working, and socializing in close quarters.

Humans were actually slow to come together in cities. For 95 percent of human history, we were hunters and gatherers. Sure, our ancestors might have had home bases to come back to, but people did that in bands, small groups presumably made up of kin. In fact, people first settled down in permanent self-sustaining clusters only ten to twelve thousand years ago. These first urbanites presumably noticed that sometimes when plants landed on the ground and were trampled into the dirt they remade themselves. At about that time in human history there were also certain animals that didn't run away. These animals could be used as various foodstuffs and cities probably seemed like a welcome relief from the daily grind of trying to find food because now it was planted right outside the back door. So, a band of people kept that home base over time, settling in. Eventually, that settlement developed and grew, making and shaping its own persona.

The growth of cites around the world went in fits and starts, but in general they came about because of the growth of the number of people on earth.[22] But John Reader, author of the book *Cities*, says that cities were in fact rather slow to grow in the historical past. According to Reader, at the start of the 1800s, when some cities had been around for six thousand years, only 10 percent of the world's population lived in cities. "By 1900," Reader writes, "city-dwellers comprised one quarter of the global population."[23] Today about half the human population lives in cities. Of course, cities have also been death traps, as archaeological remains show. There are "diseases of civilization," which include nutritional deprivation from a life of eating only one crop, and contagion from virulent diseases that found their happy place in cities, often on the backs of those placid and easily domesticated animals. But still, people came to cities and they continue to do so, thinking a city is where you find work and happiness.

Cities have their bad parts, but they also bring people together, and various inventologists have claimed that cities are where inventions happen. For example, urban studies theorist Richard Florida, who is a cheerleader for cities, claims "Cities are caldrons of creativity."[24] Florida also thinks cities explain the explosion of creativity during the Renaissance, when new technologies, great art, and new ways of thinking exploded in Western culture, especially Italy. He makes the point that this "flowing" is connected to the rise of the merchant class and capitalism.[25] His take on the Renaissance is that cities were growing in population, their economies were expanding, the rich were getting richer, and there was plenty of free money around. That money, Florida says, was used to fund the arts. Economic growth also brought about innovative ways to deal with all the money and in that Venice is the shining example. Venice had survived the plague and more people were moving to the city, where jobs were plentiful; much of the population had died, leaving plenty of employment opportunities. Add to that an economy based on trade with merchants as the upper middle class and no one family dominating everything, as they did in Florence and Ferrara and Milan. Venice was already a capitalistic economy and had, in fact, invented all sorts of money firsts, such as the first national bank in 1157, bank-to-bank transfer in 1619, the first bills of exchange around 1200 so that people didn't have to carry around gold and silver to do their trading business, and double-entry bookkeeping, the clearest sign of capitalism, in 1494. Also, there was an air of freedom of thought and action in Venice at that time because Venetians and the Venetian Republic had been notoriously antithetical to the heavy hand of the Roman Catholic Church. As Stephen Johnson explains it, what makes for new ideas is a place organized around markets, not a church or a castle, a place where there are easy mechanisms for people to exchange ideas.[26]

Although Johnson is working with hindsight here, and his model is really based on the workings of modern American culture and our consumer economy, Venice fits his scheme to a T. The geography of Venice as a small place—only 2.5 miles wide and about 3 miles long—means people live in tight quarters and run into each other all day long as they walk the streets or slide along canals in boats. Although the climate is not so temperate, life in Venice then (and even now) was largely lived outdoors. Venetians never hesitate to stop and talk with friends as they go about their business. Sometimes they have a social moment in a coffee bar in the morning or over a quick spritz on the way home, but more often these exchanges happen outside on the pedestrians-only streets and bridges. The traditional iconography of Venice is a couple riding in a gondola down the Grand Canal, but the real iconography of Venice, even today, is Venetians stopped on a pedestrian bridge, chatting.

Another way minds colluded in Venice was through their confraternities (see chapter 4), clubs that crossed the lines of class, occupation, and income—tiny mental melting pots, that is. In addition, a major aspect of Venetian trade was made up of intersecting layers of nobles financing trade ships, merchants storing and passing on those goods, and the regular citizens who did the heavy lifting from boats to warehouses and shops and back to boats headed up rivers to the rest of Europe. The social connections in Venice were fluid, familiar, and economically necessary and those types of relationships are certainly not true of American cities today. In fact, American culture eschews social familiarity in business, giving it the negative labels of cronyism or nepotism. And the very idea that you might make a sweet deal with someone because you know each other is anathema to the American ideology of individualism over the collective. Such connections and deals smack of insider trading to Americans. But other cultures, including Venetian culture then and now, operate with the understanding that relationships will be used to make capitalism and consumerism work.

Richard Florida also points to diversity in the urban population as an important ingredient in innovation and creativity. With all that trading, with ships coming and going, not to mention Crusaders passing through, Venice was remarkably diverse, a place that mingled East and West like no other city of the time. Even today, Venice is diverse. Walk along its streets and hear a plethora of languages from the visiting tourists and the ex-pats who choose to live there.

All that city closeness and diversity, inventologists suggest, breeds innovation. It is also helped along because various occupations tend to coalesce in cites because of common resources of some kind. And people doing the same work often like to live in the vicinity of other people of their profession. Think of the historic neighborhoods in New York City, like printer's row or the garment district. We see less of this today because of our ability to drive to another place, or take some sort of transport, or the ability to work online, and yet there is Silicon Valley. There's no real reason why all the tech people are there, but it's certainly part of human nature to be with "your own kind," however you define that. In general, it's a treat to be around people who speak your own occupational language and interact with people who appreciate your work because they really know what goes into it. When businesses coalesce, people share ideas. And when businesses are right down the street, they also observe each other every day as they walk by. It would seem that occupations clustered in a city might foster innovations through competition and through friendly chatting at the local bar about what you might be making. Under the rubric of "adjacent possibilities," cities also mean inventors have all sorts of other walks of life to borrow from in solving a problem or developing a product. In Venice, some of the two hundred confraternities helped bring members of an occupation together, such as cobblers, who all belonged to a guild that met in their *scuola* in Campo San Toma (a relief of shoes still adorns the building) or the bakers, who met at the Scuola dei Pistori in the Campo San Maurizio. Cities also bring with them an identity tagged to ideas and products, as in "Venetian beads" or "Venetian art," which not only adds a layer of authenticity to something but is also a protection against knockoffs. As such, the identity of a city, in turn, connects all the citizens. At the same time, cities can inspire a fierce loyalty that can be harnessed by the government to engage citizens in warfare against others who have a different identity and do not belong. The "others" can also be exploited or killed with impunity because they "don't belong." Cities often become powerful forces of progress, change, and destruction, and hence the focus of a country. Think of London or Paris or New York, iconic cities that represent the character of their country whether or not that depiction is accurate.

Venice was once more than a city. It was an empire. Unlike other northern Italian cities such as Florence or Milan or Ferrara that flourished during the same time, Venice was at its most powerful stage as an empire. Those other cities were governed by, and held captive to, one powerful family. Instead, Venice had a history of intellectual openness, and a system through which people of all classes could interact on a daily basis. It was also the longest-standing republic in the world, a moniker it still holds, and a major powerful European force. Today its place as a grand city is less visible, but still everyone seems to know Venice in some way. John Reader says that visiting Venice for the first time, in fact, initiated his questions about cities and pushed him to write about them. "I came away from Venice with nagging questions about its status as a city, and about the phenomenon, function and ecology of cities in general. Why do they exist? How do they work? Why do some seem so much more alive than others?"[27] That such a small, now rather quiet, city that began in antiquity can instigate a nonfiction writer's journey about such a large subject such as cities speaks to the still influential power of that place.

The various theories of invention and the words of all the various inventologists show that the process of invention and creativity can't be parsed out so easily. We seem to be "the species that invents," and we seem compelled to do so. Even those who don't consider themselves real inventors could probably describe a new sandwich combo they came up with, or an idea that has been floating in their heads for years, unexpressed, dormant, and yet still taking up a tiny bit of mental space. And most of us can be amazingly creative and inventive when faced with challenges or danger. Stand in front of your house with no keys and listen to your brain come up with five or six ways to break in. Therein lies the inventive human mind. The question might be, then, not why some people invent so much, or such amazing things, but why we all don't invent more.

That list of Venetian inventions at the beginning of the book, with its sheer length, chronological reach, and subject breadth, begs the big fundamental question: Why do people everywhere come up with new ideas and things? It also leads to a much more specific question—Why Venice? In Venice, people figured out how to build a city out of nothing, and then how to make a maritime empire out of very little. And then they created good government and a major feel for community. On top of that, they manufactured all sorts of things to sell and a monetary system to make everyone rich. As historian of Venice John Julius Norwich put it, "Venice, for its size, made a greater contribution to Western civilization than any other city in Europe or anywhere else."[28] Venetian inventions, discoveries, and creations are important because they became commonplace, were broadly accepted, and changed not only Venetian life and culture, but also the world.

CHAPTER TWO

Venetians Created Themselves

We should bear in mind that what was being colonized was water and mud.

> —E. Crouzet-Pavan, *Venice Reconsidered*
> 2000

AS THE CRANE lifted our small boat into the air, swung it out over the water, and then gently set us down in the busy Giudecca Canal on the south edge of Venice, I felt only apprehension. I had walked all over Venice many times and been on lots of public boats, but today was my first rowing lesson. And this would be no sit-down-and-pull sort of thing but an exercise in balance and strength, neither of which I possessed. Venetian rowing is opposite of most other rowing—you stand facing forward and leverage the boat through the water with one very long oar. My instructor, Paola, a member of the Bucintoro Rowing Society and a small woman with arms of steel, kindly suggested that I just sit there until she had rowed us into calmer waters. Moving horizontally across the canal, we bounced atop the wake of cruise ships, public water taxis, and several private motorboats. Within minutes, we had crossed that major canal and slid under a pedestrian bridge of the island also called Giudecca. Then, Paola told me it was time. I stood up, put my left foot forward, and braced my right foot on the inside of our two-person rowboat (called a *sandolo*). She passed an oar in my direction, instructed me to press it into the *fórcola*, and pointed to something that looked like a chunk of driftwood attached to the gunwale on my right. What I was about to do was particularly Venetian, so in the process of rowing I would also learn some *veneziano* words, unfamiliar vocabulary even for someone born in any Italian city other than Venice. A *fórcola* behaves somewhat like an oarlock but is no lock at all, since it's open on the back side. The trick is to guide the oar in an elegant twisting motion against the specially curved and bumpy *fórcola* in a way that dips the oar into the water and then swings it back in an arc. In that motion the boat moves forward. But if you aren't careful, or experienced, that same powerful stroke can flip the oar out of the *fórcola* and drag it backwards in the water. But done right, the sweep of the oar and the movement of the boat are an elegant dance between boat and water that is the very identity of Venice.

I wanted to learn Venetian rowing because historical accounts suggest that this forward-facing standing style probably first brought people into the lagoon. And yet it is also very evident today. If you take a moment to stand at the edges of the city or even gaze for a while at the Grand Canal and pay attention to the non-motorized traffic, you'll see its powerful pace as rowers move past. This is the stroke that fishermen use to bring in their catch and the simple way *gondolieri* singularly glide their boats through inner-city canals. In this movement is also a thread of the social history of Venice. Stand-up rowing still takes people across the lagoon from island to island to visit friends or just have a day out on the water. You frequently see two-person boats go by and hear the rowers chatting with each other. It's a sight so totally Venetian.

As I adjusted to Paola's stroke and took in her instructions and encouragement, we moved beyond anything that tourists typically see of Venice. On the other side of Giudecca we slid into the larger body of the southern lagoon. On the surface, the lagoon is still, grey, and incredibly quiet. Channel markers, large wood piers that delineate where bigger boats can go, suddenly took on new meaning from the perspective of a boat powered by people rather than machines. I tracked them in rows as far as the eye could see, clearly set up like landing field lights. To the right and left were islands I had previously ignored on maps. Some of these islands are now abandoned places with a silence all their own. One has been repurposed as a luxury resort, but even with the hotel launch disembarking guests, that island seemed a refuge of quiet. Eventually, we passed another boat. Those rowers said something to us in *veneziano* and Paola responded. Maybe they commented on the day, the sun, or perhaps my rooky stance. And after that, there was nothing.

Although it was tough work, I could have rowed all day just to be out in that watery peace. It also came to me that I suddenly had a completely new understanding of a city I thought I already knew well. To really understand Venice, I realized, one needs to get out on the lagoon. I was also reliving the primal history of Venice, the time when people moved away from solid land in search of something, or running away from something, which people always do.

HUMANS ON THE MOVE

The Venetian lagoon is like another planet that humans colonized long ago and made their own. In doing so, those first Venetians were expressing a common human strategy—they were being adaptive in the evolutionary sense of the word. That kind of adaptive is not a superficial change of style or a way of talking, for example, but adaptation in the deepest sense. Over time, the humans who succeeded, who passed their genes on to the next generation, were those who figured out how to find food and stay away from becoming food for other creatures. Often, they were the ones capable of moving into new environments full of possibilities and able to exploit new resources. That kind of adaptive is the difference between life and death, the difference between holding on or dying out. And humans are really good at this.

One mark of the human species is that we have, in five million years of history (depending on where you say humans began), seemed bent on moving long distances and trying out new environments. As early African hominids, now called australopithecines, we wandered around Africa from its southern tip all the way into the area of land known as Ethiopia at the top of the continent. At least a million and a half years ago, our ancient ancestors walked out of Africa in several waves and went as far as China, Indonesia, Georgian Russia, and Europe. Then some groups of anatomically modern humans left Africa as well about 200,000 years ago, presumably in waves of slow migration, until we covered the earth. We have, in our time, walked on savannahs, forded rivers, strolled along beaches, climbed hills and mountains, slid on ice, and jumped on boats to reach new land. If we could light up a map of the earth with the movements of the human race from our last migration from Africa to the present day, that map would be a firework of paths going everywhere. We'd see populations popping up on one continent and then another, migrating en masse or in small groups, and sometimes even resting for a while. We were hungry people probably following food on the hoof, or maybe we were looking for this or fleeing from that. What makes these movements so interesting is the puzzle solving and creativity—let's say ambition—to make do in a place that is new.

Often, the decision of where to settle down and live for a while had to do with water. And it wasn't always about drinking water. Humans, animals who can be taught to swim but don't live and grow in water, seem oddly drawn to beaches, sand dunes, rocky cliffs, sunrises, and sunsets over open water. In fact, about half the world's population now lives no more than a hundred kilometers, or sixty miles, from a coastline, where people build their houses and cities close to large bodies of water.[1] These days, freshwater can be pumped or directed to people who live far from a freshwater source, yet over 50 percent of people in the world also live within three kilometers, or under two miles, from freshwater, be it a lake, stream, or spring, where they usually find more water than they ever might drink.[2] Living close to water isn't, however, just about drinking and washing, or even watering crops. Sometimes it's about moving from one place to another. For example, humans used boats to venture out in the Pacific Ocean three to six thousand years ago. They spread out to find new island homes, and some bumped into the continent of Australia and didn't venture far from the coast.[3] In other words, water as a resource and as a transportation medium for people and goods has been with us for a very long time. More surprising, water as both a resource and a challenge has been implicated in human invention ever since the first humans forded across their first stream to find new ways to live.

Venice is apparently one of those watery places that contain all the elements that make life near water ideal—freshwater comes in from rivers, animal life is abundant, boats can transport everything within and beyond the lagoon, and people come and go efficiently by rowing and sailing. Venice is not just near the water. It is of the water.

THE VENETIAN LAGOON

Lagoons are shallow bodies of low-salinity salt water set off from larger seas or oceans. Some lagoons are formed by walls of coral around atoll islands, while others are formed by water filling in the nooks and crannies along coast-lines. Looking at a coastline from high above, repeated lagoons often form a lacy edge between land and sea. This second coastal kind of lagoon is called a "barrier lagoon," meaning it has islands of sand that separate the shallow lagoon water from the much larger sea or ocean. These barrier lagoons are not bays, because they are separated from the sea, and at the same time they are not lakes, because they have an ongoing and dynamic relationship with the adjacent larger body of water. The so-called barrier islands are always somehow permeable, letting seawater come and go into the lagoon not in a rush or a wave but as a wash of tidal water that is key to the health and welfare of a lagoon. Without that ebb and flow, inside the barrier islands would be marsh, or a very large stagnant pool.

The Venetian lagoon is long and skinny, cupped by a raggedy C-shaped coastline on three sides and capped on the fourth, eastern side by long, skinny islands, or *lidi,* that hold most of the water in. Horatio Brown wrote in 1893 that the Venetian lagoon was like a bent bow with the barrier island as the wooden part of the bow and the coastline on the west as the string.[4] Just on the other side of those *lidi* is the much larger saltwater Adriatic Sea, which separates Italy from Slovenia, Croatia, Montenegro, Albania, and Greece off to the east. Those long, skinny barrier islands between the Adriatic Sea and the Venetian lagoon are broken into sections, so the salt water easily comes in and goes out, seeping around the barriers through inlets as the sea tide rises and falls.[5] But still, those islands take the brunt of any major wave action the Adriatic might brew up. At its southern end, the Adriatic opens into the Mediterranean, which in turn opens into the Atlantic Ocean through the Strait of Gibraltar between North Africa and Spain. In other words, the water of the lagoon is part of a much larger system of water.

Only about 11 percent of the Venetian lagoon is open water; the rest is sections of mud flats or salt marshes, or extremely shallow areas that ebb and flow with the tides.[6] The lagoon is also full of all kinds of water. On the landward side, freshwater starts as Alpine snow and then melts and flows south through mountains and across the flat plains of northern Italy, called the Po Valley, until it gushes or trickles out into the lagoon.[7] Near the shoreline, the water is brackish and filled with silt and slime. That water becomes saltier across the lagoon as it mixes with inflow from the Adriatic. There are other natural forces that also swirl the waters of this particular lagoon. The current of the upper Adriatic flows in a counterclockwise fashion, circling west at its northernmost point near the city of Trieste, then pushing down the calf side of the Italian boot. As the tides and the currents rush south down that eastern side of Italy, they can become a singular force. At times, they powerfully push Adriatic water into the relatively narrow spaces between the barrier islands of the Venetian lagoon and overcome what would normally be a gentle push and pull of the tides. These currents also deposit sand along the filigree of the northern Italian coastline.[8] So, on one side of the Venetian lagoon, rivers dump silt into the lagoon, while on the other side the Adriatic brings sand. Then there is the wind. The bora blows its freezing breath over the northern Adriatic, and the sirocco rushes hotly up from Africa. Both can affect the tides and result in flooding for Venetian lagoon dwellers.

All in all, the Venetian lagoon is 192 square feet of water, making it the largest wetland in the Mediterranean basin. Calling it a wetland, rather than just a lagoon, makes sense, because on average, the Venetian lagoon is only about five feet deep. In most places, if you wanted to jump from a boat you could stand on the mucky bottom with your head out of the water. On the other hand, if there was one of those channels next to your boat, you might need to tread water and swim. There are also areas over ninety feet deep and ones less than five feet and those depths change over time due to shifts in tides, currents, and the whims of humans digging channels. More broadly, areas are classified as *laguna viva* (live lagoon), the deepest part that is also close to barrier islands and the Adriatic, or *laguna morta* (dead lagoon) closer to the mainland that contains more freshwater from rivers than salt water, and is less subject to Adriatic input. This so-called dead part is mostly marsh or swamp, and today houses commercial fish farms.[9]

Besides all that water, the lagoon is dotted with islands. The main city of Venice is actually a tiny archipelago of 118 even tinier islands smack in the middle of the lagoon. Walking through the city and skipping over bridges takes you from one of these unnoticed islands to another. Beyond the city center, there are at least thirty other sizable islands in the lagoon. While most of them are inhabited, there are at least fifteen others that were once inhabited but have been given back to nature. There are also outcroppings that come and go as the waters recede or fill in, places that are more swampy than hard land. When people decided to settle here, they had to figure out ways to live on the comings and goings of these islands, and they also had to track a very complex water system. In that sense, life on the Venetian lagoon challenged the first people there to innovate and invent just to survive. Sure, they could have moved somewhere else, but that would have meant being back on the mainland to deal with Romans and strangers, or taking a boat out into the Adriatic and rowing or sailing around for a while. In that sense, lagoon living apparently made more sense. But it came with unforeseen challenges.

The lagoon experiences two tides a day and has been classified as micro-tidal by hydrologists (people who study and engineer water), because the change in tides is actually not that great. It's not like being at the beach in Florida and quickly moving your towel as the waves lap your feet or going out later for a walk on the beach and seeing a pile of shells left behind as the tide receded. Instead, the tidal movements in the Venetian lagoon are usually subtle. Spend a lot of time there walking the same way each day, and the local canals become your backyard. You'll eventually notice when the water is "up" or "down" by the ring of algae on the cement walls lining the canals, by the fact that moored boats seem to be higher up or lower down, or that the steps leading into the canal are exposed or flooded. But sometimes, especially in autumn, the tides are extravagantly noticeable. The combination of tide, rainwater, and river flow is bringing in more water than much of the city can tolerate, and flooding is likely to occur in low-lying places. In a few hours, the water recedes because the tide has gone out. Although these tides surprise tourists and wreak havoc with building foundations, they are essential. Early Venetians understood this. During the thousand years of the republic they never made any grand moves to stop the ebb and flow of the tides, nor did they try to reduce the overall size of the lagoon with landfill.[10] They washed the lagoon clean and also created a natural topography that included reedy pads of exposed mud, spongy marshes, and channels among and across the lagoon. As archaeologists have pointed out, at the start there were plenty of fish to catch in the lagoon, lots of birds to eat, certainly mounds of salt to dry and use for seasoning those fish and birds, but very little to suggest a great city might be built there.[11] Even today, water is ever present in Venice. Water is in your view at every turn, as you skip up and over bridge after bridge, or when you have to stop and turn back at the end of a passageway leading nowhere but to the water. Venice is a city defined and delineated by boats and boat traffic (see chapter 3). It is circumscribed by water buses rather than autobuses, a place where people have to hail water taxis instead of calling an Uber.

It's important to note that Venice is not floating at all but anchored in place. Within all that water are small land masses that reach all the way down into the bed of the lagoon. These are natural formations, although some of the islands have been extended by human hands with landfill, garbage, dried mud, or something else. Archaeologists have found evidence that the first people did this sort of adaptation very early on; we look at properties built on landfill and believe this to be a contemporary alteration of the landscape, but Venetians were doing this back in 500 A.D. Then, they had to figure out how to construct homes on those shifting bits of land. Because the lagoonal islands are primarily areas of dirt, rock, sand, clay, and marshy bits, they do not make the best substrate for buildings because they don't actually have the potential to support permanent structures. Instead, buildings in Venice are anchored in place by a grid of pilings that goes deep into the lagoon bed until they hit hard ground. Imagine gluing hundreds of tightly packed tall straws to the bottom of your bathtub. Make sure they reach an inch or two above the top of the tub. Then put a tray on top of what is now the flat surface you have created with those upright stalks. Add a dollhouse to the tray because there is so much stability in that forest of straws and turn on the water and fill the tub. You now have Venice, and that dollhouse is certainly not floating or going anywhere. Underneath the Church of Santa Maria della Salute, for example, there are over a million of these hammered-in pilings, and that church is very heavy but not all that big. What makes these buildings so different is that they are not really stilt houses raised on pilings above the water, as they are across the world. Instead Venetians invented a whole new kind of way to construct an edifice in water-based geography. The buildings are actually sort of resting on real live ground but then also supported by a forest of pilings and a raft of brick or marble. How they came to figure this out is anybody's guess, but this very early innovation made Venice what it is today. And regardless of the innovations in construction materials over the last 1,500 years, Venetian buildings are made exactly the same way as they were back then simply because the method works.

Contrary to common lore, Venice is not really sinking "into the sea" (meaning, I guess, sliding out into the Adriatic) or "in on itself" (as in disappearing under water). It's true that between 1930 and 1970 various industrial projects on the mainland included sucking out groundwater from under the city. That action escalated the subsidence, a caving in of the land, under the central and western parts of the Venetian lagoon. But that sinking has slowed as the industrial projects failed and rusted away.[12] If the sucking of groundwater had continued, the city might have indeed imploded as the bottom of the lagoon lost its geologic infrastructure. That sort of subsidence-caused sinking, however, is still happening in the outermost edges of the lagoon that are framed by land, and there is no way to stop it.[13] The current aquatic threat to the city is more about the tides, about global warming and rising seas, and about the many centuries of human footprint that have altered how a lagoon is supposed to operate and evolve.

LIFE ON THE LAGOON

From a human's point of view, a lagoon is just about the most perfect place to live. There are the rivers to drink from and use as transport. The dangers of an open ocean, with its high waves, deep dark waters, and sweeping winds, are absent.[14] A lagoon also teems with fish, crustaceans, and mollusks, so there's food even if you can't grow anything. But a lagoon also means you need to figure out how to take advantage of all those resources.

As Venice's origin story goes, before there were people in the Venetian lagoon, there were people living on the land surrounding it. On the mainland next to Venice, the current buildings and network of roads, not to mention the airport, distract from picturing what this area must have looked like back then. It was simply a wet place that archaeologists say was teeming with life in the form of birds, snakes, fish, shellfish, and even the European polecat.[15] We don't know exactly when humans got to northern Italy, but they might have been in the area since our species, *Homo sapiens*, appeared in Europe as Neanderthals (simply a variant of what humans were to become; Neanderthals' traditional species name is *Homo sapiens neanderthalensis*). These almost modern humans came from Africa, as did all waves of humans, but it's unclear exactly why they migrated or what route they took. There are sites with Neanderthal remains in southern Italy that date to about forty-five thousand years ago, although these people look a lot more modern than Neanderthals.[16] As hunters and gatherers, they were surely wandering around and probably walked through northern Italy. But there is no actual evidence, meaning fossilized bones, that old in the Veneto (the region denoting the mainland area near the lagoon). Perhaps the marshy edge of the lagoon decomposed everything too quickly for it to fossilize. As such, it's impossible to say if some sort of population stayed in the area that long ago, or if people came from somewhere else to live in the Po Valley much later. We don't actually know when, or how, or by whom this part of the world was first populated.

There are real problems with tracing the origins of a people anywhere. The story of humans, as we define ourselves, is instructive. Based on fossil evidence and DNA comparisons, data that has involved the work of generations of paleontologists, the history of the human species is clearly not a straight line but instead a branching tree that grew out of Africa and spread globally. There was so much movement, migration, possible conquering, and apparent mating even between subspecies of humans (here I mean Neanderthals with modern humans) that what we have is a muddle rather than some nice clean portrait of our early ancestors. But in demonstrating that history, anthropologists have also had to dispel long-held myths about how unique humans are and sidestep various prejudices, historical and contemporary. Their analyses have a strong footing, because they are based on real evidence in fossilized bone and bits of DNA and not on myths and legends or some sort of collective memory.

Unfortunately, tracing such a history is much harder where there is little real evidence of what transpired—no fossils to compare, no DNA to trace, no written record of any kind—as in the origins of the Veneto and Venice. Instead, historians have to sort through what others have decided is the right story and grab on to any bits of real evidence. The bigger problem is that a community, social group, culture, or state always weave a tale of their origins designed to serve the body telling the story.[17] Sometimes the aim is to provide a historical glue that binds a group and gives people an identity, or makes them feel better than others because their origin story is more noble or sacred. At other times, these stories are decidedly political. A government might want its people to feel or act a certain way, and there's nothing like an origin story to keep people in line or make them move in a certain direction.

The story of the origins of Venice is fraught with these kinds of myths and legends passed down and repeated by Venetians and historians of Venice, because in truth no one knows how this city was actually created, or by whom.

WHERE VENETIANS CAME FROM

Long ago, northern Italy, southern and Eastern Europe, and Asia Minor collectively made up a vast landscape of all sorts of people. They were agriculturalists and herders who lived in villages and cities, and in those people are surely the ancestors of Venetians. But where exactly did they come from? There is a legend, for example, that the first modern people in northeast Italy came from Troy, in a part of Asia Minor that today would be Turkey. The idea is that Trojans left that city and traveled west, went above the top of the Adriatic, then down and across into the Po River Valley of Italy, ending up close to the Venetian lagoon. The Trojan-Venetian fantasy was all about giving Venetians the same characteristics ascribed to mighty Trojans. By their very nature, the story went, Venetians were born of a fierce and battle-weary stock imbued with a sense of independence and nobility as Trojans. Coming from Trojan ancestry would also suggest Venetians were not tainted by the various warring tribes that later came through that part of Europe in Roman times, or even tainted by Roman blood. In another story, Venetian ancestors came from Gaul, that is, parts of today's France, western Germany, Belgium, Switzerland, and other parts of Italy, carrying with them what is considered a Celtic outlook (as opposed to Roman). Both origin stories share the common hope that Venetians were sons and daughters of proud people not the least bit connected with barbarians and conquering Romans.[18] As such, these legends set them apart as "different" and "special" from others and gave a history to later Venetians who wanted to believe they were predestined to be people set apart and strong enough to build their own republic.[19]

Relying on historical evidence rather than legend, we do know that people called the Euganei inhabited what is now the Veneto region of northeast Italy (the mainland west and north of the Venetian lagoon) and built the grand city of Ateste sometime in the 10th–9th century B.C. Objects of both clay and metal from that time, including a bronze bucket decorated with scenes of daily life before the Romans came to northern Italy, have been unearthed from a necropolis, a major burial site, at the contemporary city of Este (Ateste).[20] More charming are the metal safety pins placed with the dead. Then came the Paleoveneti about 6-7 centuries B.C., although it's hard to know where these people originally came from, and when. Among the possibilities, scholars suggest they might have also come from Turkey by way of today's Slavic nations, from somewhere in Europe (Brittany is one theory), or northeastern Europe.[21] As for exactly when, they might have come about the 12th century B.C., or not. They might, in fact, also have been of Celtic origin, or simply hung out with Celts, who showed up later in the region and spread apparently everywhere.

In any case, the so-called Veneti built cities in the Veneto—Padua, Verona, and Vicenza—that stand today. Apparently, these early Veneti had their heyday about the 4th century B.C., and they learned a lot from the Greeks or brought this knowledge with them, including the ability to write.[22] But when conflict came about a hundred years later, the story goes, the Veneti sided with the Romans, who had arrived to conquer their land, and they turned their backs on the Celts who already surrounded them. As a result of this choice, Veneti cities and land, not to mention the citizens themselves, became dominated by Rome. By the 1st century A.D., the Veneti were speaking Latin rather than their own Veneti language. They received another identity shift in 330 A.D. when the Roman emperor Constantine included that part of Italy into the eastern arm of the Roman Empire that was to be ruled from Constantinople (ancient Istanbul, Turkey). Historically, that part of the Roman Empire has been known as Byzantium, but it was actually the Roman Empire with a twist. Byzantine citizens thought themselves Roman, but they were also decidedly Greek and worshiped in the Greek Orthodox Church from the 1st century A.D. In fact, that east-west difference was made real when the western part of the Roman Empire fell and Byzantium remained strong. And so, the early Veneti were themselves, but also citizens of Rome, with some Byzantine thrown in.[23] In other words, this origin story has many twists and turns, and who knows the truth of where the Veneti came from and what genes and belief systems they carried. But their bits of Roman heritage, forced or not, came with some real evidence.

For example, with Romans came good roads. The via Annia once ran seventy miles and connected the city of Adria (or Atria), which lies between the Po and the Adige, two rivers in northeast Italy, to the large and bustling Roman city of Acquileia at the top of the Adriatic. Acquileia is still inhabited today by a few thousand people, but back then it was one of the largest cities in Europe, and it was the Roman capital in that part of the world. More startling, another road connected Aquileia to Genoa on the western coast of Italy, a three-hundred-mile journey. In other words, in the 1st century A.D. you could ride a horse or pull a cart full of goods from coast to coast in northern Italy, and you might want to if you had goods to sell.

Archaeologists have a hard time pinning down much of the history of this time and place simply because they are dealing with a lot of soggy land, where material evidence more than two thousand years old has pretty much sunk out of sight. One exception is the town of Altino (Altinum), a town today that still sits, greatly reduced to about a hundred people and an archaeology museum, at the mouth of the Sile, the same river that tumbles through Treviso as it heads for the Venetian lagoon. In 2008, during a contemporary drought, geographers made aerial photographs of the area using various wavelengths of visible and infrared light. They were able to see clearly the ancient Roman city plan that now lies about three feet below the surface. There were temples, roads, public squares, and a theater, all of them now sinking ever more deeply into the mud.[24] There was also a canal on the city plan; since people had lived in Altino since the 6th century B.C., they obviously knew how to dredge a canal and get themselves into the lagoon. But when did they go into the lagoon and why did they do it?

VENETIAN ORIGIN LEGENDS

The various stories about where Venetians specifically, as opposed to northern Italians in general, came from were invented much later to give Venetians a history that verified them as unique, capable, strong, fearless, and independent. Later, we'll see how those values helped create Venice and foster invention and creativity, but at first, they needed a divine intervention to make that description credible. In Venetian lore, it was God, in the form of St. Mark, that founded Venice. This bit of history is, in fact, a two-part legend spanning centuries. When Rome was still in charge of northern Italy, the time when this story begins, St. Mark (of Matthew, Mark, Luke, and John), happened to be traveling from the Roman city of Acquileia at the top of the Venetian lagoon to Rome, or maybe somewhere else. He was sailing, or rowing, across the lagoon, when he landed, by chance or destiny, on a small island in the middle of the lagoon that was later to become Rialto. This tiny islet would one day be the commercial center of Venice, but when this story was imagined it was just a small exposed bit of land in an archipelago of islands in the middle of a lagoon. As St. Mark's boat beached there, an angel appeared and said to him in Latin, "*Pax tibi, Marce, evangelista meus. Hic requiescat corpus tuum*" (May peace be with you Mark, my evangelist. Here your body will rest.) This voice from God marked the place where a great town would grow. In other words, the founding of Venice on the islands in the center of the lagoon was divine destiny. And it was St. Mark and his symbolic lion that founded the city.[25] This story is a fiction made up centuries later to give religious verification for the establishment and growth of Venice, and as a way to unite people under a patron saint and a mascot, the winged lion.

Added to that legend is another story about the enthusiasm of Venetians to underscore their belief that St. Mark belonged to Venice. No one knows if this story is real either, but it has been bandied about for centuries. The tale begins with the very dead body of St. Mark, long interred in a Coptic church in Alexandria, Egypt. That part makes sense, since St. Mark spent most of his life in Alexandria and died there in 68 A.D. (or so it is rumored). In 829 two Venetian merchants traveled to Alexandria, where they enlisted the help of two Greeks, a monk and a priest, to steal the corporeal remains of the saint from the church where he had been interred for over seven hundred years. They smuggled them onto a Venetian ship and hid the body in a basket, covering it with pork so that Muslim ship inspectors wouldn't dig down too deeply and find the dried-up holy body. The relic arrived in Venice after magically helping to battle the raging storms of the Mediterranean. By some twist of whatever, the relic did not go to the main church of Venice, San Pietro di Castello, on the eastern side of the city that was still fractured into islets. Instead, it was placed in the palace of the leader of Venice, the doge, or *dux*. Four years later, the story goes, St. Mark's body was removed once again from its resting place and buried in the little personal chapel next to the duke's palace. That chapel was rebuilt into what we know as the Basilica San Marco. Putting the body of St. Mark into the duke's church made a firm connection between civil rule and religious righteousness. Yes, Venice had a previous holy symbol, St. Theodorus, but he was nothing compared with a high-ranking writer of the Gospels. What makes this particular story so meaningful as a legend is that St. Mark had predicted his Venetian resting place centuries before. All this St. Mark stuff is clearly propaganda, but it worked well as a unifying identity for the developing city and its citizens. A decade before the arrival of St. Mark's body, the seat of what had been a loosely formed organization of lagoon islands that had their headquarters on the barrier island of Malamocco, south of the Lido. But they moved to Civitas Rivalti, that cluster of islands in the middle of the lagoon. The arrival of St. Marks's body and its insertion into the doge's church solidified the power and glory of the burgeoning Venetian state (more on this in chapter 4). It also gave the city and the populous the symbol of the lion, then plastered all over the place as a symbol of communal membership. The lion of St. Mark was, in fact, eventually used as a stamp of ownership by the later Venetian Republic as it extended its power on the mainland and throughout the Adriatic.

There is another "official" story of the founding of the city of Venice, and this one comes with an exact date and time: noon on March 25, 421 A.D. In this version, a couple of men from Padua laid some stones on a bare spot where the Rialto Market developed six hundred years later.[26] It's unclear why the Paduans chose the spot that eventually became the commercial hub of Venice. At this spot rose the Rialto Market in 1097 and the Venetian fish market. Back in 421, however, it was nothing. Even the Grand Canal was not much of anything. Life on the Rialtine islands was instead oriented along various canals that went horizontally from San Pietro di Castello in the east to Rivalto (today's Rialto) in the west.[27] Only much later did the Grand Canal become the central waterway. The Paduan origin of Venice appeared in the 14th century, more than a thousand years after people first inhabited the lagoon. That would be like historians, or storytellers, writing for the very first time about the American Revolution four hundred years from now and expecting anyone to believe them, even with the date of July 4, 1776 and a signed piece of paper that actually shows that date. As in the story of St. Mark once landing on the same spot and declaring it his burial place, this story was made up centuries later to alter history and give gravitas to the founding of Venice. The time and date, the story itself, are fictitious; there is no written record of the stone laying event, and no witnesses who wrote anything about it. But histories, cultures, and people like to hang their myths on exact dates because months and days seem to make their storytelling more credible.[28]

By far the most repeated story of the founding of the city involves barbarian hordes overrunning the mainland around the lagoon. Attila the Hun arrived in 453 A.D., the Visigoths came around 500, and the Lombards in 568. These hordes were said to pillage the countryside, burn everything in sight, murder and rape, and completely destroy towns. Many have suggested, or simply repeated this meme, that the first people to move into the lagoon were running from the destruction wrought by these bandits. There is also one written record, although it was written five hundred years or so after the various hordes showed up. Giovanni Diacono (John the Deacon), who was the first real historian of Venice, wrote that around the year 1000 the Veneti people left the mainland and ran into the lagoon because of a Lombard invasion about 550. In this version, locals escaped to the lagoon not only for safety reasons but also for religious reasons. The Lombards were Christians like the Veneti, but the Veneti were scared and so carried their religious treasures with them into the lagoon. As such, Diacono suggested that their "escape" was actually a religious miracle.[29] Well, probably not.[30] There were, in fact, any number of "barbarian" peoples (read people looking for goods) coming through northern Italy in those centuries, but it is hard to base the founding of lagoonal life, let alone the city of Venice, on their repeated appearance. Taking into account the need to make Venetian ancestry noble and pure, the image of more culturally sophisticated people running from vile barbarians makes for good theater. On the other hand, if those barbarians had no knowledge of big bodies of water, and hadn't arrived with a few boats at the ready, the fleeing Veneti would have had the upper hand. This story of repeated mass migration into the lagoon as a place to hide is told over and over, with each writer adding his or her own bits of speculation, creating a visual of people running and screaming into the water, pushing each other out of the way to board boats, landing on mud flats, and exhaustedly ripping twigs off scrabbly trees to construct primitive shelters. In 1893 Horatio Brown wrote, "It was difficult for the refugees, who were largely agriculturists, to adapt themselves to the new conditions of life on the waters, where fishing was the principal source of livelihood, and where they, doubtless, missed the luxury of their mainland towns."[31] It's the word "doubtless" that highlights the insertion of Brown's own feelings about ancient life on the lagoon. He also claims, based on nothing, "They returned in large numbers to their ruined cities."[32] Maybe they did and maybe they didn't.

An alternative story, that people moved into the lagoon long before Attila and continued to move out there over the centuries, was perhaps just too dull to account for the birth of a city that would become a republic. Historian Edward Muir also points out that the Huns weren't settlers but nomads, and although they might have sacked towns and villages, they would have moved on and the escapees could easily have moved back.[33] The idea of "waves" of these "hordes" makes it seem that day after day, week after week, unkempt people from the north or west repeatedly overcame Veneti towns and burned everything down. In fact, the dates show a century between the Huns and the Lombards. If you were a small child when the Huns came, you were surely dead when the Lombards arrived. And if your family went into the lagoon, chances are they came back and rebuilt their houses and reinhabited their cities and kept on with crops and livestock. Or maybe they stayed because the fishing was so good.

Archaeological evidence, conducted totally underwater or on partially submerged sites, suggests a more permanent presence of humans in the lagoon. One site places Roman villas with beautiful mosaic floors on an island in the northern lagoon around 1000 A.D.[34] There was also a strong and active Roman presence in the lagoon by then, as evident in piles of ceramic jars of Roman age that were scuttled and apparently used to shore up islands.[35] Bits of Roman bricks have also been found in at least eighteen places around the lagoon (underwater), a sure sign of solid building.[36] It makes sense that Romans would want to come and go through the lagoon and put up buildings on islands, because it was their territory; the lagoon was also within easy reach of their thriving cities of Altino and Acquileia on the mainland. There are even recovered planks of a cloth-covered boat, as evidenced in stitch holes, uncovered near the mouth of the Adige River where it empties into the lagoon. This boat dates from 2000 to 1000 B.C. and is identified as Roman or pre-Roman.[37] But a continuing story of Romans and non-Roman people gradually inhabiting and utilizing the lagoon is pretty boring stuff to include in an origin story.

The idea that the first Venetians ran into the lagoon out of fear and built makeshift houses for themselves (because they weren't capable of building anything else even after living in Roman cities on the mainland) has also been carried through the centuries by a letter written in 537 A.D. by the Roman intellectual Flavius Cassiodorus. A statesman whose lofty aim was to preserve documents in the 6th century, Cassiodorus wrote a letter to the early lagoon people commenting on their primitive lives at the time: "You live in houses like those of sea birds…by weaving together flexible reeds you manage to increase the solidity of the ground in the lagoon and you are not afraid to put such a fragile defense against the surging tides of the seas."[38]

From that, everyone has had a picture of early Venetians as humble, scared peasants barely scraping by.[39] Cassiodorus also instilled even more character into his story when he continued: "The inhabitants have only one concept of plenty: that of filling their bellies with fish. Poverty and wealth, therefore, are on equal terms. One kind of food sustains everyone. The same kind of dwelling shelters all. No one can envy his neighbor's home; and living in this moderate style they escape that vice [of envy] to which the rest of the world is susceptible."

Where he got these notions no one knows.

The fantasy of the lagoon as a refuge from fear and annihilation, a place where independent people were cut off from the world and therefore had to rule themselves (and did so with equality), has been rewritten in history books and told on the street corners of Venice for centuries. It has been told so often that the story feels like it's based on eye witnesses and recorded in old documents at the time. Instead, these original stories served later Venice to help make itself and its sense of community, as they so often do in human groups (see chapter 4).

On the other hand, Venetians didn't appear out of nowhere.

THE ARCHAEOLOGY OF EARLY VENICE

In truth, as writer Jan Morris put it, "They became Venetian by fits and starts."[40] That large body of water had attracted people forever. The lagoon offered fish, relative peace, and refuge. Some probably made homes on islands poking out of the water. The idea of the lagoon as a place of refuge or safety has probably been exaggerated for dramatic effect, and until recently there was no hard evidence of what actually happened, except for that trail of Roman detritus found in the water.[41] How could there be anything at all left by the first real Venetians when their lands were so scattered, they had such a small footprint in the first place, and those footprints are often underwater? In fact, no kind of urban archaeology—archaeology conducted on dry land—happened anywhere in the Venetian lagoon until the 1960s. That work began with a dig at Torcello, a small island north of central Venice thought to be a major population center. There's also a story, a myth really, that has yet again skewed assumptions about how and where people began living in the lagoon. This story concerns only the people living in the mainland town of Altino. When Attila the Hun arrived in Altino in 472, the people responded by fasting for three days. Then a voice from heaven told someone to climb up into a tower and look about and the bishop of Altino did extactly that. From up high, the bishop saw a bunch of islands in the lagoon, as if no one had ever rowed a boat out there before, even Romans. With this "discovery," the scared Veneti of Altino made haste to Torcell making Torcello the real first Venice. The Torcello story is repeated often and is printed in all the guidebooks. And indeed, in the 1960s, archaeologists found evidence of a tile walkway seven meters long (twenty-two feet) that can be dated to the 2nd century A.D. Over that walkway is a layer with some small wood framed houses built on tile and stone footings and a pebble walkway that dates from the 6th century. Those houses also had cooking hearths. But these early dates for this first archaeological attempt have been disputed, because it turns out that the archaeologists had dug on a spot of reclaimed land (which means way back then the early Venetians were already fiddling with their islands).[42] And yet, there are walls inside the standing church on Torcello that date from the 5th century. It does appear that in 639, the Catholic dioceses moved from Altino to Torcello, and it became the ecclesiastical center for all the various communities in the lagoon for the next thousand years. Archaeologists have also found evidence of a very early baptistery, window glass, coins, and human burials next to the small and very lovely Basilica Santa Maria Assunta on Torcello, where you can pray today. Those items all date to the 7th and 8th centuries. But that doesn't mean Torcello was the early epicenter of Venice, only one part of it.[43]

It wasn't until the 1990s that more successful attempts to understand the origins of Venice were conducted. Researchers were able to use more sophisticated archaeological technologies and techniques for working in places where one might be digging below the current water level. They also adapted their approach and acted like Venetians by doing their work from boats. Armed with shovels and trowels, they rowed small boats close into places with low levels of water.

Their discoveries are revealing. For example, urban archaeologist Albert Ammerman and his team found parts of a wooden boat 2.8 meters below today's ground level on the island San Francesco del Deserto, a small island that lies northeast of what we now consider Venice. The archaeologists dug out part of the hull, one rib of the frame, and some wood pegs made from a mix and match of tree species. These materials come from 425–550 A.D., just when barbarians were supposedly pushing people into the water.[44] The boat is small, only two meters wide, suggesting it could carry only a person or two. Maybe this boat was an escape vessel, but it could also have belonged to someone already living on San Francesco del Deserto, a person who had opted for the lagoon life long before the Huns. That idea is supported by other bits and pieces the archaeologists found as they dug down even further. Nine inches in the mud below the boat were two flat pieces of oak lying horizontally and a bunch of wood poles from several types of trees. One of the flat pieces of wood had a big X carved into it, and the poles were upright, spaced tightly together, and driven into the layers below. The scientists speculate that the poles and flat pieces represent a wharf or canal side structure.[45] These material remains were in marshland, and it's possible that whoever left them there had been living on the island for some time. Although most books on the history of Venice emphasize the "mass" migration of refugees running into the water to escape barbarian hordes, this evidence, and other bits of archaeology, show that there was a strong and permanent human presence already in the lagoon before any of the barbarians showed up.

The same archaeological team found further evidence of permanence in the Church of San Lorenzo in the Castello region of the main part of Venice. At the time of this excavation, archaeological techniques had advanced beyond marking off large squares and digging down inch by inch, layer by layer. Instead, borrowing from geological surveys, Ammerman and his crew drilled slender tubes into the ground with special equipment. Those tubes sliced down cleanly without disturbing the surrounding materials, and they came back up with a long vertical sample of that spot with all the layers kept in place. Those layers, called core samples, can also be accurately dated with accelerator mass spectrometry. As Ammerman says of this technique, "The parallel here is with modern surgical biopsy."[46] The present-day Church of San Lorenzo dates to the 1600s. But under that church is another much older structure. About ten feet below the current church floor set in the 1600s is a series of upright poles denoting some sort of building. And below those poles there are bits of brick, mortar, tile, slag, and pottery suggestive of permanent building materials. They also found some actual construction made out of matted reeds. Whatever this building was meant to be, it had been built on what is called, in *veneziano*, a *barena*, the word for an exposed island that sometime gets covered by water.[47] These materials were dated about 550 A.D., so they show that people were building permanent buildings very early on and taking into account fluctuations of the lagoon.[48] In other words, back in 500 A.D. they didn't just throw up lean-tos for quick coverage but took time to make sturdy and lasting buildings.[49]

Ammerman's group has also pulled up core samples around the middle of the wide open and very public area called the Piazzetta in front of the Ducal Palace in Piazza San Marco. These core samples date to the middle of the 7th century and they show that people were using this part of Venice regularly in the 600s.[50] People were also doing something across the way at the site where the Civic Library of Venice, the Marciana, stands. Going deep within a storeroom, the archaeologists hand-dug deep cores that were spotted with pottery and mortar and pieces of wood and reed from about 750–800 A.D.[51]

These well-dated archaeological sites undermine the old stories claiming that the first Venetians were temporarily attached to their islands. In fact, very early on people were living nicely in the center of the lagoon on the very islands that would become the center of Venice. And they built there to stay.[52] What archaeological digs also show is that the nascent Venetians were manipulating their environment and making new land all over the place. For example, on the island of Torcello, they added about 1.8 meters of dirt to the land under the church over its early five-hundred-year history. People also piled up 2.5 meters of additional landfill under the various iterations of the Church of San Lorenzo over six centuries.[53] They brought in lots of building materials and mountains of fill and molded their landscape boatload by boatload. The same studies show that the early Venetians considered the daily fluctuations of tides in addition to annual fluctuations, and made sure to keep their houses about one meter (approximately three feet) above that mark.[54] Venice, in other words, was not necessarily at the mercy of rising water or sinking land, and lagoon dwellers smartly started manipulating their environment a very long time ago. Studies of mean sea level over Venetian history also show that anyone driving a bunch of pilings through shaky land to reach a good base would only have to have gone down about nine feet to reach a decent hard substrate that could hold up a building.[55] In other words, the founding of Venice was not such a fragile proposition. And there was not one central location for the founding of that life—those first Venetians were spread across many islands.

THE ORIGINS OF VENETIAN CULTURE

Venetians established a civilization and a culture in a body of water. By 466 A.D. there was an assembly at Grado in the northern lagoon, where tribunes, or local leaders, were elected to govern the twelve established communities scattered across the lagoon. In other words, by the late 400s, there was a governing mechanism in place that united everyone in the lagoon into one larger community. Alvise Zorzi, a Venetian historian and a Venetian, says 466 A.D. is actually the best founding date for Venice and the establishment of civil life in the lagoon. Note that this date has little to do with barbarians at the gate.[56] As historian Elisabeth Crouzet-Pavan writes, "In the absence of any preexisting ancient site, any organizing central kernel, any inherited plan, the urban fabric was gradually knitted together out of scraps, scarce bits of land that loomed up out of the water."[57]

In a sense, Venetians treated the lagoon as if it had been habitable land rather than a large pool of water, and that was their first spark of genius, and one they share with other water-based communities around the world. But in doing so, Venice stood alone and apart from other cities in their area. Italian city-states such as Florence, Milan, and Mantua were typically bounded by walls that kept out enemies. Sometimes they were established on hilltops as fortresses. Rulers of those city-states had to live within palaces that were not so much homes as castles. Venetians, in contrast, had the lagoon; the water acted like a very large moat.[58] Indeed, the expanse of water meant they could see anyone coming, because intruders were fully exposed for quite some time before they reached land. The shallows and sunken islands also meant intruders had to be very good seamen. They might need to back a boat off a *barena* and find a deeper channel for their attack. Or they might have to figure out why their deep-bottomed boat was stuck. Sure, Venetians had to hammer a forest into the lagoon and ferry materials from somewhere else to build their houses, yet there was great freedom in the site. As architectural historian Robert Goy writes, "The architecture of the Venetian lagoons in all periods is based on function, practicality and on beauty, but never on considerations of security."[59]

And this is what even casual visitors see today as they ride the train or bus over the causeway that connects the mainland with modern-day Venice. No matter the method of modern transportation, looking out the window as you cross the lagoon to Venice brings back the eerie isolation that has formed the physical and psychological atmosphere of this city. Like a beehive hung alone in a tree or an apartment building left after a nuclear holocaust, Venice became a socially self-contained city that ran on its own time and in its own space.

The early Venetians adapted the lagoon islands for daily human life and in doing so, they created themselves, their first and perhaps most important invention.

CHAPTER THREE

The Seven Seas

Sweet dreams are made of this.
Who am I to disagree?
I travel the world and the Seven Seas.
Everybody's looking for something.

> —Annie Lenox and Dave Stewart,
> "Sweet Dreams"
> Eurythmics, 1983

For water, too, is choral, in more ways than one. It is the same water that carried the Crusaders, the merchants, St. Mark's relics, Turks, every kind of cargo, military, or pleasure vessel; above all, it reflected everybody who ever lived, not to mention stayed, in this city, everybody who ever strolled or waded its streets in the way you do now. Small wonder that it looks muddy green in the daytime and pitch black at night, rivaling the firmament.

> —Joseph Brodsky, *Watermark;*
> *An Essay on Venice*
> 1992

SOMETIMES IT TAKES a very long walk to see the world. In this case, the world is a 15th-century Venetian map, and the walk begins at the door of a museum in the Piazza San Marco. The curious must pass through the grand ballroom and various salons of a Napoleonic palace, and cross room

after room full of huge sea battle paintings, armor, or coins toward a gloriously huge room with a high curved ceiling and intricately designed parquet
floors. At the far end of this room is a much smaller, even tiny, after-room.
Few tourists have the stamina, at this point, to enter. Finally, there it is: a world
map, or mappamundi,[1] as it was called back when it was finished in 1459. It's
mounted in a gilded frame (actually two gilded frames) and is about two feet
by two feet square. Painted by Venetian monk Fra Mauro, the map belongs
to the Biblioteca Marciana, the library of Venice, and is the oldest medieval
mappamundi in existence. It's not only beautiful but also accurate. There is an
image of this map set next to a NASA photograph of the Earth; it's startling
to see that a 560-year-old map is pretty much spot on, except that North and
South America and Australia are missing. But to be fair, no Westerners knew
about those continents back then. This mappamundi is also upside down,
with south at the top. The orientation is confusing until you locate Italy
(a tiny upside-down boot) and move out from there. The eye then recog-
nizes the Mediterranean Sea, the Atlantic Ocean, Africa, the Middle East, the
Indian Ocean, India, China, on and on like a reverse telescope moving back
to take in the world as it was known in the middle 1400s. All the coastlines
are frilly, sort of like the edge of a pie crust, and all the seas and oceans are
deep blue and scored with jaunty white lines that denote waves. The map
is dotted with castles and rivers and mountain ranges. There are all kinds of
boats in its seas and oceans and they add a sense of movement and adventure
to the flat surface. These boats sail here and there and even around the Cape
of Good Hope before, it should be noted, anyone had actually done that.
The three thousand or so bits of text on Fra Mauro's map are almost impossible
to decipher because they are written in Venetian vernacular, the dialect of the
region around Venice. To be sure, Venetian is a living language—this is
the language that gondoliers speak and is the local sound that fills bars and
floats above the tourist masses and confuses even native speakers of Italian.
Translations of these texts find an interesting mind at work over this map,
someone who wanted to not only draw the world but influence it as well,
a surprising goal for a 15th-century monk. Fra Mauro's map isn't only the
oldest surviving map of the world; it's also the first map to show a way to
sail around the Cape of Good Hope and connect Europe with Asia by sea.
It's also the first map to include the island of Java, the first Western map to
depict Japan, the first map to show that the Indian Ocean was an ocean and

not an inland sea, and the first map of the world based on the reality of its time rather than faith, religion, or some other sort of belief system we might now dismiss as quaint. Fra Mauro's map is a cartographic masterpiece based on reliable sources, and it catapulted the Western world toward globalization.

More than anything, Fra Mauro's map is a testament to Venice's relationship with water. It speaks to the need for good navigation, an understanding of geography far from home, and the essential Venetian occupation with commercial trade everywhere ships could go. In fact, the great Roman geographer Pliny the Elder considered the "Seven Seas" to be the area at the southern end of the Venetian Lagoon.[2] To Pliny, the sailors of that area were the finest, meaning they sailed the complex ins and outs of the lagoon, navigating through saltwater marshes, around the bits of low islands revealing their landscapes only during low tide, and out into the open Adriatic. Admittedly, other Romans used the term "Seven Seas" to include the Adriatic, Baltic, Mediterranean, Black, Caspian, Red, and Arabian Seas, while today we assume the phrase "Seven Seas" refers to the seven now known oceans. But that phrase was in use long before the Pacific was "discovered" and the Indian Ocean was still considered an inland sea. Pliny's definition is significant not just because of his influence as a geographer but also because it highlights the accepted understanding that Venetian sailors were the best. That label was recognized early in history, and it applies today.

Spend time sitting by the Grand Canal and keep track of the parade of boats, all kinds of boats. Watch gondoliers maneuver their odd craft around corners and up to the docks to pick up tourists. Stand at the edge of the city, stare out at the lagoon proper, and track the kinds of boats that pass by—rowed, motorized, and sail-driven boats. Hang out on some small footbridge over a back canal and see locals putt-putting around in their small launches or rowing alone or in pairs while standing up. The ease with which Venetians navigate their way, and the fact that people of every age and gender are piloting these boats, demonstrate that these skills have not, in any way, been lost. Being on the water is a deep identity for Venetians. As a rowing teacher in Venice once explained to me, Venetian kids just seem to know naturally how to get into and out of boats, row standing up, handle an excessively long oar, make tight turns, and tie a boat to a morning post. In fact, Venetians have always had to be expert navigators, accomplished geographers, great boatbuilders, and soldiers on the sea. Their very livelihood as merchant traders depended on these skills. Venetians went far beyond any other people living on or near water in taking advantage of their environment to establish trade networks across the known world, becoming rich for their efforts. In that waterborne historical ambiance is a long list of firsts and inventions. Those accomplishments began with Venetians knowing where they were going, and why. They also encompass a group of Venetian explorers who were compelled to see the world. All that sailing and rowing on the Seven Seas and far beyond necessitated expertise in shipbuilding and innovation in nautical military technology that would allow them to defend their city or take over other places.

THE MAPMAKERS OF VENICE

Maps don't just show us where to go. Cartography has always reflected culture and politics.[3] For example, every schoolchild learns the outlines of his or her country, and when that outline changes over time, as they always do, kids relearn that map. Throughout history, maps have also been used in government policy (think voting districts or school redistricting), civic identity (think of the phrase "Venice is a fish," which reflects both its actual shape on a map and its fishing history), and as instruments of war, empire, and polity. Today, we are used to relying on satellite photographs and Google Maps to give us two-dimensional renderings of our world. We know that the earth is round, with the Arctic at the north, and full of seven continents and lots of water. We might think this image is static, without prejudice, and forever, but all maps have subtext. For example, current printed maps of the world intended for sale in the United States place America on the left center and somehow make that country seems bigger than it actually is. Australia, about the same size as the United States, is placed on the lower left, below the grand landmass of Asia, making it almost hidden. In other words, maps reflect the person, country, or culture that produces them. These days, when we employ maps to get around, it's easy to ignore the fact that maps are political and cultural, and often incendiary. They chronicle changes in the world as new places are added and others are broken into parts or sewn together into new countries. More important, maps can change the world.

Maps have been particularly important for Venice because of the city's trading economy and the fact that it's a bunch of islands set in water.[4] It was essential for this small place to realize visually where it sat within the larger world. Venice's trading economy went abroad in all directions, meaning maps were basic tools of economic trade. In 1460, Venice was the first European city and government to survey and map its own territories, and they did so for military and administrative reasons.[5] Accurate maps of the mainland west of Venice were also critical because the government needed to manage the course of rivers running through the Po Valley that dumped silt into the lagoon.[6] To maintain the lagoon, hydraulic engineers had to divert those rivers by scouting origins and courses and then diverting rivers into other rivers to bypass the lagoon and push all that silt directly into the Adriatic.[7]

The first great Venetian cartographer was Genoa-born Pietro Vesconte. We know of Vesconte's work, and know that it's his, because in 1311 Vesconte signed and dated a map of the eastern Mediterranean; this map is the oldest surviving signed "portolan chart." *Portolano* is the Italian word for a mariner's directions, which long ago meant written directions, line by line, for boats under sail or oar power.[8] For some odd reason, modern historians started calling all maps made specifically for water navigation "portolan," which in fact they are not. Instead of written instructions, they are the visual representations of where a sailor might go. But portolan maps are navigational directions of a sort, hence the contemporary mix-up.

In any case, these so-called portolan maps are not like anything we think of as maps today. Sure, they are composed of hand-drawn blotches of land and craggy coastlines, but the vast expanse on these maps is open water. There is no top or bottom to these charts because they are meant to be rotated as needed by ship captains.[9] What marks them as particularly "portolan" is the busy network of intersecting lines that overlay everything. First, there are the faint horizontal and vertical lines that lend a certain precision to land and sea. At first glance, they look like miscalculated longitude and latitude lines. Actually, these lines were simply grids to aid the cartographers when drawing points of reference.[10] Once the landmasses and seas were accounted for, Vesconte and other portolan cartographers took a ruler and spent hours superimposing radiating straight lines across water and land that intersected with each other.[11] The compass had been around in Europe since the 12th century, and sailors used the compass and astronomical coordinates to get about.[12] By the time of Vesconte, there was a tradition of drawing a compass rose on various points on nautical charts and then extending between eight and thirty-two equidistant radiating lines from the compass rose center out as far as possible to the end of the map. The result is a complex web of lines, a mathematically precise spiderweb overlaying everything. The other odd part is how Vesconte's maps, and other portolans as well, show just about nothing inland. There are zillions of place names along coasts, including the names of settlements and river mouths, but the world of the 1300s simply fades away not far from the beach.

Looking at portolans now, the web of lines is both disconcerting and confusing, and the empty interior suggests geographic ignorance. That, of course, is not accurate. As with all maps, portolans had a significant function, especially for the times. They were the most practical of diagrams, and although hand-drawn and rather precious, they were usually aboard ship.[13] Such maps were purportedly used to help a sea captain find his way, say, out of the Mediterranean and down the west coast of Africa. At other times, they were used to help a ship get back on course using a mathematical table that translated the lines and what to do next. Along with dead reckoning, these miscategorized portolan charts were the driving directions of the Middle Ages.[14]

Vesconte drew various atlases, portfolios of several maps that were intended to be together in a folder or cover. We are now familiar with world atlases that are so large you have to put them on a table before flipping the pages. Back then, an atlas might have as few as four maps or portolans. Vesconte had a reputation for accuracy, and he must have received his geographic information from mariners who had traveled around and made notes or remembered what they'd seen.[15] Because of that reputation, Vesconte was enlisted to contribute his navigational charts for a very political purpose. In 1306, Venetian nobleman Marino Sanudo[16] was on a mission from God, literally. There had already been ten Crusades aiming to bring a Christian presence to the Holy Land, but that was not enough for Sanudo. So, he wrote a "secret book" (*Liber Secretorum Fidelium Crucis Super Terrae Sanctae Recuperatione et Conservatione*) to convince the pope to authorize a new Crusade to, once again, bring Christianity to the Holy Land. The irony was that there was constant tension between Venice, that is the republic, and the pope, so an appeal from a Venetian probably had little persuasive power, even with great maps by Vesconte. In the end, there was no eleventh Crusade. But the Vesconte world map included in Sanudo's document was world changing nonetheless. This map made it clear in drawings that most of the world was not Christian, and that was a motivating, and probably scary, punch line.[17]

By the 14th century, Venice had the reputation of being a place that encouraged trade, manufacture, and creativity, and that had little conflict with the Church. Because of this, many businesses either moved to Venice or were initiated there. In the 1400s there were several great cartographers working in Venice, and they produced some of the most magnificent and most significant maps of the age, or any age. Andrea Bianco drew, wrote, and signed a nautical atlas in 1436 that was a combination of portolan charts for navigators and a mappamundi of the world that was basically pictorial. Unlike useful portolans, mappaemundi were considered works of art to be part of atlases pored over in living rooms or hung on walls to impress people.[18] These maps were not really about directions or navigation; they were artistically painted and splattered with all kinds of information about places and people.[19] Bianco's world map preceded Fra Mauro's, and it was faithful to what the Greek mathematician and astronomer Ptolemy had written about the geography of the known world in the 2nd century. Ptolemy's book *Geographia* had been translated into Latin in 1406, and it was still the standard for depicting the world twelve hundred years later, even though many lands had been "discovered."[20] Bianco was a Venetian sailor, and during a stay in the Port of London he drew these maps. He had firsthand geographic knowledge and a fine hand. Once back in Venice, Bianco actually helped Fra Mauro draw his world-changing mappamundi.

In 1452, Venetian cartographer Giovanni Leandro made another mappamundi. Several survive today, and they are masterpieces of both art and cartography. Drawn and painted on velum (cured animal skin), his worldview was typical of the time in that it was intended to give a solid geographic foundation to Christianity.[21] In fact, mappaemundi in general, as opposed to portolan charts, were seen as a reflection of the hand of God because God created the world.[22] As such, Leandro placed Jerusalem in the center and a depiction of Paradise at the top. He also included symbols for the four evangelists in the four corners. This world map was also typical in its secular description of the known world: three continents—Asia, Africa, and Europe—all surrounded by a rim of ocean. Charmingly, Leandro's world also had a set of rings surrounding the map that can be used to pinpoint the date for ninety-five years' worth of Easters. The rings also depict various astronomical and astrological moments yet to occur.

Leandro's lovely maps were and are overshadowed by Fra Mauro's ground-breaking work that appeared only seven years later.[23] His map is famous not only because of the "firsts" mentioned at the beginning of this chapter but also because Mauro took advantage of some very Venetian information to construct his map. Mauro traveled in his youth as both a merchant and a soldier, but once he joined the Calmadolese friars in 1409, he was monastery-bound on the island of Murano.[24] But still, Mauro apparently availed himself of every written or spoken account of the world he could find. He had a well-known text of a fellow Venetian to start with—the traveler and trader Marco Polo. Mauro was able to read Polo's travelogue (written by someone else, see below) and make a stab at getting the Far East in relatively accurate perspective. Polo's tales of twenty-four years traveling to, from, and around the court of Kublai Khan were full of place names, descriptions, and routes of travel, but no maps. Mauro was also helped by two other Venetians, the words of explorer Niccolò de' Conti[25] (see below) and the established cartographer Andrea Bianco. Mauro had the advantage of simply being in Venice, the then center of interpersonal information exchange where sailors came and went, many of whom must have had conversations with the monk about what they had seen. Mauro referenced these voices as "many have said" in various texts on the map. Although this reference might seem shaky in our modern times of grand university libraries and endless ways to cite information for posterity, such notations were credible in Mauro's day. By then Venice was, in historian Angelo Cattaneo's words, "the most important entrepôt in the world."[26] Venetian merchants, and the state, were very interested in information about other people and places, though a skeptic might say they were gathering information about consumers, like an ancient Amazon or Google. And they were. But all that sailing, trading, and chat made for lots of geographic and cultural descriptions from abroad.[27] In fact, historian of medieval maps Evelyn Edson calls mappaemundi "encyclopedias," not just directions or illustrative drawings of continents and oceans.[28]

Mauro distilled available Western knowledge to discern the shape of the world and where it was possible to reach by ship. He gained reasonable accounts of bodies of water, coastlines, and the interiors of continents. Then, he gathered helpers in his workshop on Murano and began to fulfill a commission from the king of Portugal to draw a map of the world. The copy hanging today in the Venetian library is not the original sold to King Alfonso but a copy apparently made for the Venetian Republic and that one hung for years in the monastery on Murano.[29] It might not have been considered significant then, but the Portuguese would soon be moving about the world, conquering lands and goods and, in doing so, opening up the world for themselves and other people, such as the Dutch. Fra Mauro's map, therefore, was integral in eventually removing his native city from her own trading dominance.

It's clear that Fra Mauro had a singular vision. Earlier mappemundi were based more on hearsay, religious faith, and celestial fantasy than on geography. Their job was to inculcate citizens to be good Christians. Surprisingly, Fra Mauro didn't seem to care about that, and he indicated his indifference by not putting Jerusalem in the center of the map. Also, he oriented the map "upside down" as described in Ptolemy's *Geographia* and as shown on Arabic world maps that came before. He also pulled the Garden of Eden completely out of the world, where it usually sat in other versions, and placed it on the lower left corner, outside the planisphere proper.[30]

What's on the map is even more significant than what's not. There are over three thousand texts that not only explain place names but also have descriptions of peoples and cultures. These texts are also full of opinion and argument, because Fra Mauro was determined to drag the West into a more global perspective. And that he did. Bouncing in the waves of this map's deep blue waters are many kinds of boats relevant to the sailors of various countries. Below Africa, there is a thin line of water connecting the Atlantic Ocean to the Indian Ocean. Before this map, everyone in Western culture believed the Indian Ocean was an inland sea and that there was no water route around the African continent. But Fra Mauro believed the traders and sailors, the words of those who had actually gone east, that it was indeed possible to sail around the Cape of Good Hope and quickly head for the spice islands and the countries that produce silk and gems.[31] He emphasized this prospect by drawing in all the various trade routes throughout the Indian Ocean ports, enticing merchants to take the water route rather than the tedious and long journey by land.

Historian Marianne O'Doherty calls Fra Mauro's map not just a map but a sign of various connections, an "agent of great change between oral tradition and written cosmography, between medieval and modern and between periods of Arab and European dominance in the Indian Ocean world."[32] Mauro's map was also a map of the people, and an agent of change in that. All previous mappaemundi had place names and explanatory text, but this one was written in Venetian vernacular. It's true that Mauro's particular monastery was devoted to translating manuscripts into Venetian, so the choice of that language was probably normal for him. Still, to have this map in the local language shows how central Venice and its language were at the time. Sited in a major European city involved in trade, the Venetian language had become the lingua franca of its time, and was certainly the language of trade.[33] He also personally commented that writing in Venetian was the way to have more than just a few intellectuals read his map. So, what we see as an odd choice for writing in what we now think of as an obscure language was, in fact, totally understandable. Mauro was clearly not interested in impressing with language. And yet he used geographically or culturally local names for places, emphasizing their difference from Western translations but at the same time giving equal credibility and stature to that "foreignness" as an anthropologist would.[34] He wanted to bring the West to meet the East. Because of this map, in 1487 the Portuguese sailor Bartolomeu Dias went all the way down the coast of West Africa and looped around the Cape of Good Hope before turning back because his crew rebelled. Because of Mauro's map, Vasco de Gama and his crew rounded the tip of Africa and sailed across the Indian Ocean to India in 1497.[35]

These journeys by the Portuguese were also helped by a series of portolan charts drawn in 1468 by Venetian Grazioso Benincasa, who charted the Cape Verde Islands and the coast of West Africa all the way to Liberia. Benincasa also drew the first full map of Ireland, a portolan covered in those radiating lines, with the landmass edged in green. That prophetic color detail has now been surpassed by the commercial success of this map and the atlas that contains it. In 2014 the *Irish Times* reported that the Benincasa atlas with the included portolan of Ireland would be sold at auction and was expected to fetch over three million euros. As the auction house commented at the sale, the atlas is still in its ancient Venetian goatskin binding and "the colours and gilding remain as fresh and vibrant as the day it was finished."[36]

But it was Fra Mauro's radical map that showed the way to sailors and merchants and sparked their courage and greed to venture forth and find a sea route to the East. As William Boelhower notes, "[Mauro's] mappamundi offers an exemplary cartographic homology of the dawning moment of the new oceanic order."[37] This newly conceived oceanic order meant there was an efficient, less time-consuming water route to the East that didn't require one to mount camels. Mauro's map was the first document to initiate and encourage globalism, and it laid the groundwork for what another historian calls the "aggressive attitude of the West toward the rest of the word."[38, 39]

There are few maps of Venice itself from these early centuries, and historians have suggested that for Venetians, maps were about being cosmopolitan, not about being Venetian; about turning outward, not inward. But in the late 1400s, German banker Anton Kolb funded artist Jacopo de' Barbari to create a "bird's-eye view" of Venice.[40] The finished product, an enormous 4.4 x 9.25 feet, stands as the largest image ever made by woodcut. The secret is that it was printed with six separate woodcut blocks of pear wood. The prints were were then stitched together.[41] There are several surviving prints of de' Barbari's map, and the original woodblocks are on display in the Correr Museum in Venice. The enormity of the map's size and its complicated and accurate detail, and its orientation from above, are best appreciated by standing in front of those woodblocks. The map itself is startling for the attention to every building, every church, every campo, every single brick in Venice, and it's fun to identify known places that have not changed since de' Barbari's time (which is just about everything).[42] That includes 10,312 chimneys, 114 churches, 47 convents, 103 bell towers, and 253 bridges (147 fewer than there are today).[43] The woodcut blocks themselves also speak to the technical skill of de' Barbari and his ability to see from above what Venice looked like, although the map is not drawn to scale. Historians speculate that de' Barbari employed several surveyors to climb the many bell towers (*campanile*) of Venice and report back to him.

De' Barbari's map is clearly an homage to Venice and her power.[44] It's not about the lagoon, not about Venice's place in Europe, but about Venetian power. The very fact of a perspective from the sky might have meant, as some have suggested, that de' Barbari intended the view to be the same as God's: God was looking down upon Venice in particular and presumably giving approval. This map was also quite different from other urban maps that came before because it was based on "science," that is, actual surveying and attention to proportion. And it was different in that it included every single house, shop, church, and bit of garden.[45] Recently, during a show of European craft masters at the Fondazione Giorgio Cini in Venice, the Parisian embroidery house Maison Lesage invited exhibition participants to help them finish a map of Venice made on a large cloth panel executed by hand with silk embroidery thread and beads.[46] This collectively hand-sewn map was a perfect execution of de' Barbari's map of Venice. There were all the buildings of Venice, in the three dimensions of tiny beads, and the green gardens in looped French knots with tiny green seed beads. And viewers and embroiderers were engaged in a bird's-eye view of the city.

Although de' Barbari produced this one amazing map, he was no cartographer and there is no evidence of any other maps created by his hand. Instead, de' Barbari made a living as a painter, and his small painting, *Still-Life with Partridge and Gauntlets* is the earliest still life painted in Europe. This work is also trompe l'oeil, a painting of photographic realism and a bit of an optical illusion. There's the perfectly realized dead bird nailed to a perfectly painted wooden wall with the bolt of a crossbow and the perfect metal gloves of the hunter hanging there as well. Hanging in a hunting lodge, the figures in this painting might be mistaken for real life.

The world was biting at Venice's heels in the first half of the 16th century, and the cartographers knew it. In 1526 Venetian cartographer Pietro Coppo drew a four-volume, twenty-two-map atlas of the world; a shorter version of fifteen woodcut maps was published a few years later, and this work is considered the first modern atlas. In less than a century between Fra Mauro's map and Coppo's atlas, the East had become accessible to sailors and merchants, and the Americas were on the map. The world had become both larger and smaller, and other countries such as Portugal and Spain were overtaking the Venetians in sailing to faraway places in search of goods. Once the mighty power of Venice was threatened by the feasibility of sea trade routes to the East, Venetian cartographers began to make a different sort of map. Politically and culturally, de' Barbari's can be considered one of these—a map focusing on the city and glorifying Venice, not a trade route map. As urban archaeologist Erkia Mattio, who works in Venice, has said, "Venetians saw themselves as surrounded by Europe." And with that perspective they began to feel threatened, and in need of some self-satisfaction and self-glorification. A map from 1528 by Benedetto Bordone makes this point clear. Bordone was a cartographer and engraver from Padua, part of the Venetian Republic at that time. Bordone was also known for his miniatures, and he was a successful commercial seller of maps.[47] Bordone was the first person to draw the world as a squat oval, three hundred years later, it is the basis for an equal-area elliptical projection of the globe. But his greatest contribution to cartography was a three-volume atlas, *Isolario*, or "Book of Islands," which was the first ever island atlas.[48] But it wasn't just islands. Instead, this atlas consisted of eighty-four folios of island and city maps, seas and oceans included. It's the map of Venice proper that stands out. On this map, the city of Venice appears as a central figure of small islands encircled by land on the left. That land then bends around Venice at the north and south like protective arms and the embrace is completed on the right by the barrier islands. These islands are pictured as more substantial than they really are. As a result, the lagoon waters were pictured as incidental, and the land surface of the city of Venice was exaggerated. In general, Bordone's map makes Venice look like a powerful place that is totally protected from outside forces. That perspective was probably also influenced by the Battle of Agnello (1509), in which the French army invaded Venetian territory near the city of Bergamo and from there proceeded to gain what was then thought of as inviolate Venetian land holdings on the mainland. Having these properties conquered had a deep psychological effect on Venice, so it's no wonder Bordone drew a map of the city that made it look much like a fort with a moat and encircling wall of land.[49] Bordone's atlas also included a quick sketch of Japan (a very inaccurate one) based on Marco Polo's description although Polo never made it to Japan. That map, including its inaccuracies, was the first Western detailed cartographic image of Japan.

Maps were also prominent in Venetian homes in the 1500s.[50] The idea was to show off how cultured and cosmopolitan one might be. Owning an atlas was also a plus and was made more feasible when civil engineer Giacomo Gastaldi (from Piedmont but working in Venice) used copper plate engraving to print the first pocket-sized atlas in 1548.[51] Apparently, this flaunting of maps in the home did not include maps of Venice because that would be all too familiar. And that makes sense. Even today, we can all appreciate an antique map of our city, but no one would take a road map from the auto club, press it out, and frame it. Maps were also used by the state to emphasize that Venice was outward facing, cosmopolitan, and a leader on the world stage. Maps of the world, and some of Marco Polo's journeys, were once frescoed on the walls of the Palazzo Ducale, the offices of the state.[52] In the same way, maps were used by Venetian general Antonio Barbaro in the late 1600s to decorate the Church of Santa Maria del Giglio. Incised in marble panels at the base of the church's exterior are reliefs of all the towns outside Venice where Barbaro led campaigns, showing off his accomplishments.

Mapmakers from the 1700s onward used a strictly straight-on aerial view. This view was a good trick since no one had ever flown over Venice at that point. These maps were also not as politically skewed, and are thus more reliable. The website The Venice Atlas visually makes this point. Digital cartographers decided to study maps of Venice and how they changed over the centuries. The website video blinks with one historical map of Venice after another overlaying in quick succession. This video shows that from about 1700 onward, the fishlike outline of Venice flops about a bit but essentially remains the same.[53]

Printing had a major effect on the consumer market for maps. Once printing became a major industry in Venice, the city published more maps than any other place in Europe.[54] There were some sixty thousand maps in circulation around Europe in 1500, and that number rose to 1.3 million by 1600. The printing press was the reason for such a massive increase.[55] Cartographic skills and techniques included copperplate etching, a process more fragile than woodblock printing but one that allowed for more finesse, especially in lettering. In the late 1500s there were between five hundred and six hundred copper plates of maps circulating in Venice. Borrowing, renting, and purchasing these plates were commonplace. This thriving business was only brought to its knees by the plague of 1575, when approximately forty-six thousand people died in Venice.[56]

As exploration of the world expanded during the Age of Discovery, places quickly appeared on the maps made by Venetians. For example, Vincenzo Coronelli, master globemaker, cartographer, and official "Cosmographer of the Republic of Venice," was the first to draw accurately the source of the Nile, and his map of the Great Lakes region of the United States was the most precise of the time.[57] Franciscan friar Coronelli founded the first geographical society, l'Accademia Cosmografica degli Argonauti, in 1705. But it's the Coronelli globes that stand as his greatest contribution for both their accuracy and their beauty.[58] There are several in Venice, including a pair in the Correr Museum, standing next to each other in a small room, not protected by a glass case but waiting for exploration. One is celestial, the other earthly, and up close they amaze in their detail and artistry. They also speak of other worlds, the desire for knowing, for adventure, and of the heart of the explorer. Focusing on the tiny boot of Italy, even closer on the dot that is Venice, and then spinning one of those globes, if only we were allowed to do so, that city would be lost in a blur of expanding possibilities, a world beyond what was then envisioned.

With their background of seafaring and trade, and their feeling that they lived at the center of the universe, some Venetians were prompted to move out from home to travel the world. In doing so, they found new places, learned about different cultures beyond their own, and expanded the art of trading goods into one of the most successful businesses of all time, one that runs the world today.[59]

VENETIAN EXPLORERS AND
WHAT THEY DISCOVERED

Marco. Polo. Marco. Polo. We all know this swimming pool game, a water-based blindman's bluff, where you close your eyes and try to find your mates by following their voices as they move about. This game is actually based on a very real Venetian who traveled to China in 1271. One version of the game's history says Marco Polo, as a young boy, wandered off asleep on his horse as his uncles' caravan headed for China. Waking up, Marco heard voices calling his name and he answered with "Polo." The other version says people decided Marco Polo had no idea where he was going, so they invented the game and gave it his name to honor Polo's lack of a sense of direction. Nonetheless, this pool game has made the name Marco Polo familiar.[60]

Marco Polo didn't get famous all on his own. In fact, it was his father, Niccolò, and his uncle Maffeo, not Marco, who went to China on a trading expedition before Marco was born. This means Venetians not named Marco Polo were the first Westerners to visit the Far East.[61] That trip made the Polo brothers rich, and after they returned to Venice and stayed for a while they set off again, taking seventeen-year-old Marco in tow.[62] The length of these travels, even where they went, was not all that rare for merchant traders who traveled for profit. In fact, the Polos had a house in Soldaia in the Crimea on the Black Sea, where many Venetians had trading outposts. The Polo family also had long-standing roots in Constantinople. What makes Marco's fame so great and so long lasting is his book, *The Travels of Marco Polo*, or as it was translated into Italian, *Il Milione*. It was well written, widely read, translated into several languages, and basically a hit. It's still read today. But it's not as if Marco Polo sat down and wrote his adventures about the twenty-four years he spent living in the Far East. Instead, Polo came back to Venice and took his place in

Venetian society, a forty-one-year-old merchant trader. A year later he joined in the fight against Genoa, the oft rival of Venice, and even paid to arm a ship which he then commanded. His ship was captured by the Genoese and Polo landed in jail, although some historians have suggested that he might have been under a kind of house arrest in a private home rather than in a cell. There, Polo told his life story to his jail mate, Rustichello da Pisa, who wrote it all down in Old French, the language of the literate. Apparently, that manuscript was copied, and translated into several languages, and it spread across Europe. Some historians think the work was not so much a conversation between Polo and da Pisa as a collaboration. In any case, the book was wildly popular, and so it received all sorts of criticisms, especially from fellow Venetians who were skeptical of what Polo claimed he'd seen, and modern historians who are still arguing about its authenticity.[63] No matter what, Marco Polo's book is the first "travelogue" and a book that changed the world. "It is no exaggeration," historian John Larner writes, "To say that never before or since has one man given such an immense body of new geographic knowledge to the West."[64] In that sense, Marco Polo invented travel writing and the popularization of travel adventure. More seriously, Polo's book brought the non-Western world to Europe, even if some Europeans doubted what they saw as fantastical tales. The book is also a merchant's book, full of goods that could be bought and sold, items that would do well on the consumer market, and the size of markets in every town the Polos visited. It even included the prices of everything and the profit margin for each good.[65] The book was also a geographical and anthropological tour de force. For example, Polo reported there were 7,448 islands in the China Sea and 12,700 in the Indian Sea (before anyone knew it was an ocean). He described habits and ceremonies, ways of dress and language, and just about everything else. He was the first Westerner to write about Japan, although he didn't go there. He and his uncles traveled farther and for a longer time than any geographer of the time. Their travels were the impetus for Columbus, Magellan, and the great cartographer Fra Mauro, among others. Polo was the first great Western explorer, geographer, and anthropologist, and clearly the first great travel writer.

Polo is universally known, yet two other early Venetian explorers have been ignored or doubted. As told by contemporary Venetian writer Andrea di Robilant in his book *Irresistible North*, Nicolò and Antonio Zen spent two decades in the North Sea around 1380.[66] The travel of the Zen brothers was then documented by a great-great-great grandson who made a narrative out of their letters and redrew their maps and published the account in 1558. According to that account, the Zen brothers explored the Shetland Islands, the Faroe Islands, the Orkney Islands, Greenland, Iceland, and made it to Nova Scotia, a hundred years before Giovanni Caboto (see below) planted the Venetian flag in Newfoundland. Clearly, the Zen brothers were all over the North Atlantic in the 14th century. But their explorations have also been subjected to various layers of doubt by other explorers, cartographers, and geographic historians. Some believe the Zen brothers never went anywhere, while others believe they probably "discovered" Canada. Their redrawn map of the North Atlantic was such a mess that it took di Robilant on his own journey to the area to figure it out. Unlike Marco Polo, there is no coherent travelogue, no tales handed down for centuries, just the idea that they must have been exploring far from home in an area unknown to most Europeans. They were involved in the life there, party to various battles, and made note of where they were and what they saw. And just like Polo, they were merchant traders in search of new goods and new markets.

Another Venetian was also a great traveler across land and sea and he, too, began as a merchant but was clearly waylaid by other things. There is no available information about the background of Niccolò de' Conti, only that he was Venetian and he began a twenty-five-year trip through Asia in 1419. De' Conti was more an anthropologist even before there was the discipline of anthropology because he stayed a long time and dug in deep, and he kept journals. For example, he made the effort to learn Arabic while staying in Syria and then learned Persian in Iran. Besides China, he also visited Vietnam, Malaysia, and Burma. He spent a year in Ceylon and Sumatra, where he learned about cannibalism, and nine months in Java. He was the first to describe the Sunda and Spice Islands in Indonesia. De' Conti also reported that the Venetian gold ducat, a currency of the city, was in circulation in India, which underscored the strength and long arm of Venetian trade.[67] In India, he noted both the wealth of gemstones and the practice of suttee, when a wife threw herself on the burning funeral pyre of her husband.[68] He went elephant hunting and wrote about tattoos. And he intrigued Venetians with his description of huge ships—junks, that is—that were four times the size of the usual European ships of the time. These ships, de' Conti said, were apparently built in self-contained sections so that when one part was damaged the ships could continue to sail.

De' Conti traveled about in local boats, to his credit as an explorer and so-called anthropologist, but being Venetian, this part of the adventure was probably nothing to him.[69] He returned to Venice in 1444. We know about his travels because the pope made de' Conti relate his travels (published as *Viaggi Fatti da Vinetia, alla Tana, in Persia, in India, et in Constantinopoli*) to Poggio Braccioline, the papal secretary, to atone for becoming a Muslim even though de' Conti said he was forced to convert.[70]

That combination of merchant trader and explorer was also apparent in Alvise Cadamosto, who worked for Prince Henry the Navigator, of Portugal. Cadamosto was an experienced sailor. Funded by Prince Henry, in 1455 and 1456 Cadamosto sailed down the west coast of Africa and discovered the uninhabited Cape Verde Islands and the mouth of the Gambia River. On that journey, Cadamosto made what is considered the first European observation of the Southern Cross in the night sky. He was also looking to gain from the slave trade. Cadamosto spent much of his life on the water and the rest of it trading people and goods. His book *Navagazioni* was all about his travels for Prince Henry the Navigator. His detailed descriptions of people and places in West Africa are probably the first of that time.[71]

Merchant and statesman Giosafat Barbaro also traveled extensively in the mid-1400s to Georgia, Poland, Russia, Germany, and around the Black Sea. In 1472 he became the Venetian ambassador to Persia. One of Barbaro's wackier moments came in 1437 when he conducted an archaeological dig aiming to find gems and gold in what turned out to be an ancient mound of kitchen garbage.

Giovanni Caboto, or John Cabot—a name every American schoolchild learns as the "discoverer" of North America—also worked for a foreign government. In his case, it was Britain. Cabot was born in Genoa but grew up in Venice and was a Venetian citizen. He had a hard time finding work but appealed to England's Henry VII for a "patent" to explore the Atlantic and find a northern passage to Asia. That patent was granted, and in 1497 he and a crew left Bristol, England, and headed west. Caboto hit Newfoundland and, not knowing that the Vikings had been there first, and thinking Canada was Asia empty of people, planted both the Venetian and British flags to claim the land.[72] Caboto was so convinced that he had made it to Asia that he made a second voyage to North America in 1498. This time he went farther south, presumably looking for some sign of the people and riches that the East were supposed to provide. Some suggest he went as far south as Delaware, or even

North Carolina. Of course, by that time Columbus had already landed in the West Indies.[73] Caboto then discovered the Grand Banks of Newfoundland, big news to fishermen, but he found no signs of Asia at all.[74]

Such grand explorations were at their height when Ferdinand Magellan, a Portuguese sailor, was hired by the king of Spain to look for an easier passage to the Spice Islands by going west, as Caboto had, but this time using a southern route. A Venetian scholar, Antonio Pigafetta, was on board one of Magellan's five ships that left Spain in 1519. In written accounts, Pigafetta is charmingly referred to as the "gentleman passenger." He was no sailor, but as a Venetian he was certainly familiar with riding around in boats. Apparently, Pigafetta's only job was to take notes of the trip; in doing so, he became the first person to ever write a journal of a long sea trip. This was no logbook or notes by a first mate of the condition of the sea and the direction of the ship. Instead, Pigafetta noted down everything he saw, including the behavior of various island peoples they encountered. And he drew maps.[75] Handwritten copies of his book *Relazione del Primo Viaggio Intorno al Mondo* were passed out to monarchs throughout Europe, and the volume was eventually printed.

That we have Pigafetta's account of Magellan's famous and world-changing voyage is just pure luck. Of the five boats that set out with Magellan, only one made it back to port three years later. Also, of the 140 men who began the trip, only eighteen survived. Magellan himself was killed in the Philippines. But Pigafetta was one of the lucky eighteen in the last remaining ship, and he just kept taking notes. As a result, everything we know about the first circumnavigation of the globe is from the notebooks of an adventurous Venetian scholar.[76] Pigafetta was there when the first Europeans navigated through the Straits of Magellan at the southern tip of South America, and still on board at the end of the trip, making him among the first people to circumnavigate the globe. It is also Pigafetta who famously said that the Pacific Ocean was so large that surely no one else would ever cross it again. Of course, he had some right to say this, because he and the Magellan crew were the first Europeans to see the Pacific Ocean. He was wrong about that prediction, but his detailed descriptions of other lands and their people are invaluable.[77]

The written works of Barbaro, Cadamosto, Polo, de' Conti, and Pigafetta, among many others, were included in the first ever collection of travel writing put together by Venetian geography enthusiast Giovanni Battista Ramusio (*Delle Navigationi et Viaggi*).[78] What made the collection so remarkable was not only that it contained these firsthand reports written by explorers but also that Ramusio produced three full volumes defined by geographic location. The first came out in 1550 and the last was printed after his death.[79] Ramusio was such a fan of foreign lands that he started the first school of geography in his own home.

Although these adventurers were widely traveled and their observations have been handed down through the centuries, in general Venetians were less interested in conquering places and sailing around for adventure than were people of other Mediterranean countries such as Spain and Portugal. In fact, Venice sent only one fleet a year to North Africa and none into the Indian Ocean when the route around Africa was clear. Instead, this merchant maritime republic held fast to routes they already knew in the eastern Mediterranean and the Adriatic. More than anywhere, Alexandria was their portal to the East, to India and the Spice Islands where they successfully sought trade.[80] So there was no real impetus to move beyond those boundaries or find new routes to the East.[81]

But then there was Giovanni Belzoni. A painting of this barber from Padua shows a very large man with a very large beard wearing a turban and flowing robes, an outfit intended to mark his adventures as an Egyptologist. After leaving Italy and his barbering profession, Belzoni lived in England where he made a living as the bottom strong man in a human pyramid. At some point in 1815, while traveling around with the human pyramid, he drew up plans for a hydraulic machine that might help Egyptians reclaim land. Belzoni then went to Alexandria to shop the device, but his machine was rejected. Having nothing better to do, Belzoni stayed in Egypt and became involved in figuring out how to haul a giant stone head from Thebes to the Nile for transport to the British Museum. While waiting for a boat to take on the big head, Belzoni went exploring as far as Aswan and the temple of Ramses II "discovered" four years before but still encased in hard sand. Belzoni, a "can do" kind of guy, put a local crew to work removing the sand but didn't pay them, and the project soon ended. He continued to travel about, often bringing his British wife, Sarah, with him, picking up Egyptian treasure including bronze vases, life-size statues, a carved sarcophagus of Sethos I (it now rests in a museum in London), and an obelisk that was helpful for deciphering hieroglyphics. Besides Belzoni's more commercial pursuits removing and selling priceless antiquities, he also made some major discoveries. For example, the paintings in the tomb of Sethos I are among the most glorious and colorful ever uncovered in the Valley of the Kings. He also found the tomb of the pharaoh Ay in the Valley of the Kings and carved the words DISCOVERED BY BELZONI 1816 in the stone lintel over the entrance. He did the same when entering the inner rooms of the pyramid of Khephren at Giza in 1818 even though someone else had already been inside. Belzoni became so famous in his time that the British made a coin with his likeness, and some claim that Belzoni was the model for avaricious archaeologist Indiana Jones.[82] That comparison is based on the fact that Belzoni's trips to North Africa were not just about adventure, or even knowledge, but also about a race to uncover, confiscate, and then sell priceless treasures from the tombs of pharaohs.[83] But that's not all. Belzoni also wrote a long narrative of his travels and observations, noting places of archaeological interest and local customs, habits, and attitudes. Altogether, and putting his adventures into the context of the day, Belzoni was one of the first accomplished Egyptologists and certainly the first Venetian Egyptologist.

HOW LIFE ON THE WATER INVENTED THE FACTORY LINE

Frederic Lane, historian of Venetian ships and shipbuilders, has made the point that Venice was not just a maritime republic; it was, and still is, a nation of boatmen (and boat women).[84] The intimate relationship between Venetians and water can be seen in the hundreds of types of boats that have enlivened the city's history over the many centuries, and many of those types are still a vibrant part of Venetian life today.[85] For example, when Venice was actively trading goods up the rivers of the Po Valley they used flat-bottom barges to transport the load. Today, descendants of those barges are alive and well motoring down the Grand Canal delivering everything from washing machines to fruit. A lifetime on the water has also produced, by necessity, feats of civil engineering, such as rerouting rivers to avoid the build up of silt, and the invention of the first seawalls on the barrier islands to the east of Venice proper. There had been wood barricades there to keep the Adriatic at bay, for the same reason that rivers were a problem—silt. Those wooden walls were, however, inadequate, so new ones were built of stone. Construction of these seawalls, called *murazzi*, started in 1716 and took thirty-seven years to complete. They remain today, fourteen feet above sea level, forty-six feet wide at the base.[86] From the land side, the *murazzi* on the island of Pelestrina look like simple walls along the roadside, but a quick walk up stone steps to the top reveals that the Adriatic is right there, blasting its aggressive waves and howling winds against these walls that still hold. Historian Alberto Cattaneo says the *murrazzi* are "the crowning glory of the Venetian era,"[87] and so it feels standing on the top looking at the vast expanse of the Adriatic, and then turning around and seeing peaceful and protected Venice far off inside the lagoon.

Other significant inventions from Venice can also be attributed to all that water. In the 1600s Venetian physician Santorio Santorio built the wind gauge and the water current meter, both proving essential in Venice and around the world. Other creative minds turned to Venice as a naval power often under siege, directing their inventiveness to the art of war. In 1378 the Venetian navy was the first to mount gun powder weapons on ships during a war with the city of Chioggia to the lagunal south. Two hundred years later, Venetians were the first to place guns one right next to the other on a ship's deck and fire broadside at pirates and enemies during the Battle of Lopanto. In 1531 that maneuver had been perfected by Venetian engineer Niccolò Tartaglia who used mathematics to figure out the path of cannon-balls. His calculations initiated the science of ballistics. Sixty-four years after Tartaglia, Galileo Galilei, professor of mathematics at the University of Padua, invented the sector, a military compass designed to solve easily any mathematical problems in gunnery, ballistics, surveying, and navigation.[88] All these calculations were enhanced and made possible by the instruments invented by Venetian mathematician and astronomer Bernardo Facini, who constructed a logarithmic calculator in 1714, and many other tools used to efficiently measure angles and such. Facini is also famous for constructing a very complex astronomical clock, a *planisferologio* housed in the Vatican Museum today.[89] In 1738 Venetian admiral Angelo Emo, veteran of many a naval battle, invented the floating battery, a raft carrying a pyramid of cannonballs that effected a powerful broadside as the raft crashed into pirate vessels in shallow waters.[90]

In a moment of genius that likely looked like craziness, Leonardo da Vinci drew up plans for a "diving suit" in 1500 to help Venetians in their water wars. His design suggested that someone could wear a skin of leather with a sort of floating bell on the surface that would supply air into a face mask. The idea was that an army of soldiers could don these outfits and surprise the enemy by walking across the bottom of the lagoon and sink ships by boring holes in their hulls, or maybe by using explosives. Although his idea was brushed aside as too kooky, a modern reconstruction and water testing in 2003 showed that da Vinci's idea would have worked just fine. In 1609, Galileo was more successful with his revamp of the refracting telescope when he presented it to the doge and his council. Together, so the story goes, this group went up into the *campanile* and looked toward the Adriatic. It was obvious that the telescope would be a very nice asset since it would allow Venetians to spot the enemy as soon as they entered the lagoon.

But the most influential invention born from Venetian waters is actually the site where boats are built. Contemporary maps of Venice show that the eastern end of the city is dominated by two things—a very large public garden and an even larger blue square surrounded by a wall with a watery arm leading out into the lagoon as if pointing toward the Adriatic. In fact, this complex is a shipyard that has been around since 1104. Today, it houses a contingent of the Italian navy, some giant warehouse spaces for art exhibitions, a café, and who knows what else, since most of the buildings are now closed off from the public. Taken together, the entire area is called l'Arsenale in Venetian (or "the Arsenal" in English) after the Arabic *dar as-sina'ah* ("house where things are made" or "workshop"). Long after Venetians appropriated this Arabic phrase for their shipyard, Italians took it on and used the word for a place where munitions are held, hence our current understanding of "arsenal" as a place of guns, and a lot of them.[91]

The primary entry gate to this structure is iconic for Venice. A canal leads north from the much larger and heavily trafficked Giudecca canal toward a brick complex with two stone towers at the entrance that make it look like a watery castle entrance. There, a high pointy wooden pedestrian bridge crosses the canal right in front of an ungated opening into the complex. Today's main walk-in street door is over that wooden bridge in the small *campo*, or city square, on the left, where several very large white marble lions sit or lie about. Back at the water entrance, the canal that leads into the Arsenal eventually empties into the lagoon at the north. This canal was once at the heart of production and is called the Canale delle Galeazze because ships called galleys were pulled through the canal for outfitting. A sharp water right after the main gate leads to a massive basin of water

(Darsena Grande), the center of the shipyard, a place for building, repairing, and launching many a Venetian ship.

All in all, the Arsenal looks like a fort with its three miles of high brick walls topped with serrated sections encircling the complex. From any angle, the Arsenal is imposing, as it's meant to be. This structure began as a place to store naval goods, including munitions, used by the Venetian merchant fleet that sailed along the Adriatic and across the Mediterranean on trade routes.[92] At that point, in the 12th century, the complex was only about eight acres, but it was still the largest, and first, industrial complex in Europe; its function and sheer mass predate the Industrial Revolution by six hundred years. Over time, and a switch from merchant ships to warships, the Arsenal grew to the sixty acres that it encloses today (that's about 15 percent of the land of a tightly packed city). Naval historian of Venice Frederic C. Lane has described how the Venetian state was in charge of large merchant ships, trading vessels known as round ships often built in private shipyards and not the Arsenal.[93] But by the mid-1400s, the state was building and servicing warships in the Arsenal, a move brought on by the threat of the impressive Turkish fleet.[94] At that point, the government of Venice moved from only building and operating merchant ships that they rented out to traders to becoming a builder of ships of war. Previous to this switch, the Arsenal was more a warehouse and repair shop, but it quickly reoriented to become a state shipyard pointedly directed toward building huge ships that could defend Venetian interests all over the Mediterranean. Venetian shipwrights were especially devoted to the galley, a large but lithe ship that was designed as a rowed boat but soon had sails in addition to man power.[95]

But the most significant contribution that the Arsenal gave to posterity was the very system that manufacturers use today—the factory line was first initiated there in 1320 and was still operating five hundred years later during the mid-18th century when the Industrial Revolution began. This system is used today across the world. The process came about because Venice was a port that made its living on the sea, and their efficiency and success also put them at risk from pirates, rival cities, and countries. Venetians knew what it took to build ships in their water-based republic (that is, how to bring in all the resources from the mainland, especially lumber) because their livelihood and survival depended on it. Once at war, they had to make that process faster. Venetian ship designers had, from the outset, adopted the keel-and-ribs construction of ships, instead of the more traditional way of building the hull first and then fitting in the keel and ribs. This design made it possible to piece together the outline of a ship, like a prefabricated house, then pull it alongside the warehouses organized by type. Starting at one point in the line of warehouses, the put-together hull could be pulled down an interior canal past buildings where ropes, sails, shipboard food, and the like could be easily tossed on board. As such, this complex of warehouses full of parts served the republic well when warships were needed in a hurry—they could be assembled and launched right on site.[96] At the end of this assembly line, a fully fitted galley could simply be rowed out of the Arsenal and into the lagoon.[97]

There was also a chain of command in the yard that functioned well and was a model for how factories operate today. Given that the Arsenal was a state institution, there was managerial oversight. It began with a panel of three nobles, aristocratic civil servants required to serve, called the *Patroni all'Arsenale*, with an experienced admiral in charge of daily operations. Eventually the government added three *provveditori* once the Arsenal took its place as a grand shipyard. Sometimes, these six oversight positions are translated as "The Lords of the Arsenal," but they came with all sorts of obligations, such as living close by and being present during operations. The state could also perform the magic trick of turning a merchant vessel into a warship when needed, and all Venetian merchant sailors knew how to be armed merchant marines. In other words, trade and war on water were highly regulated and owned by the Venetian government.

In general, the Arsenal was run like a tight ship. Shipbuilding and outfitting were concentrated within and right outside the Arsenal walls. In fact, the community area around the complex was inhabited by the craftsmen and women who worked inside, like a factory town of the much later Industrial Revolution.[98] Workers at the Arsenal, especially carpenters and caulkers, were considered highly skilled and they were well paid and proud of their part in keeping the city a maritime republic.[99] At one time, there were between two and three thousand individuals, called *Arsenalotti,* employed at the Arsenal. There were also employee benefits. Included in the state budget for the Arsenal was a vast quantity of wine, apparently a major perk, or simply an expectation, like a smoke break, for workers at the Arsenal.[100] Skilled workers also received pensions.[101]

Remarkably, the Arsenal has not been torn down to make way for hotels or become a camping ground for weekend trippers, nor is it bustling with tourists. Instead, the complex, even in its current half-empty state, stands as one of the most powerful faces of the Republic of Venice, part of the city's identity in the past and now. Venetians have always thought of themselves as a cooperative collective born of water and necessity, and there it is, enclosed in a brick wall.[102] The Arsenal and how it was built, why, and how efficiently it ran, and the fact that it's still there, are a grand testament to the lasting identity of Venice as a maritime republic, an identity Venice has not been able, or more probably has never had the desire, to shake. There is, in fact, a very real and very current contract between Venice and the sea. It began in 1177 and is renewed each year during the Feast of the Ascension (*Festa della Sensa*), a major holy day in early June in Venice. The annually repeated ritual is called "the Marriage of the Sea" (*Sposalizio del Mare*). It began when Pope Alexander III thanked Venice for support during a war against the Holy Roman emperor. The pope stood in a boat with the doge, took a ring off his finger, and gave it to the doge to throw. That ritual morphed from a thank-you gift into something completely Venetian: although the pope part has been forgotten, the marriage ceremony is repeated each year with a parade of boats rowing to the point where the Adriatic meets the lagoon to watch the mayor of Venice toss a gold ring into the water.

CHAPTER FOUR

A Most Serene Community

The basic difference between Venice and the other Italian city-states was the greater unity and solidarity of allegiance at Venice.
—Frederic C. Lane, *Venice; A Maritime Republic*
1973

Nonetheless, Venice does offer a central story, one that acts as a constant in the ever-changing complexity of its history. It is a story of Venetian stability.
—John Martin and Denis Romano,
Venice Reconsidered
2000

Because of the scarcity of space, people exist here in cellular proximity to one another, and life evolves with the immanent logic of gossip.
—Joseph Brodsky, *Watermark; An Essay on Venice*
1992

You have to be born in Venice to understand why anyone lives here.
—Spoken by a Venetian character in the TV
production of the Donna Leon mysteries,
which take place in modern Venice
2018

SEVERAL YEARS AGO, while spending a month in Venice, I took my young daughter by the hand and we walked three minutes to the Rialto Market to buy some parsley. I was making spaghetti carbonara that night, and we needed that little bit of greenery to make the perfect dish. We asked our usual vegetable seller, who happened to be magnificently beautiful (we had privately bestowed on her the title Madonna del'Orto, (the name of a famous church in Venice which means Madonna of the Vegetable Garden), for a sprig of parsley. Since I was always practicing my Italian, I added that we were making spaghetti carbonara and needed the parsley for that. To my surprise, our Madonna began to scold me. *"No. No. No. No prezzemolo* (parsley)," she said emphatically, wagging her index finger back and forth at me as if I had committed a really bad sin. "This is Venice," she continued. "When you are in Venice you make carbonara as it should be done. As it is done in Venice. *No prezzemolo.*" Instead, she instructed, I had to go buy these tiny containers of very thick cream and put that in the recipe, which I did because, well, I was in Venice and wanted to do the right thing, the Venetian thing. A few days later I was telling this story to an American woman who had retired to Venice with her Venetian husband, and she nodded sagely and explained, "Ah yes, this is the Venetian Republic. It always has been and always will be. No matter what the rest of Italy thinks or what the tourists think, this is the Venetian Republic."

The real Venetian republic might have officially ended in 1797, but this city is still a strong community with the closely held identity of a clan. And that Venetian identity is played out even in a recipe for spaghetti carbonara.

WHAT MAKES A COMMUNITY AND WHY WE THINK COMMUNITY IS A GOOD THING

The modern Western definition of a community is a group of compatible people living in the same place and having something in common. As anthropologist Tim Ingold suggests, it can be a group of people joining to form a life together, an inter-personal combination that requires attention to work efficiently and well.[1] A community can also be defined geographically. Small towns and villages are obvious communities with many shared beliefs and knowledge of each other. Under this rubric, community is of small size, physically close, and interpersonally interactive. The contemporary notion of a community also includes groups that are bound by a shared interest. Hobbies such as knitting or scrapbooking, or service to organizations, are good examples. These days, the internet has also enabled this type of community to become both global and huge. Why do people feel drawn into allegiance with one or more communities?

Joining together is an evolutionary adaptation. Anthropologists speculate that early humans lived in small kin-based groups hunting and gathering in Africa, and we stayed that way for more than 95 percent of our history. Grouped together, our ancestors were less likely to be eaten by other animals and more likely to learn from others about a food source or get help caring for dependent human children. At a fundamental level, humans are a social species that depend on others for staying alive. The very fact that solitary living today increases the risk of mental illness and lowers life span and quality speaks to the deep roots for human sociality.[2]

Within that evolutionary model of community is also the drive for identity. Anthropologist Ruth Benedict understood the idea of identity better than anyone else. In her classic 1934 book *Patterns of Culture,* she wrote, "No man ever looks at the world with pristine eyes. He sees it edited by a definite set of customs and institutions and ways of thinking."[3] Today we anthropologists call that cultural indoctrination a belief system, something that subconsciously forms our thoughts about everything, especially who we are. That identity is crucial because it forms how we see ourselves and how we fit into the larger scheme of life. There are also the identities that others impose on us, and those are the ones that cause so much trouble, both when they are true and even more so when they are untrue. In Western culture, we assume freedom can be measured, in part, by the ability to define one's own identity and this closely held belief underscores the right to an identity. Community, and identification with that community, then, are natural and necessary human attributes.

These concepts of community and identity have played out in Venice for over the past twelve hundred years, and the Venetian perspective tells us a lot about how we think of the costs and benefits of the community today. Venice was one of the most tightly knit communities of its time—or any time—and it was a community with an unusually strong and stable identity. Venetians also figured out ways to enforce (and manipulate) that sense of being Venetian, and in doing so they invented our modern concept of community and our modern dream of what it means to have a stable and useful identity.

THE COMMUNITY OF VENICE

Venice is a naturally born community. Although Venice began as a group of 144 islands, there were eventually pedestrian bridges connecting these islands.[4] And in any case, people got from place to place by boat. Venice is also geographically a small town, one that could never really expand outward (although there has always been a lot of landfill going on). Its 102 acres (160 square miles) can be walked east to west in two or so hours and north to south in an hour. It is also a tightly packed labyrinth of houses, pedestrian streets, and canals, outlined on all sides by an expanse of water. There is nowhere to escape, no place to run, nowhere to hide, and no suburb to move to.

By definition, this city is also about people interacting and talking with others, or watching others. As Italian writer Luigi Barzini has said about Italian cities in general, this is expected:

> There are usually café tables strategically placed in such a way that nothing of importance will escape the leisurely drinker of *espresso* or *aperitivo*. Reserved old ladies peer unseen at the spectacle through the wooden slates of green-painted blinds....There are free benches of little walls in the sun for the elderly, there are balconies along the façades of all houses, as convenient as boxes at the theater. You can place an armchair there, or stand leaning on your elbows, and see the days, the years or your whole life go by, as you watch a cavalcade with thousands of characters and hundreds of subplots.[5]

Venice is Barzini's description on steroids. There is no place to sit and do nothing without someone or everyone knowing what you are doing unless you are inside your residence or sitting in your palazzo's walled garden. One historian suggests that in the past there was, in fact, little distinction in Venice between public life and what we would today call private life. Privacy has always been hard to come by there.[6] Instead, Venice is a place made for one-on-one close encounters all day, every day. It's a maze of slender streets lined by buildings several stories high that often cut out the sun, so close to each other that neighbors on the upper floors can toss an onion back and forth. This way of life is anathema to places like America where everything, including the way city centers and suburbs are designed, is about efficiently getting in and out without talking to anybody. In Venice, even if you know no one, even if your language skills are nonexistent or minimal, you can still be part of the show.

Back when Venice had a population of over 100,000, it was also a place of social chaos not often seen in modern cities today. The streets were crowded, and neighborhoods were abuzz with human activity. Since many businesses were run out of homes, or in buildings where people lived above their shops, workshops, and warehouses, there was a constant coming and going of everyone. Apartments of non-nobles were often small, and people entered the homes of others with ease.[7] Historian Monica Chojnacki writes that Venice was a place with "permeable" social boundaries, and she means social and interpersonal boundaries.[8]

Maps show Venice as divided into six communities, or *sestieri* (San Marco, Dorsoduro, Santa Croce, San Polo, Cannaregio, and Castello) but Venetians also recognize the local parish as their neighborhood. There are about seventy parishes in Venice. While the family was of ultimate importance in other Italian city-states, the parish was also an important identifier.[9] The parishes had their food markets, stores, workshops, and a *campo* with a well where everyone met up every day.[10] And the parish had a communal baking oven where women gathered together. It seems that these parishes were not so much about religion but in the service of society—that is, they made for an identity.

Even more interesting, there was never any real separation of the rich and poor in Venice, and that too can be seen today, with palazzi sitting right next to humble edifices.[11] Rich and poor also had to work together every day, with the poor servicing the rich, the rich needing the poor as employees, and everyone directly or indirectly involved in the mercantile business that was Venice.[12] Most of all, the population density meant no one could get away from anyone else. Nobles walked the same streets and rode on the same canals as the middle class and the poor; nothing physical separated people off from each other. The pace of trade also made for a hustle and bustle of life, as a sort of "city-world" within a very small space.[13]

The government of Venice also enclosed the city in a bubble of protective oversight that fostered togetherness and loyalty. For example, from 810 A.D. onward, the very critical hydraulics necessary to maintain the lagoon and the city were under the auspices of the city government.[14] In 1224 the government established a separate ministry for monitoring the conditions of the various channels dug in the lagoon. Three hundred years later, in 1501 they established a zoning board, the *Magistrato all'Acque*, for hydraulic issues. Also, Venice seems to have been the first city with free, unlimited, and clean public water for all citizens. Evidence of this early community service can still be seen across Venice. Many of the *campi* still sport a squat stone or marble wellhead that anchors the center of the *campo*. There were once over 6,500 of these wellheads across the city in the late 1800s, but only 230 remain. They are now artifacts, locked, sealed, and no longer functional, because as of June 23, 1884, fresh water arrived via aqueduct from the Brenta River. But these wells were once a service to the community, and an important one in terms of health and social life. Each wellhead tops a brick shaft that leads to a huge underground cistern filled with various layers and types of sand. Rainwater and building runoff entered the grates set out from the well.[15] The water flow then passed through the various sand layers, filtering out all debris and seeping toward the shaft, where it was drawn up in buckets. The importance of these wells is illustrated by the decorations, religious symbolism, and family crests that adorn many of them. These wells were also the real hub of local life in the parish community. Back then they served as community watering holes, and today they still serve as meet-up places in a geographically confusing city. Even when there was no rain and water had to be ferried in by barge from the mainland, the state protected citizens by regulating the price of water.

Governmental concern for the general safety of citizens was also evident in the first-ever city ordinance mandating street lighting, as mentioned in chapter 2. That's why today every single street in Venice is lit at night and why it's an incredibly safe city.[16] Citizen safety was also addressed after a spate of murders in the city, and in 1450 a new law required that anyone out and about after 3:00 A.M. had to hold a lantern aloft.[17] In 1284 Venetians also enacted the first child labor laws to protect children from working in the more dangerous areas of glass production on Murano. Those laws were revamped in 1396 and extended to exclude children under the age of thirteen from certain other dangerous trades. Although today we abhor children working where their lives and health might be at risk, the Venetian laws were ahead of their time, the only child labor laws in Europe until the middle 19th century. In 1291, the city ordered glassblowers to move to Murano because of the threat of fire in such a contained space. The move was the first forced exodus of a particular trade to protect a municipality. And in 1443, the republic proclaimed it would provide an attorney for anyone who could not afford one, making Venice the first city or state to provide public defenders. Presumably, this act reflects the general Venetian attitude of social fairness.

All around the city today are slabs of stone carved with long lists of dos and don'ts. They are titled *Il Serenisimo Prencipe Fa' Saper* and were put there over time to make clear to whoever lived nearby that there were common social rules they must obey.[18] They covered items such as gambling, the state of canals (as in don't throw anything into them), and the rules of business. Punishments were also explained. Other stone placards still in place clarify the standard rules, according to the republic, for such things as the amount of flour to be used in making bread or the correct way to size a fish for sale. In general, but certainly not for everyone, these policies made for an efficient and pleasant place to live—the rules were clear, understood, and applied equally—and thus it was a community that functioned well.[19] The same sort of signs, but not carved in stone, decorate the city today. But these days, the proclamations tell visitors, not citizen, that there is no siting in Piazza San Marco and don't feed the pigeons, for example. Spending time in Venice now, it is also clear that the public visual proclamation is alive and well. Recently, after the crash of a cruise ship into another tourist boat, it took only one day for the appearance of a poster, as if by magic, on walls and buildings all over town, announcing a protest march the following week. Even more surprising, the day after the protest, eight to ten thousand marchers strong, those posters were covered over by other posters thanking the participants. That act must be the only example of an organization—in this case, the *No Grandi Navi* (No Large Cruise Ships) movement—delivering a thank-you note to followers, especially under the public eye. Other posters about events, art shows, plays, and concerts are pasted up on dedicated walls, city trash cans, and on boards that enclose construction. These ever-changing posters are so much a part of Venice today that there is really no reason to look up "what's on in Venice" when you arrive because the list is right there on every street. This kind of visual announcement works, of course, because everyone is walking everywhere rather than looking out a car window.

THE IDENTITY OF VENICE

Venice, as a community and a body politic, has also worked hard to form an identity that sets Venetians apart from others. Over the centuries, Venice played the game of maintaining an exclusive community with a special identity; at times this was confusing because they also depended on global trade. For example, the merchant galleys were owned by the state and only Venetians could be on board, but they traveled throughout the Adriatic and the Eastern the Mediterranean and as far as England and Flanders buying goods from and interacting with other cultures. In its early days, Venice ruled that foreigners could not live in the city, an act that was repealed presumably because it threatened Venice's trading network and might have seemed overly restrictive given Venice's global reach. In general, foreigners were suspect because loyalty to Venice was paramount and these "outsiders" presumably had other allegiances.[20] Foreign dress was encouraged, and that seems like an inclusive and liberal policy, but as Gary Wills points out, that apparel also announced a person's foreignness.[21] Venetians were also not allowed to sell their boats to outsiders unless that boat was worn out or in bad shape.[22] And they held their manufacturing secrets close. Nowhere was that xenophobia more apparent than in the glassmaking industry. By law, glassblowers were not allowed to share their methods with non-Venetians, and glass workers who left Murano for factories on the mainland were drummed out of the guild. Some artisans were even threatened with death in absentia if they left the city permanently or shared techniques with those outside Venice.[23]

Venetians also conspicuously imprinted their brand across the Mediterranean. The lion of St. Mark, a fierce beast with wings and a book in his paw, is the proud symbol of Venice. When Venetians aggressively captured ports on the Adriatic and elsewhere, they permanently emblazoned that lion of St. Mark on stone bridges, building, bell towers, and forts, and those lions remain today as a reminder of the reach of the Venetian Empire.[24] At the same time, Venice had no real interest in conquering anyone or taking their land. They only wanted safe ports for their trade ships, and they wanted to make sure the populace knew who was in charge. That symbol has lasted like no other brand. The lion of St. Mark is plastered all over the city on buildings, doorknobs, jewelry, and tourist stuff. That leonine figure is, of course, also the center of the emblematic deep red and gold Venetian flag which waves over museums and private homes. Branding on such a large scale might be familiar today in commercial marketing, but this small, powerful city used that marketing technique starting in the Middle Ages, and it has had a lasting effect. Maybe the gondola is the symbol of Venice for tourists today, but that lion, seen everywhere across the city, is the real logo of Venetians. It carries the former power of the republic and the pride of modern Venetians in its very paws.

Venice reinforced its identity in its public buildings such as the Palazzo Ducale, the Mint, and the libraries and various iconic bridges. But the most identifying space in Venice is Piazza San Marco, which used to have an orchard, was once full of market stalls, and today is trampled by tourists. The piazza remains the largest square in Venice, and it has been used over the centuries not only to impress visitors but also to let the people of Venice know exactly who they are, the inheritors of all that beauty and history. The bell tower, or *campanile*, was first built in the 9th century as a watchtower. It reached its current height in the 12th century thanks to Nicolò Starantonio Barattiero who had to figure out how to bring material up the bell tower and finish the job. In 1160 he put together a series of wood boxes and pulleys that accomplished the task, inventing the first elevator. A loggia was added in 1549, and the tower was completely rebuilt when it fell over in 1902. This major symbol of Venice was also used to coordinate citizens over the years. Venice ran on time because the daily chiming of the *marangona*, the bell in the bell tower, opened and closed the workday with a clang. Other bells announced noon or an execution. That sound was eventually enhanced by the city clock on the north side of the piazza. Although its major face looks like a simple astronomical clock, it was the most complicated one of its time. This clock also kept accurate time with real numbers and soon ushered Venice into regulated "scientific" time for all citizens.[25] When Luigi di Lucia restored that clock in 1858, he added numbers on drums that flip forward every five minutes, thus inventing the first digital clock and the first public digital city clock. And it keeps accurate time today.

This special identity of Venice was underscored by its various myths (as the origin myth in chapter 2 explains) over the one thousand years of its independent existence.[26] First came the myth that Venice was an "ideal republic, a strong maritime empire, and an independent state in which the Venetian nobles were devoted to the ideals of civic humanism and the commercial virtues of sobriety, hard work, and self-sacrifice."[27] That eventually devolved into the Venice of low morals focused on fun times. The opportunity for visitors to exploit Venice at that point was elaborated by foreigners who were shocked by Venice's wealth and power and then reveled in, and romanticized, its economic decline and apparent decadence in the 17th and 18th centuries. They showed up in Venice during their Grand Tour, taking in the art and music but also engaging in the gambling and loose morals that seemed to make Venice a "wild" place. The fact that there was less money around to maintain grand buildings simply added to the aura of decadence about the city. No matter the century, Venetians lived by the myth that they were a special people, destined to own the lagoon and destined to be a sovereign nation called The Most Serene Republic (*La Serenissima*).

Now we are living in a Venice that appears to exist only for tourists and in their view as a city existing only in the past. That myth is exemplified by a gondola ride down the Grand Canal. But the current myth lived by tourists in Venice is also laid bare by the scores of Venetians going to work and school, pushing through tourists simply to get on with their lives.

Venice has also kept its identity through the centuries with various rituals, processions, and events.[28] Some were and are religious, such as the day celebrating the ascension of Mary into heaven. Others honor the Virgin for sparing the city from the plague, even though at least one third of the inhabitants were dead after the plague of 1576 and there were more plagues to come. The close contact between Venetians that fostered such social cohesion and conviviality, along with the free flow of creative ideas, also fostered the spread of disease in time of plague. Other rituals include the *Festa della Sensa,* when the doge, now the mayor, throws a gold ring into the sea to symbolize the marriage of Venice and the water.[29] Carnival used to be a big party. It started in the 12th century and eventually attracted much of Europe with six weeks of sexual freedom in the 18th century, but that all died out along with the republic in 1797 when Venice joined formally with the rest of Italy, making the peninsula not just a group of city-states but a nation. Today's revived tourist-enticing version is there solely to bring money into the city, and this carnival lasts about a month. These rituals make for superficial but universally iconic images that shout "Venice" even to people who have never been there—the gondola, the carnival mask, the powdered wig, fancy dress, and that fierce lion. On the more local scene, Venice also holds fun and free neighborhood festivals; they pop up all summer long across the city. Some festivals celebrate the harvest of wine on an outlying island, while others are all about food and music. And there are any number of regattas that praise antique watercraft. Venice was also the first city to hold a regatta and the first women's regatta, in 1569, and those races still occur regularly in Venice.

Venice has also preserved its sense of self through language. *Veneziano* (also known as Venet and Venetian) is not a dialect of Italian but an actively distinct, living language that stands on its own.[30] It now contains bits of Latin, Italian, and English but is still decidedly Venetian, that is, incomprehensible to everyone but those from the Veneto, the region that includes Venice. This language was once the lingua franca of the trading world (supposedly half of the Greek spoken today is *veneziano*[31]), and one might think it has faded from use since then. But it lived on as the language of the Austrian navy until 1918 because that navy was composed of Venetian sailors who worked for an inland country that had occupied Venice from 1798 to 1805 and for a year in 1814. Venetian is still spoken all over Venice and other cities and towns of the Veneto, and at various enclaves around the world where Venetians emigrated, including parts of Croatia, Brazil, Argentina, and Mexico. At least four Venetian words have become international. *Ci sciavo* (I am your servant or slave) elided into *ciao* for hello and good-bye.[32] Another is *ghetto*, a word that initially meant the former brass foundry area where Venetian Jews were sequestered in 1516. This word has been appropriated in our modern times to denote a neighborhood in a city that is ethnically insular and overcome with poverty and all its ills. We assume *ballot* is a French word but it's Venetian. *Marionette* is also not French but Venetian, and it refers to giant wooden dolls called *Marione* crafted for the *Festa delle Marie* which celebrates the return of twelve virgins who had been kidnapped for their jewelry in 844. During the festival, enterprising artisans crafted smaller versions for sale and they were called *marionettes*. The word *sequin* is also Venetian and is based on the shiny Venetian coins called *zecchini* which were widely used in Europe. Also, according to Lodovico Pizzati, author of a Venetian-English dictionary, Venetian has seen a resurgence because of texting.[33] There is all kinds of eliding that happens in Venetian which makes it the perfect shortcut language for texting.[34] Those from the Veneto now appreciate their language for its precise meaning, often elided words, and the fact that it comes close to being a secret langue that can only be understood by family and trusted friends.[35]

CROSSCUTTING CLASSES AND GENDERS

Although history books often focus on the monied and noble, the citizenry of Venice was, out of economic necessity and geographic familiarity, tightly interwoven. As historian Gary Wills puts it, "Venetians of all classes threaded their way through each other's lives and living spaces."[36] At its height, Venice had about 150,000 citizens (today it has 50,000). It's not hard to imagine the city as a beehive, with everyone buzzing about, everyone with a role but all of them connected for a singular purpose. That purpose was trade and money, and in that sense, Venice was unusual. The social and economic classes interacted on an hourly basis because they were entwined in trade. There was lots of unpleasantness, some oppression, and surely outright crime in the process,[37] but Venice might have been the first city to fully understand that for a place to run smoothly, everybody of every socioeconomic class needed to join together for a common goal of prosperity. In that sense, Venice might be the first modern socialist economy.

But Venice was certainly not a classless society.[38] Ninety percent of the population were *popolani*, a group that included shopkeepers, dock and ship workers, domestic servants, and artisans. The *popolani* also included the very poor, prostitutes, and beggars. Higher-ranked individuals were *cittadini*, with subcategories of *cittadini originarii*, whose ancestors were born in the city, and the rest from lineages arriving from somewhere else.[39] The *cittiadini* comprised only about 5 to 8 percent of the city, but they pretty much ran the government as civil servants. The nobles, only 4 percent of the population, claimed their status from families with long histories of living in Venice. These people were not born of kings, and no one had any real title beyond *cavaliere* (knight). Unlike patricians in other places, Venetian nobles built their wealth by trading so they directly interacted with manufacturers, sailors, shipbuilders, dock workers, and those who took their trade goods upriver to the rest of Europe.[40] Members of this class usually, but not always, had money, but what they all had was governing power because all nobles were required to work in the government for free. Every young noble was automatically inducted into the ruling body called the Great Council (*Maggior Consiglio*) at age twenty-five, and the doge and other powerful councilors came from their numbers. As part of the ruling class of Venice, the nobles also held positions outside Venice as ambassadors and *provveditori*—governors of ports, islands, and regions of land on the mainland.[41] But being noble was a rather unclear designation, so in 1297 they voted among themselves to make their status official. This move was called the *Serrata*, or "closing," and what was closed was the governing body, the Great Council. After the

Serrata, only sons of "true" nobles would be allowed to serve in the Great Council, but that denomination was still a vague concept. Since there were no royal lines in Venice, and no king had knighted anyone, how could this so-called noble class be delineated? They drew the lines themselves in 1315 when they passed a law that put all the names of 142 original families (although plenty of non-nobles also had those last names) into a book known as the Gold Book (*Libro d'Oro*) and then claimed these were the only true Venetian nobles.[42] Only their sons would thereafter be registered as nobles and admitted to the Great Council. To be listed in that book, someone had to prove their status with documents of paternity and maternity.[43] This act of regulated exclusivity, some suggest, was the beginning of the end of Venice as a great power because it froze their classes and thereby made an issue of what had been an easygoing and flexible society. That social flexibility actually encouraged inventiveness in business and trade, making Venice a meritocracy rather than a place where only the upper class could succeed. After the *Serrata*, there was no way for non-nobles to improve their class designation even if they had all the money in the world. Economic historian Diego Puga claims that with the *Serrata*, Venice lost the "institutional dynamism" that was responsible for much of its success and the efficient and collegial running of the community.[44] The *Serrata* also turned a place that favored community or parish over family into a place that elevated one class based on heritage, and not just money, since many of the nobles were broke.[45] Nobles also became rather afraid of their actions and power. In an unusually insightful move, they reigned themselves in with sumptuary laws that controlled displays of wealth and laws about excessive gambling,[46] and would even be the root of why there is a law requiring that all gondolas be uniformly painted black.[47]

Class identity for Venetians was a tricky, and often fluid, subject. For example, it was close to impossible to become a Venetian citizen if you came from somewhere else. Citizenship was only granted after someone had lived in the city for twenty-five years. And without citizenship papers, it was difficult to be part of the merchant life. For example, in 1305 the state declared that only full citizens could be part of long-distance trade.[48] That decree excluded Greeks, Jews, and members of the various Venetian port colonies. This policy shifted with war, plague, and whenever man power was needed to keep the trade networks running smoothly.[49]

THE ROLE OF CONFRATERNITIES AND GUILDS

Another city-wide system of confraternities (*scuole*) and trade guilds (*arte*) cut across lines of money and class. Although there were lay religious confraternities in Florence starting in 1230, the Venetian version was different. In 1260 the first confraternity in Venice, called a *Scuola Grande*, was established as a civic organization where nobles were not allowed. This *scuola*, and the eight other *Scuole Grande* that followed were not based on trades or occupations. Instead, they had religious affiliations with a specific church and they met in a building associated with the church to plan charitable works.[50] There were also 925 lesser confraternities called *Scuole Piccole*. These *scuole* provided a social network among members, an identity beyond class or parish, and, most importantly, a place to meet up. But more than that, these organizations performed extensive community service. They gave aid to the poor, housing to the homeless, took care of the sick, dying, or elderly. At one point there were 120 almshouses in Venice run by the *scuole*, some small houses donated by members and some purposely built large complexes.[51] Because of these services, the *scuole* were highly regarded in Venice and considered an integral part of community life.[52] They joined the ritual promenades on feast days, proudly holding up their banners.

Although there were associations for several crafts and trades in Venice as early as the 11th century, the trade guilds, called *arte*, began to flourish in the 1300s. These guilds were highly specific, for example, gold beaters (gold leaf makers), wool workers, hat and glove makers, blacksmiths, bakers, painters, and the like. Each guild was composed of workers from every part of the specific craft, from janitor to master artist, so they functioned as a great leveling field within each trade.[53] The guild of gondola builders was established in 1607, making it the first, and only, guild dedicated to this particular Venetian boat. These occupation-based guilds acted as trade unions; they made up rules and regulations for learning a trade and ensured techniques were handed down only orally to preserve their exclusivity.[54] From this practice came the first codified patent system in the world and introduced the concept of intellectual property rights in 1474. But in the case of the Venetian guild system, those rights were more about the guild collective rather than any individual. In other words, patents began as corporate patents.[55] The state upheld the guilds in other ways. In 1272 it offered wool workers a free house and the option to operate free of taxes for ten years. This monetary bait can only be considered the first tax incentive offered by a government to aid economic growth. Each of these practical trade guilds was attached to a *scuola* for religious practice and charity work. One could belong to several *scuole* at the same time, as well as an *arti,* which were guilds based on particular craft skills.[56] A Venetian man's identity was therefore tied up in these overlapping networks that crossed the city.[57]

Venice diluted its class system by encouraging and relying on these nongovernmental charitable organizations and trade guilds. From the Middle Ages onward, the Venetian *scuole* did the kind of community support that is now provided by governments. Instead of taxation supporting these endeavors, the non-noble citizens did the work, and by doing so, they made a responsible, interconnected, and caring community. The nobles might have run the government, but the rest of Venice ran the city and made it a decent place to live, no matter one's status or wealth. In that system, Venice presaged what our local communities are trying to do now. Journalist Thomas Friedman recently wrote about a small town in Minnesota that has successfully incorporated a large population of immigrants from all over the world. The town's success, Friedman writes, has three essentials that made it happen—jobs, embracing outsiders to do those jobs, and "leaders without authority" who check their politics at the door and form networks that "spearhead economic and societal change."[58] Venice had all three of these essentials, but especially the third, with many networks of people and organizations that stepped in and made the place better for everyone. When Napoleon closed those confraternities in 1806, he destroyed those networks and thereby plunged the city into a societal mess. He took away a social fabric that was the identity and glue of the city, and which is missed even today.

WOMEN IN VENICE

Women in Venice had surprising freedoms, especially non-noble women, when compared with their contemporaries throughout other European societies. If they had family money, these women could own businesses and property, and they were accepted in the public life of the marketplace. Venetian women could also defend themselves in court. Unusual for most confraternities in Italy, Venetian women were allowed to join as auxiliary members of the *Scuole Piccole*. They could elect their leaders, but they were still not full members.[59] Venetian women worked as farmhands, fishermen, and herbalists, among many occupations.[60] Some even owned glass factories.

Noblewomen pretty much stayed at home, where their job was to obey their husbands and have children, but they also interacted with everyone who came to the palazzo.[61] They had complete control over their children and made parental decisions without counsel from their husbands. And they had the right over their dowries; a dowry was returned to them, intact, if their husband died, giving Venetian noblewomen a place of power in the family.[62] Women could make loans to relatives, for example. But the lives of unmarried noblewomen were often undermined by dowry inflation.[63] Some families could only afford a dowry for one daughter, so the others were either sent to a convent or allowed to remain spinsters and live with a brother or with other unmarried women. Their behavior, however, had to be as proper as that of a cloistered nun.[64]

These freedoms also apparently encouraged some women to find their own voices. Many noblewomen could read and write, but that was also true of many lower-class women, who presumably learned from their family members.[65] Sometimes they used that ability as a weapon within the family; women could write their wills and exclude husbands if they chose. They could also speak out. In 1487 Casandra Fedele, at the age of twenty-two, delivered a speech at her cousin's graduation from the University of Padua. This speech was then published in Venice and Germany, and Fedele went on to correspond with intellectuals of the day. She was also asked to talk in front of the Senate about education for women. Ironically, and presumably as a sign of the times, Fedele stopped writing when she married. In the mid-1500s, poetess Gaspara Stampa, who was born in Padua but grew up in Venice, became renowned for her 311 poems, and she is still regarded as one of the greatest Italian poets of all time.

Venetian women were also instrumental in the foundations of feminism. In the late 1400s, Laura Cereta became one of the most famous intellectuals of the day. Her writings, which held work and female friendships in high esteem, were not acceptable to many, especially men, but she didn't care and attacked them in print. Modesta da Pozzo de Forzi (or Zorzi in Venetian) wrote poetry and other works in the middle 1500s, but her most signifi-cant work, *The Worth of Women* (which includes *Giustizia delle donne* and *Il merito delle donne*), was tellingly published in 1600 after her death. Those works made the case for a natural superiority of women and they scathingly opposed the way men treated women.[66] Together, these particular writings can be considered the impetus of modern feminism.[67] Lucrezia Marinella took Forzi's ideas even further when she responded to antifemale pamphlets in 1601 and a book by Giuseppe Passi that was unkind about women. Her book, *The Nobility and Excellence of Women and the Defects and Deficiencies of Men*, was the first time a woman argued with a man in print. Marinella's thesis was also that women were superior to men and they should stand up and own their intelligence.[68] Cloistered nun Archangela Tarabotti pub-lished another anti-male treatise in 1652, *La semplicità ingannata*, an anti-male denunciation of fathers who banished their daughters to convents rather than give them dowries or allow them to live as unmarried spinsters. As historian Patricia Labalme writes, Tarabotti called these fathers "pimps and procurers, highway robbers, ministers of Satan, sewers of sin" for spending money on themselves rather than dowries for their daughters. She was furious about being confined to a convent by a system that treated women as parcels to be thrown away, stripping them of free will in the process.[69]

Venetian women were also accomplished writers and editors. Elizabetta Caminèr Turra became director of the magazine *Nuovo Giornale Enciclope-dico* in 1777.[70] This publication full of book reviews and philosophical essays had been founded by her father, and it was highly successful; the biweekly magazine was sold in bookshops in nineteen cities across Italy.[71] Pushing her editorial role further, Caminèr Turra was a voice of protest that targeted the rich, the powerful, and all kinds of authority. She managed the magazine through various iterations for twenty-eight years and covered what were considered "female" issues, such as childcare, but her feminist stance was much broader than seen in any other publication of the time. According to historian Catherine Sama, Caminèr Turra addressed the vaccination of children, the issue of dowry, whether or not women should marry at all, and female lives in other cultures and through history. She was adamant that women had a right to a "life of the mind."[72] In that sense, her magazine was the precursor to contemporary feminist publications such as *Ms.* magazine, and she was just as analytical, informed, and fierce.[73] And when no one would publish her groundbreaking magazine at one point, she turned around and opened a print shop. Three years later, Gioseffa Cornoldi Caminer founded the first magazine specifically for women, *La Donna Galante ed Erudita* (The Elegant and Educated Woman), a clear statement that a woman could be both high class and educated.

Other Venetian women with groundbreaking credentials include Giustina Renier Michiel, who published the first translation of Shakespeare from English to Italian in 1798.[74] Although much more modern in time, the fact that Venetian Adele Della Vida Levi founded the first Italian kindergarten in 1859 and promoted education free of economic and religious discrimination speaks to the acceptance of allowing all classes to school.

These accomplishments could only have come about in a place where women were historically accepted as active participants in the daily life of Venice. As a sign of the acceptance of women as part of the population, when the Café Florian coffeehouse opened in Piazza San Marco in 1720, it was also the first coffeehouse in Europe to allow women. All this can be summed up by a book published in Venice in 1740 that declared women should study. Learing, was a right and something owed to their families.[75] It was written by a Sienese priest named Giovanni Niccolò Bandiera; he published it anonymously, presumably worried about the flack.

But being a woman in Venice was not always so great. There were numerous prostitutes and courtesans in the city, especially between 1350 and 1450.[76] They catered to a large resident population and the many foreigners. Prostitution was validated when in 1358 the state made it legal by suggesting there was a "need." The government put in various controls while also opening a state-run brothel at Rialto.[77] Some prostitutes earned great sums and had near-celebrity-level reputations, but the much larger number led difficult lives. By 1509 there were 11,654 prostitutes in Venice (about 10 percent of the population), and we know this because the government kept track.

Venice also had a large number of nuns. Noble families often could only afford one extravagant dowry, so the other daughters had to become one of those quiet spinsters or join a convent as a lay sister or become cloistered.[78] Former prostitutes and other destitute women also ended up in convents. At the height of sequestering unmarried and destitute women, Venice had fifty-nine convents, but only eight remain today.

THE GHETTO

A chapter on community and firsts in Venice must consider the Venetian Ghetto.[79] The only first in this arena, however, is the use of the word "ghetto"; Jews had already been sequestered in Frankfurt, Germany. When Venice locked up its Jews in 1516 in an area of the city, they used the name already applied to that area, Geto Nuovo (New Ghetto). There are many ideas about where the name "Geto" came from, but it is surely connected to the brass foundry that was originally sited close by in the *Geto Vecchio* (old place of the brass foundry); *geto* in that early context probably referred to the jets, or *geti*, of molten brass that were poured out to make cannons.[80] Or it might have been connected to the process of unloading copper slag from that foundry onto the closest island, then called Geto Nuovo.[81] Also, documents from the time show that the word *gèto* also applied to the taxes that were collected and used to clean canals and public spaces. It seems that the original *gèto*, pronounced "get-o" in Venetian, was transformed into the modern spelling when German Jews arrived in Venice and added the *h* to preserve the soft *g*, as it would be written in German.[82] *Ghetto* is now, of course, used disparagingly about an area of a city where one ethnic group clusters. The original Venetian word has also taken on a brutally negative connotation of a place that is not desirable, that is poverty stricken, unclean, and infused with crime and hopelessness.

For good or ill, Venice has always had a conflicted relationship with whomever they see as outsiders. All sorts of foreigners arrived in Venice because of its focus on trade, in addition to refugees from Venetian colonies, and those long-term residents made major contributions to the city. For example, there were so many Greeks in the city by 1498 that they formed a sort of Greek nation, and the city granted them ownership of a parcel of land to build a church and cemetery. That autonomous Greek area of Venice, called the Hellenic Institute, includes the Greek Orthodox church, an icon museum, a research institute and a library. It can be visited today.[83] Many others arrived because of Venice's open door to foreigners and its sense of cultural freedom. But nothing shows ambivalence toward "foreigners" more dramatically than the way the Venetian government treated Jews. The Venetian Ghetto was established by governmental decree on March 29, 1516. It was not the first time Jews had been moved away from the general population and made to live in a separate area, but Venice was the first Italian city to do so, and other Italian cities soon followed.[84]

The Venetian ghetto today is a small corner of one of the nicest districts of the city. The main part of the ghetto, the Ghetto Nuovo, has a large central *campo* with benches, trees, and a Jewish museum focusing on the history of Jews in Venice. There are also the Holocaust memorials. Perhaps the most arresting vision in the *campo* is the run of rusted barbed wire on top of one wall. It was left there after the German occupation of Venice during World War II and the subsequent deportation of 200 to 250 Venetian Jews to Auschwitz-Birkenau for extermination. Only eight of those Venetians came home.[85] This *campo* has three exits, all over bridges to neighboring islets. Areas on two of the neighboring islands were eventually incorporated as areas solely for Jews. The openness of the *campo* and the exits is misleading. Holes that held the iron hinges used for locked gates still festoon the back walls of houses. And looking around, one realizes that the whole area is surrounded by canals that were ideal ways to section off the area from the rest of the city.

The history of this particular ghetto underscores the Venetian push and pull with people they saw as foreigners. In this case, Judaism was a "foreign religion" traditionally at odds with Christianity even though early Christians had been Jews.[86] Jews had lived in Venice for centuries, but in 1396 they were ousted to the town of Mestre on the mainland. Given their role in the economics of the city, they were soon allowed back, but they could stay only fifteen consecutive nights anywhere in Venice. A war in 1509 brought many Jewish refugees to the city; that population pressure initiated the negotiations in 1516, and that was considered a compromise. The state gave the Jews a place within the city where they could rent, but not own, housing but that area would be walled and the gates locked at night.[87] "From the outside it looked like a fortress," writes historian Patricia Fortini Brown.[88] At first, this plot of land was considered a good thing because it meant Venice was "allowing" Jews to live in the city permanently. It was also better than other nations at the time, such as England, which had banished Jews completely. And in 1290, Jews were involved in the process, making this seem like a reasonable compromise.[89] But that compromise was tainted by the fact that gates were locked and surrounding canals were patrolled at night by Christian guards, whom the Jewish community was forced to pay. Jews were also required to wear a visible identification of a yellow cap, scarf, and circle sewn on their clothes. And they paid rents one third more than Christians, bringing an inflated income to Christian landlords who mostly neglected their properties.

In spite of this sequestering, Jews were integrated, albiet in restricted ways, into the city economy. They could conduct international trade, sold used clothing, were moneylenders and pawnbrokers. These occupations meant Jews were financially involved with the aristocratic population that often ran out of money from gambling or other pleasures. Also, it was a sin at that time for Christians to make loans so they had to turn to the Jewish money lender (as in the *Merchant of Venice*). Jewish doctors were highly respected, and doctors were allowed out at night to tend the Christian sick. And while Jews had their own stores within the ghetto, the area became crowded and run down. With the ghetto forbidden to expand physically, renters built up, adding stories to houses sometimes refered to as the first skyscrapers.[90] They also internally subdivided the buildings, adding between floors by lowering ceilings.[91] That crowding, and landlords who did not care, made the ghetto a target for the various plagues that hit the city.

The ghetto remained locked at night for three hundred years. When Napoleon conquered Venice and ended the republic in 1797, he declared that Venetian Jews could live and work anywhere they wanted. That lasted 150 years. Then came the Nazi occupation and death camps.

Anthropologists say identity is everything, but that statement does not usually carry the rider that identity can also get you harmed and killed, that identity can bring genocide. For all Venice's progressive approach to democracy, liberty, and freedom of thought, they fell short when it came to outsiders unless those outsiders had something to offer in the world of trade. When you take people of a religion and lock them behind walls, restrict their movements and their occupations, and mark their clothes with bright colors, they seem like a "population," a particular "culture," so the Jews were identified as a people separate and apart based solely on their religious beliefs, even though they were as European as anyone else in Venice.

POLITICS AND GOVERNMENT

When John Quincy Adams was looking for examples of democracy as he helped give birth to the United States, Venice was the only standing republic in the world.[92] Adams was impressed by the interlocking oversight of various governmental bodies and the fact that the doge had no real power and was always watched. Adams even quoted the liberal and erudite Venetian friar Paolo Sarpi, who, in defiance of the pope, had been the first statesman to promote the separation of church and state in Venice in 1606. Adams also wrote to Thomas Jefferson in 1820 that he hoped the United States could model Sarpi's words. From that correspondence the United States became a country founded on the Venetian principle of the separation of church and state.

In Venice, that separation seeped into everything. Unlike a monarchy, the office of doge, the highest state office, was not sanctioned or directed by God. The religious part of the dogeship grew with the centuries. It became part of the identity and celebrations of the city, but that was less about giving the doge godly credibility and more about telling the pope that he had no power at all in Venice. That move was a reaction to pestering by the papacy, and the more Rome pestered, the more religiously accented the office of doge became. The popes wanted to control the republic and of course, the Venetians would never allow that to happen, so they passed laws to constrain the church. In the 13th century, the Venetian Senate ruled that any will bequeathing money to the church had to be reviewed and approved by them. In 1604 the Senate passed a law stating that if a proposed building was religious—a church, monastery, or convent—the Senate had to approve of the plans. No priests or friars could hold public office or behave in a political manner, yet many did if their behavior honored the state and not the church. Supported by priest and Venetian statesman Paolo Sarpi, the government also passed a law unlike any legislation found in another Catholic place: clerics could be tried for crimes just like anyone else.

The battle between Venice and the Roman Catholic Church is still alive and kicking today. In 2017, the *Times* (London) ran a story with the headline VENETIANS CONDEMN CHURCH FOR LETTING ITS PROPERTY TO TOURISTS with the provocative subheading VENETIANS SAY THE CHURCH IS HELPING TO TURN THE CITY INTO A TOURIST TRAP. At issue was the monastery of Santa Fosca where the church was renting former monk cells as hotel rooms. Venetians reacted with a mocking flash mob in a busy *campo*, cardboard houses used as props, and a large banner that read, YOU ARE DESTROYING VENICE FOR MONEY. Venetian protesters were also pissed about the sign on the monastery door that read, YOU WILL SLEEP LIKE GOD, a blasphemous branding move if ever there was one.

The underlying philosophy of the Venetian government, and Venetians, was distrust of individual power. Venetian rule sat with a group of people, some wealthy and some not, who considered themselves aristocrats but they were also men in competition with each other in the trading sphere. And so, the Venetian government was run by various committees, some of which spent all their time watching over other committees. This first clear example of a system of governmental checks and balances began in 675 A.D., when the people of the lagoon started to manage themselves as a collective. That structure, born of suspicion and competition, also kept power in the hands of many. It employed all those *cittadini* who served as civil servants, and it was imbued with a system of checks and balances.

Historian Frederic Lane also suggested that the Venetian governmental structure can be considered a pyramid rather than the separate but supposedly equal parts of government as seen in most democracies today.[93] At the bottom of the pyramid was the Great Council (*Maggior Consiglio*), a body that included every male noble over the age of twenty-five who applied and could prove his high-born ancestry. In 1255 the Great Council became unwieldy because of population growth, so a new body was formed—the Senate (*Consiglio dei Pregadi*). That body started with 60 members but grew to 275 over time.[94] The senators did just about everything: they served on powerful committees, passed laws, developed many departments that are familiar today—water projects, food safety, and education, for example. They also served on the committees that passed laws trying to keep their class intact and exclusive. And the senate named state attorneys, decided on ambassadors, and appointed the heads of various branches of government.

The structure outlined above only gives a general idea because the details of the Venetian government system are mind-boggling in their various layers, stemming from oversight among committees and the fact that Venice was a self-ruling republic for eleven hundred years. There was, for example, the Council of Forty (*Quarantina*) which appeared sometime between 1207 and 1222. It ran the justice system and enjoyed a broad swath of power. The Council of Ten (*Eccelso Consiglio dei Diece*) was initiated in 1310 and included the doge, his six usual advisors, and three others.[95] This council started its mandate as a group of judges but then extended their reach to anything that could vaguely be connected to acts of purported treason or bad behavior and be declared an issue of state security. Sentences of imprisonment, torture, and public whipping or execution could be laid at the feet of this council. But the main job of the Council of Ten was especially aimed at keeping the aristocracy in line, something the ruling patriciate strongly felt was necessary for a successful republic to function. In that sense, Venice was an oligarchy with a conscience, something highly unusual today.

The apex of the Venetian government was the office of doge, or *dux* in Latin, which simply means "leader." The first doge was elected in 697, the last in 1797, with a total of 120 doges during those eleven hundred years. Doges swore to promote the "honor and profit" of Venice.[96] To make sure the dogeship could not be bought or corrupted, from the 1500s onwards the Great Council used a complicated and secret system of election that involved ten separate steps.[97] The steps began with placing brass balls (plus thirty gilded ones) into a container. These balls—*balote* in Venetian from which we get the word *ballot*—denoted everyone in the Great Council; that number ranged from one to two thousand over the lifetime of the republic. By definition, they were all aristocrats. The councilors passed by the box and a randomly chosen young boy reached into the box using a pair of wooden hands and gave each delegate a ball. If that ball was gold, the particular noble went to the next level of selection, and everyone else had to leave making that level a quick study of an elimination round. The next round narrowed the possibilities to twenty-one, and from there a series of votes and elimination and they came up with a list of 41, none of which could have come up in the other steps. Finally, with all these levels of increasing and decreasing, it took a majority of twenty-five votes to elect the doge.[98] The process was so secret and so very convoluted, and the rules of the process also changed over time, that no one, no matter their power, could have controlled it.[99] Versions of this method were then copied, more or less, by France and the United States in the 1700s.[100]

Today, politicians, especially in the United States, refuse to address the issue of buying a political office or the role of outside lobbies and corporations aimed at influencing the election.[101] Venice was more paranoid about any whiff of centralized power, so they essentially emasculated the doge with the election process, his duties, and his permitted behavior. The doge was a sort of "prisoner of the state" in the Palazzo Ducale.[102] He was not allowed to open or read his mail, go out of the palace just for fun, talk to his family without a witness, see ambassadors privately, or even make his apartments in the palace more elaborate than they already were.[103] When the doge died, there was a major audit of his whole record, including his finances. If there were discrepancies, the doge's family had to pay giant fines. Families, in general, were not exactly thrilled when a relative became the doge, because they were hobbled financially as well during his lifetime (another difference from modern democracies). Doges often ended up broke because their salaries did not cover what it took to support all the trappings of the office which they had to pay for personally.

The real leadership in Venice was not the doge but the combination of the doge, his six advisors representing the six districts of Venice (called the *Minor Consiglio,* or Lesser Council), and three leaders of the judiciary.[104] But even these positions were shuffled about to protect against bribery, collusion, personal gain, or lasting influence. The six advisors were elected by the Great Council and served only eight months. And they were not so much advisors as "watchdogs" over the doge.[105] The six councilors also sat the doge down once a year and read him his oath of office to remind him of his lack of power and influence just in case the doge had forgotten.[106]

Venice also had a healthy paranoia about outsiders, especially the Vatican and other European governments. That paranoia, and the employment of a great information gathering system, made it possible for Venetian agents to routinely intercept foreign dispatches. Spying, in other words. In 1506, Giovanni Soro was employed by the Council of Ten to decipher code, making the Council of Ten the first Secret Service to use code breaking.[107] They even gave Soros his own office and a small team of helpers. Soro was the first cryptologist in Western history; he invented encryption and deciphering others' codes, and he is deemed the father of modern cryptology. In one story, Pope Clement VII sent Soros two encrypted messages and asked him to decode them in the service of the Vatican. Soros claimed he could not figure them out, but that was probably a lie and just a political move to install a false sense of security in the Vatican. If Soros couldn't break their code, no one could, so the Vatican had no need to invent difficult codes. Those might interfere with the political maneuvers of the Republic of Venice, a city that always liked to be one up on the pope.

We know an unusual amount about the daily goings-on of the Venetian government, at least for the thirty-seven years of its operation from 1496 to 1533, because Venetian nobleman Marino Sanudo, son of the geographer of the same name who appears in chapter 2, took copious notes as he sat in the Great Council and the Senate. He then published those notes in fifty-eight volumes, surely the first, most detailed, and longest daily chronicle of a republic. Sanudo was writing everything down before there was such a thing as meeting minutes. And this work is certainly not just about statecraft. Sanudo comments on every current event in Venice and beyond, gives historical context when useful, and deftly describes whatever is happening around town. His writing is lively and interesting, if one has the energy to plow throw it all. Venetians were well known for keeping lists of everything, accounting for trade good transactions and household accounts; the halls of the *Archivio di Stato* (Archives of the State) with their miles and miles of shelving stacked to the ceiling with all these papers, groan under the history of Venetian note taking (and what a boon to historians). Sanudo's "diaries" stand out in all this blizzard of accounts, lists, and legal documents, because his notes were daily observations (they really served no purpose except for posterity) of the inner sanctum of the Venetian government over decades. Although Sanudo was repeatedly disappointed not to be given the official post of city historian, his volumes are still one of the staples of Venetian history.[108]

What stands out the most is the length of the stable Venetian government—eleven hundred years, the longest-standing republic in history. During those eleven hundred years, the republic was never invaded, never conquered. There was no standing army keeping would-be invaders away, just the reputation of the might of the Venetian Republic and the fact that everyone was dependent on their trade. It took economic collapse and Napoleon to change all that. But how did the Venetian Republic survive so long? Today's modern republics are not even half the age of the Venetian Republic, and thus there are lessons to be learned: bits of socialism, working nobles who interacted with everyone, a concern for public welfare, a separation of church and state, and oversight across the government, not to mention a healthy paranoia about corruption. These varied beliefs and behaviors are the infrastructure that made the city a place of active invention and free creative thought. It was also a city with a solid identity and a collective purpose. In 1849 the Austrians were trying to take Venice and part of their plan was to attack from the air, the first time this had ever been attempted. From the mainland, the Austrians launched a fleet of balloons loaded with bombs containing tar, gunpowder, and shotgun pellets.[109] As curious Venetians watched, the balloons floated toward them, then off course and away from the city. The Venetians, in their usual spirited manner, cheered and laughed as the dangerous artillery ascended into the atmosphere.

CHAPTER FIVE

The Art of Medicine and the Idea of Public Health

Measure what can be measured, and make measurable what cannot be measured.

— Galileo Galilei

Medicine was impotent against plague.

— Eugenia Tognotti
2013

Liberty, stability, longevity, social tolerance, and economic prosperity infused the history of Venice, and never more effectively than when the city was in hot water (so to speak) in times of crisis.

— Jane L. Steven Crawshaw,
The Plague Hospitals: Public Health for the City in Early Modern Venice
2012

IT'S ABOUT A four-minute boat ride from the long barrier island called Lido that delineates the easternmost side of the Venetian Lagoon to one of the tiny specks of land that dot the watery landscape. That ride is only available on two weekends a year, at the end of the summer, when volunteers meet visitors and give guided tours around the interior. The rest of the time the island is left to wind and rain and archaeologists, and its own complicated history as the first plague island in history, called Lazzaretto Vecchio. Long ago, when plague was raging, a similar ride to this island directly northwest from the city center couldn't possibly have taken much longer than it does now even though the boat back then would have been rowed rather than relying on a marine engine. And the passengers would not have been tourists and locals out for an off-city interesting day trip but plague victims being exiled to an island within spitting distance of Venice to curb contagion. And they would have been banished to Lazzaretto Vecchio as a place to die or perhaps to beat the odds and live. Even so, the two rides of today and long ago still have something in common; the sick and now the well can both appreciate the sense of isolation and containment that Lazzaretto Vecchio presents. Approaching the island today, what stands out is the high perimeter of brick that comes down to the water's edge. Some of it is purposefully built wall while much of it is formed by the outsides of buildings with inset windows and tile roofs that take up three sides of the island, not unlike an offshore prison. No one lives there now and on shore it has an atmosphere both peaceful and depressing. Inside the various buildings that housed the sick, the vastness of the quarantine undertaking that began in the 1400s becomes clear. The isolation is marked by views of the lagoon and Venice out of windows or over the walls from an upstairs porch. But this facility was never meant to be so threatening or so sad. It was built not as a place of punishment but to treat and sequester those infected with the plague. And in that action, Venice became the first city in the world to quarantine sick citizens as a way to protect the health of others. That act of public oversight established an organized and strategic community health plan for the first time in history. And they did it without a clear understanding of what plague was, where it came from, or how it was transmitted.

Centuries later, the Venetian Republic also fostered a major revolution in medicine by giving professors at the University of Padua, geniuses such as Galileo Galilei, William Harvey, Gabrielle Fallopio, and the lesser-known Santorio Santorio, the freedom to think about, experiment with, and reframe everything that was known at the time about illness and health. Those physicians, anatomists, scientists, and philosophers of science didn't just make the scientific and medical revolutions happen. They also created how we view and practice medicine today. Venetians changed our very concept of how the human body works, where disease comes from, how treatments can be tested, and why public health is critical to modern society.

VENETIANS AND CONTAGION

Venice was not the first, or the only, place to feel the cold hand of plague. For example, the Black Death, also known as the bubonic plague that produced buboes or pustules all over the body, killed hundreds of millions of people worldwide and many other medieval plagues swept from Asia to Europe, felling populations in their paths. Plague, a contagious disease that killed en masse through a bacterial infection, is not with us today because we have antibiotics.[1] Some have considered HIV infection a plague, or Ebola or dengue fever plagues, but in the strict definition, these are not plagues because they do not infect and carry away a large swath of the population. These modern infectious diseases are better named epidemics or pandemics, because even in our times of modern medicine we have now seen that they can spread far and wide and they can come in viral form. And of course, they can be just as deadly and frightening as the plague on an individual basis, especially before medical experts fully understand their method of spread.

Disease also comes with a belief system that guides the identification of an illness and its treatment. Across human history and among cultures, various stories have been rolled out to explain the causes of disease and death. Illness has been explained as a curse from the ancestors, or evil spirits, or payback for some bad deed. Such belief systems underscore the idea that sickness and death are moral punishments.[2] A recent example is the HIV virus and its spread among homosexual men in North America, as if God suddenly disapproved of homosexuality and decided to send down a fatal disease. That belief was discounted as HIV spread among heterosexuals and was handed down from mothers to babies. Before the discovery of bacteria and viruses, many suggested that illness was not so much a personal failing as it was the result of stars at work in a negative way, or illness floated in on the air of certain seasons. Underlying these belief systems has always been the easily recognizable fact that many illnesses hopped from one person to the next, sometimes rapidly and fatally. And it was the fact of contagion, the idea that the spread could come by contact, air, water, food, anything, that can incite collective fear. The Apollo 11 astronauts spent twenty-one days in quarantine because they had gone to the moon, and who knew what lingered there. And the question that guided that post-lunar quarantine is the same one that guided the original fear of the plague—what exactly was causing this particular contagious disease?

Although many believe that Louis Pasteur, and somewhat later Robert Koch, were the first to understand that small particles we now colloquially call "germs" were responsible for spreading many diseases, these men were certainly not the first to think about contagion. In 1362 Ibn al-Kathib, an Arab intellectual and physician, wrote about contagion after seeing what plague had done. He pinned it to goods and ships coming into foreign ports and commented on the fact that not everyone died, only some and that suggested the possibility of immunity. His thinking was heretical at that time, especially in Muslim culture. And long before Pasteur, Padua professor and anatomist Girolamo Fracastoro proposed in his book *De Contagione et Contagiosis Morbis* (1546) that disease was passed along by tiny particles. He called them *fomites*, a word that has been translated from the Latin as "spores" but actually means "tinder," as in to make a fire, and that's what Fracastoro meant—these *fomites* started the raging fire of infection. He also made the enlightened suggestions that these spores could move from person to person in bodily fluids, they could survive for a very long time, and while the spores themselves were not infected, they were the causal factor of an infection. Fracastoro's initial insights into contagion came from studying the then rampant disease syphilis. In what seems an odd pairing today, the doctor first wrote about the disease in an epic poem called "*Syphilis sive morbus gallicus*," which translates as "Syphilis or The French Disease," published in three volumes in 1530. Poetry and disease, let alone syphilis, don't seem like a good match to us today but back then medicine and anatomy were tied up with philosophy, theoretical musings, and poems about love gone bad. Fracastoro had the honor of giving that specific disease the name syphilis based on his epic hero, Syphilus, who had contracted the disease after he insulted the god Apollo.[3] Fracastoro offered some cures for syphilis that might have actually worked, such as doses of mercury. He was also the first person to describe typhus, but not in a poem.

Venetians surely recognized that the plague was passed around, and that it arrived from outside the city on incoming ships, so very early they had a vague idea of transmission but no clear idea of how it actually happened (thus the generalized scramble of quarantine). The plagues that hit Venice came in three types, and often they arrived together. One rested in the lungs, causing pneumonia. The infection went from person to person; a mere cough could spread this kind of death. Bubonic or Black Plague, the second kind, was only transmitted by flea bites. The name Black Death comes from the purplish-black pustules, blood hemorrhages, that form under the skin. The bacterium that caused Black Plague, *Yersinia pestis*, infected fleas, causing them to search for blood meals in hosts such as black rats and people.[4] The fleas bit the ubiquitous rats on Venetian merchant ships, and the sailors as well, and then the ships carried the infection into the city.[5] At the dock, the rats jumped ship, taking the bacterium with them; a rat bite could easily transmit the disease. The diseased fleas were also carried into town on goods for sale in Venice or aimed at transfer to other European cities. Also, once on land the fleas had an additional new population of potential hosts beyond rats and people—local dogs and cats. Between the rats, fleas, pets, and sick sailors who could be bitten again by a flea that passed on the disease, *Yersinia pestis* had a field day of rapid infection and spread in Venice, and everywhere else.[6] When the infection entered someone's bloodstream, it became septicemic plague, the third kind.

The fact that the city was tightly packed with people didn't help. We know the numbers because Venetian authorities kept records of everything, and there are long lists of the dead and dying; Venice was the first city to take statistics of people and things seriously.[7] The city experienced approximately twenty-two waves of plague between 1361 and 1528, but the worst plagues occurred in 1347–1349; 1462; 1485; 1506; 1575–1577; and 1630–1632. When the Black Death of 1347 killed about half the population of Europe, it cost Venice about 60 percent of its population.[8] That would be about 50,000 people in a year and a half. And then it struck again in 1575 and killed one third of the 50,000 residents of Venice—over 16,500 people. The population then recovered to about 190,000 people, but the plague of 1630 took a third of the city again. It just never stopped, until it did, which was after 1630. Bubonic plague never recurred after that in Venice and was less frequent across Europe. The decrease of incidence happened, in part, because the brown rat survived well in colder European climates and in crowded cities replaced the black rat. The brown variety was also not such a great climber and was unable to jump on and off ships.

In such a tightly wound city, a place where everyone was socially and economically dependent on each other, the recurrent loss of much of the population was devastating. To some historians, these plagues and the repeated decimation of the population set Venice on its course toward much later self-immolation in 1797. They have also suggested that the waves of death made the city fragile, and in want of citizenry, and that that weakness echoed down the centuries. The only reason Venice didn't fall completely apart after each plague hit was that immigrants came from the mainland to fill job vacancies once the plague passed. But still, the absence of a plague after 1630 in Venice bears special mention. As historian Frederic Lane says, "The Venetian Republic was on the frontier [where the plague could arrive from the East], fighting off the threat by vigilance."[9] It was this vigilance that was so groundbreaking and surely instrumental in halting the spread into Venice and onward. But more importantly, the massive die-off of citizens prompted the Venetian government to enact a many-pronged defense against these diseases. Although isolating the sick, such as lepers, had been practiced in the past, Venice was the first city to have a comprehensive plan about protecting public health against infectious disease.

THE BIRTH OF QUARANTINE

From 1347 at various ports around Italy, ships had been turned away regularly for signs of plague. But on March 20, 1348, the Venetian Republic imposed the first-ever public policy on quarantine for incoming ships.[10] Venetians went into governmental action although no one was really sure what the plague was, where it came from, or how it was transmitted. But Venetian authorities believed that it arrived in a foul miasma, that is, on the back of some bad air, and so they wanted to get the infected away from others who might breathe those foul respirations.[11] The authorities could have taken the attitude that there was no way to see or stop that air, but instead they enacted a series of laws and procedures to put up a sort of curtain between its trading partners and colonies and whatever cloud of invisible contagion was bearing down. They began to inspect arriving people and ships for signs of plague and then not allowing further passage if someone has symptoms. In 1377 Venice went even further to protect its colonies. Much of the trade in Ragusa (now Dubrovnik, Croatia) on the other side of the Adriatic was owned and operated by Venice. But that trade occurred with mainland Greece, a place rampant with the plague. In an enlightened move for the times, the Venetian government imposed a policy of quarantine over the whole city of Ragusa. That quarantine included boarding ships in port and inspecting people, animals, and cargo or personal goods for signs of disease. Then, all those people, animals, and goods had to remain on ships anchored offshore for thirty days to see if everything remained disease free or died off. By the 17th and 18th centuries, Venice forbade Ragusans to have any contact with Greece.[12]

Imposed isolation began with that thirty-day rule in 1348 but was then extended to forty days in 1403. The label "quarantine" stuck because *cuarànta* means "forty days" in *veneziano* (*quaranta* in modern Italian).[13] No one knows why they started with thirty days or why it was increased to forty days, but one historian has speculated that the forty might be connected with Hippocrates, Pythagoras, or Jesus wandering in the desert.[14] Maybe it was just an arbitrary fit with some other scheme related to trade by ship. Another possibility is that Venice, and other cities that adopted quarantine later, thought that length of time separated the acute plague from some other chronic condition or infection.[15] There was also some inkling back then that the plague pest might not have died with the host. That assumption, fear, or a kind of bureaucratic thoroughness resulted in Venice initiating another kind of quarantine. From 1348, all those who died of the plague in any part of Venice had to be buried away from the city, usually on islands.

About seventy-five years after the first proclamations for quarantining outsiders, the Venetian Republic stepped up vigilance of its own citizens. Isolation, or quarantine, was a snap in Venice because all that water in the lagoon meant the city was also surrounded by separate islands set off from Venice proper and yet relatively easy to access by boat. The state could ferry over materials for buildings, then bring infected individuals and the staff and supplies to care for them. All three of the plague islands that eventually formed a complex and strategic approach for quarantining against the plague had been occupied by monks and friars of various orders, and some housed nuns. These were livable islands, places of quiet and isolation where the inhabitants would not be disturbed and where they could lead pious and quiet lives. Monastic groups could also grow vegetables and fruit and feed themselves. In other words, these already established self-contained facilities made ideal places for isolating the sick and the mentally ill. And that came later.

In 1423, the state established the first-ever plague island by taking over the islet described at the beginning of this chapter. Its original name was Santa Maria Nazaret (Saint Mary of Nazareth) and it was a convent. Once the state commandeered it for the sick, the island was rechristened Lazzaretto, perhaps as a twist on Nazareth or in honor of Lazarus who rose from the dead.[16] Based on this Venetian example of living quarters and long-term hospitals for plague victims, the rest of Europe initiated their own hospitals, then named them Lazzaretto as well. The island eventually became known as Lazzaretto Vecchio (meaning "old") when an additional island was requisitioned for quarantine and called Lazzaretto Nuovo ("new").

Archaeologists have cleared away most of the undergrowth that had overtaken Lazzaretto Vecchio. They revealed six large *campi*, or open spaces, with the remains of long brick buildings squaring off each section, a cemetery, arching trees, and corresponding brick buildings that housed the acute hospital and administrative buildings. Standing in one of those reconstructed cavernous residential buildings and imagining the many beds of sick people is sobering. Looking out at the lagoon from the windows, or from a balcony of the second story of another building, it's easy to imagine that life here would have been both peaceful and terrifying. It wasn't a prison, not intended to punish anyone, but it was indeed enforced isolation from friends and family, and it must have been crushing. Being able to see Venice in the near distance must have been even worse torture.

Recently, the graves of fifteen hundred plague victims were uncovered on Lazzaretto Vecchio, and there are surely more.[17] The skeletons are laid out in large pits, everyone lined up in one direction with what seems like an attempt at hurried respect while the bodies were piling up. But what is most striking about Lazzaretto Vecchio is the clear sense of a community, a hospital community, established to house and care for masses of victims, waves and waves of sick people disembarking, with death their most likely future. On one outside column holding up part of a building there is a series of inscriptions recently translated by a linguist. The words are notes about one patient, written about himself as he was relaxing on a stone step. Other inscriptions and drawings in red ocher on interior walls also haunt Lazzaretto Vecchio. The archaeology organization Archeove, now in charge of preserving the Lazzaretto, has also amassed a large cache of letters from plague victims. These papers sport signs of fumigation; the state assumed the plague could be transmitted by touch. The letters were handed down through generations and saved by collectors all over the world and now they represent a database of the sick and dying communicating with others about their plight. Lazzaretto Vecchio was in business as a plague quarantine island and a hospital for lepers from 1443 until 1630. It then found new life as a military installation, then as a dog pound, and now as a place for historical research and the occasional tour group.

In 1456 another island was sequestered to fight the plague, and Lazzaretto Nuovo opened officially in 1468. This island was designated for those who might be suspected of having plague when coming off a ship, or those in Venice who had family members with the plague. In other words, it was not a place to die, and certainly not a prison, but instead a facility all about waiting and watching. There is one historical written note indicating that someone was living there as far back as 1015 when it was called Vigna Murada (walled vineyard). Supposedly there were planted grapevines and some sort of wall surrounding them.[18] Today Lazzaretto Nuovo is open to tourists on weekends during the summer months. Guides familiar with the extensive archaeological work that has saved the main building and cleared off wild vegetation lead guests around and tell the island history. The island is close to the main outlet from the lagoon to the Adriatic and so it was especially useful over the centuries. It has been an outpost and stopover for Roman travelers a monastery, then a pre-plague holding facility, and then ending up as a military outpost in the hands of the French and then the Austrians. The Austrians dismantled almost every building to put together a defensive brick wall ringing the island. The archaeological salvage work was spearheaded by Italian archaeologist and now director of the island Girolamo Fazzini. A visit today requires a fifteen-minute public boat ride out into the flat greyness of the lagoon, heading northeast. The view includes reedy patches of marsh called *barene*; they pop up left and right, reminding the visitor of the abundant natural history of this part of the world. Birds stand on thin legs pecking in the *barene* at whatever looks good to eat that is burrowed in the shallow water. Fish sometimes surface, making their own little ripples, or schools of small fish wiggle in the shallower parts of the lagoon. This ride echoes what it was like to be a lagoon dweller a thousand years ago and what it is like now for fishermen and others who still make a living off this lagoon. Then it takes a polite request of the vaparetto captain to get off at Lazzaretto Nuovoa because it is not a usual stop. Normally, the public boat would slip past the plague island along a canal that divides that outpost from the garden island of Sant'Erasmo, the largest island in the lagoon, where most of the town's vegetables are grown. That particularly Venetian produce from Sant'Erasmo includes the very young artichokes with a tinge of purple (*carciofi violetti*) that are grown by the thousands on Sant'Erasmo.[19] A tourist steps out of the vaparetto onto a deserted wooden dock at Lazzaretto

Nuovo. The main entry is right ahead, and once inside the compound, the same sort of ancient peace and quiet of Lazzaretto Vecchio descends. Inside the entry there is a large grassy *campo* and two lines of aged trees that lead to the main building, the Tezon Grande, the last remaining brick building that has been excavated and restored by Archeove.[20] This building is massive, only outpaced in Venice by the Cordiere building of the Arsenale shipyard where they used to make rope. The Tezon Grande was once the warehouse for all the personal and trade goods offloaded from ships. These goods were sorted, organized, ID'd, then fumigated with smoke from piles of burning straw, tobacco, and pitch. Caretakers also tried dousing goods with smoke from herbs, berries, and gums as well as wood full of resin in an attempt to "perfume" the goods and make them usable after possible exposure to the plague.[21] These treatments represent the first attempt by any authority to decontaminate goods that were associated with contagious infection. The authorities on Lazzaretto Nuovo also burned piles of incoming trade goods, especially expensive imported silks. It was like setting fire to money. By 1493, the authorities also attempted to disinfect anything made of paper, including mail, by dipping it into vinegar.[22] A testament to the accuracy of off loading, sorting, organizing, and fumigating all this stuff is found on the walls of the Tezon Grande. These walls were used like chalkboards listing what belonged to whom, and the writing is still there.

Documentation (prints and drawings) shows that families and individuals traveling solo had their separate housing, each with a small garden. Each section that corresponded to all the passengers of a particular ship, for example, were clustered in one compound and had their own small church.[23] These compounds were closed off from other sections in case the plague broke out. A doctor examined people every day and if someone showed signs of plague, they were transferred to Lazzaretto Vecchio. If after forty days of quarantine there were no signs of disease, they were free to go. During the plague of 1555, this island was also a resting place for those from Lazzaretto Vecchio, the real plague island, if they were past the worst part and recovering from the plague.

Plagues were still somewhat in evidence when a third island, Poveglia, was put into play as a backup. Over two hundred years after those two Lazaretto hospitals were set up, yet another island and its current population was forced into service. Poveglia is made up of three small islands and is farther away from Venice than the other two plague islands. An octagon was built in 1380 on the smallest islet as part of five octagons that formed the early Venetian military defense system for the lagoon; today only the low perimeter brick rampart maintains the eight-sided shape. Another islet had been planted with fruit-bearing trees and various herbs and vegetables and it has become an overgrown orchard today; a bridge connects that orchard with the main islet. When plague victims were dropped off in 1782, there was a thriving community. The ruins of the life here remain—a bell tower, a central *campo*, a well, a lovely disintegrating villa, and steps leading down to the water. There were as many as 1,600 quarantined individuals on the island during those times of plagues, and then in 1793 and 1798 when the bodies of plague victims from two ships were brought to the island for burial. It earned the name Lazzaretto Nuovissimo ("really new"), but that name never stuck, because unlike the other plague islands, it was a non-religious community before, during, and after the plague centuries. There also appears to have been a psychiatric hospital there at one point, and it became a rest home for the elderly in 1900. The island was abandoned in 1968 and is now up for sale.[24]

Venice was also the first city to set up an active "maritime cordon" to protect the city itself from outsiders. Ships carrying plague had to signal with a flag as they entered the lagoon and there was always someone up the bell tower in Piazza San Marco watching for these flags. A small boat rowed out to the ship and collected the captain. He was then interviewed through a glassed-in window in the offices of the health department and had to prove that the ship was free of disease and receive a bill of health. Otherwise, everything and everyone was "quarantined" on a plague island for forty days.

By 1528 travelers into Venice and governmental representatives of Venice in foreign ports and cities were also required by law to report any signs of plague in the places they had visited.[25] The same rules apply today for those entering areas of disease. A clean bill of health to return to one's home country means having the correct vaccinations in hand, if they are available. Also, in the United States, the Centers for Disease Control and Prevention (CDC) does the same job established over five hundred years ago, tracking diseases as they pop up, looking for the route of transmission, and enforcing mandatory quarantines while asking questions about visits to farms and such as a traveler comes home. The CDC also calls for quarantine when they deem it prudent, but it's not, alas, in charge of other public heath issues such as sewers or garbage collection.

MENTAL HEALTH QUARANTINE

A fourth island right off the southern rim of Venice, San Servolo, also reportedly housed over two thousand plague victims in the great plague of 1630. But San Servolo is better known for its time as a mental hospital—a hospital for the insane, used to remove people with mental illness from the general population. At first, a monastery was built there in the 9th century, then it was transformed into a hospital for sick soldiers in 1716 and paid for with state money. The monks were experienced doctors, pharcmacists, and caretakers. They were also good gardeners and pharmacists; they grew whatever herbs were needed for medicine and made elixirs, creams, and poultices in their on-site apothecary shop. The monks moved from tending sick soldiers to caring for the insane in 1725. Before this, rich Venetians had provided for their mentally ill family members at home, at their country houses on the mainland, or by paying for housing in a monastery. The "mad" were also thrown in prison or kept on ships parked at the mouth of the Grand Canal. These particularly unfortunate souls either ended up back in prison or became galley slaves and rowed for the Venetian Republic. San Servolo as an insane asylum began with a few noblemen and later added the mentally ill of the middle class.

Even after the fall of the republic in 1797, San Servolo remained a hospital for the insane with some wounded soldiers thrown in. By the mid-1800s the monks had moved out and a real doctor, or alienist as they were called back then, was in charge. This physician, Prosdocimo Saliero, believed that idle hands were the devil's playground and he gave the patients something to do, hand-icrafts and such. He also had patients working as carpenters, mattress makers, bakers, printers, and the like because he thought they might forget their mental illness if they were occupied. But Dr. Saliero's great contribution to history was the fact that he used case sheets for each patient, the first evidence of any physician keeping individual medical records. These patient charts are preserved at San Servolo and they show the attention given to each subject. Salierio also monitored the patients for various biological measures to get a sense of their overall physical state. And he introduced clinical measures that hope-fully would make a difference. For example, he recommended hydrotherapy, defined as sitting in a tub and soaking in alternating hot and cold water, for the manic and demented. And Salierio followed the normal mid-1800s practice of restraining mental patients in the hope that this forced physical stability would bring about normal mentality. Basically, they tried everything at this facility. Some doctors proscribed electrotherapy, the passage of electric current to the brain with electrodes positioned on the skull. Medical anthropologists measured body parts in reaction to changes in emotion or changes in respiration under emotional moments. They also mapped the bumps on patients' skulls to predict the various types of mental illness. And they did talk therapy. The doctors also performed autopsies on the brains of deceased patients to try to gain some understanding of brain pathology as the cause of mental illness. By 1900, San Servolo and most other places saw mental illness like any other disease, one that could respond to chemical or biological treatment. That view was supported by working with elderly demented patients and those suffering mental illness from syphilis. Also, many of their patients presenting as insane were suffering from pellagra, an illness brought on by a diet dependent on unprocessed cornmeal, a staple of northern Italy especially among the poor. In the late 1800s, polenta, not pasta, was the mainstay of Venice because maize was grown on the mainland close by. But without the kind of processing done by many cultures around the world, maize is deficient in vitamin B_3 or niacin. That lack was common in the Veneto, especially among poor farmers or peasants.[26] Basically, pellagra is a disease of malnutrition, and it is with us today around the world, the result of a one-crop diet made from dry corn. Pellagra first presents as a skin disease, but it has many complications that arrive in a cascade of misery, one of which is dementia—the reason why pellagra victims ended up at San Servolo.[27] Photographs taken of these patients upon arrival and then when they were discharged show the dramatic changes that could happen when they had a more varied diet and better food, even for just a few months.[28] Pellagra victims went from madness and illness to normality simply with a dietary shift. Based on nothing, Saliero fed these patients lots of beef broth, and that may have done the trick. It wasn't until the 1930s that the connection with niacin was discovered. What stands out in this story of San Servolo as a place for the mentally ill is that it was enlightened in simple ways—the poor and undernourished were fed and housed, those with what we now call mental illness were observed, the doctors tried to understand their mental illness as a biological phenomenon, and the cures included use of locally sourced plants and herbs that were prepared right on site. Taken together, life as a mental patient in San Servolo appears to have been humane compared to other asylums at the time that were rife with neglect, mistreatment, and abuse. Yes, patients were sequestered on the island, but for most—the poor, but also the rich who were hidden away by their families—this was probably not such a bad thing. In essence, the treatment of the mentally ill and pellagra victims on San Servolo was a public health policy aimed at not just helping citizens but also understanding them. And given that patients from all classes were sequestered there together, Venice is probably the first Western health care system where rich and poor were treated equally.

PUBLIC HEALTH IN VENICE

Taken together, the use of these various islands by the state was one arm of an organized and strategic plan to maintain public health in Venice. Those plans were written and adopted by the members of the government council and financed by the republic, that is, by taxes from individual citizens. In 1485 the Venetian state established the *Magistrato alla Sanità* (Ministry of Health), a committee of three that had clear goals and lots of power. This ministry ran for over four hundred years and was in charge of keeping Venice, as a city and a population, healthy. Their broad range of duties included designing and managing the process of quarantine; managing the plague islands; oversight of physicians and barbers who also performed medical procedures; making sure the sewers continued to flow, garbage was removed, water was clean, and where dead bodies went; monitoring how the insane were treated; and attempting to control plague, smallpox, and venereal disease.[29] They also put into place various official documents including certifications that denoted health or illness or suspected illness.[30] During the various plagues, they managed the function of the *pizigamorti*, those employed to gather up and bury the dead on a daily basis.[31] Venice, by the hand of the *magistrato*, was also the second city, after Florence, to use inoculations for children to ward off diseases and the first city to provide such vaccinations to the public for free.[32] The ministry also kept an eye on prostitutes, beggars, and the homeless.

By 1721, the Venetian health ministry was considered the best in Europe.[33] Their overarching policies included detaining anyone at the border who showed any signs of illness, health inspections and written clearances, disinfection of persons and goods, and isolating people away from others in quarantine, are still with us today because they work against contagious diseases.[34]

Quarantine eliminates the possibility of more hosts, and it does so in the most matter-of-fact way—by keeping the diseased away from the disease free. Today, quarantine is a major cornerstone of public health around the world, guarding against everything from the common cold to Covid-19. Those suffering colds and other viruses are enjoined to stay home and wash their hands. Those with more serious infectious diseases are subject to national and international quarantines, although the rate at which people now move around the globe makes spotting and restricting infection difficult. But quarantine has always involved a conflict between stopping contagion and suspending someone's civil rights, such as the right to travel without inspection and health clearance. Obviously, with quarantine rules come the possibility of officials rejecting travelers or immigrants for non-health-related, prejudicial reasons.[35] But it's no surprise that the history of quarantine as a public health policy leads right back to Venice, a place based on collective community values and at the same time an open-door policy because of the economic necessity that came with trading partners and ships carrying imported goods. Public health is also still organized by the government as in medieval Venice, paid for by taxes, and aimed at protecting the collective. For example, in America the CDC and the United States Public Health Service are inheritors of the powerful, efficient, effective, and Venetian *Magistrato alla Sanità*, even if no one working at either of those organizations is aware of what we all owe to Venice's aim to protect the health of its citizens.

UNIVERSITY OF PADUA AND REVOLUTIONARY MEDICINE

In 1405 the city of Padua, just west of Venice on the mainland, became part of the Venetian Republic and remained so until the fall of the republic in 1797. Unlike Venice, Padua was not a place of trade, not a city built on money, or the seat of a maritime empire. Instead, Padua was, and still is, a city dedicated to learning, and the University of Padua (Padova) is the heart of that academic life.[36] The university was founded in 1222 when a group of scholars and professors split from the even older University of Bologna (the oldest university in Italy) because they felt their academic freedoms were in peril. In protest, they moved to Padua because that city was known for its dedication to individual freedom and equality. Usually, the pope's blessing was needed to start a university, but Padua just did it.[37] The university's motto has always been *Universa universis patavina libertas,* underscoring that the University of Padua was established on a foundation that anyone could learn anything. That motto has been followed for eight hundred years in research and teaching.[38] More importantly, that sense of academic freedom attracted and fostered some of the greatest minds in human history, and in doing so it revolutionized science and medicine and established the way medicine is practiced today.

When Venice took over Padua, the university was already famous and the republic wisely fostered that reputation rather than attempting to subsume it into Venice. In fact, there was a clear separation between Venice, the economic seat of the Venetian Republic, and Padua, the intellectual seat. It is reasonable to suggest that the Venetian state didn't want academics mucking up their well-oiled merchant empire, and surely the academics were happier on the mainland and not under the close scrutiny of the government.[39] But both cites shared the belief system of intellectual freedom and liberty compared to other Italian city-states at the time. Also, Venice might have been a Catholic city, but the rebellious and well-known disconnect between Venice and the pope was a defining factor in perpetuating and supporting freedom in that city and then in Padua. For example, Galileo Galilei taught at Padua for eighteen years and was never in trouble with the government or called a heretic. That changed when he went back to his native Tuscany. Padua was the first university in Italy where even non-Catholics could study, and for that reason, among many, the school attracted students from all over. For example, Nicolaus Copernicus, a Pole, studied medicine there in the early 1500s, and William Harvey, an Englishman, first worked out his ideas on blood circulation there when he entered medical school at Padua in 1599. It was also here that Elena Cornaro Piscopia, a Venetian noblewoman, became the first woman in the world to receive a PhD, in 1678. But that didn't happen easily for her. Piscopia was not allowed to attend lectures and she could only study at home where the learned came to her, and was not allowed to fraternize with male students. Her first attempt at a degree was rejected by the mayor of Padua, who had that authority, because he felt that her major, theology, was not seemly for a woman. Undeterred, Piscopia switched to philosophy and received her degree in 1678. She graduated at the age of thirty-two and died at thirty-eight. Some might say her death was the result of her struggle for an education.

Venice showed its support for the university with financial backing, and the members of the university returned that support with loyalty to Venice. For example, when University of Padua mathematics professor Galileo Galilei perfected his telescope, he brought it to Venice and showed it to the doge. The doge's council quickly realized the telescope's military potential—now they could easily see their enemies entering the lagoon. Venice also instituted a rule that anyone born in Padua or Venice could not be a professor because, as per usual with that government, they were paranoid about nepotism and greed infiltrating into any organizational system.[40] The academic freedom in Padua was the catalyst for the minds of geniuses, philosophers, and plodders. Together those scholars started the scientific revolution and changed the very face of medicine.

At the start of the 16th century, European notions of health and disease were based on the writing of Hippocrates, a Greek physician born in 460 B.C., and Galen, also of Greek origins but practicing as a Roman citizen in the 1st century. Both felt that disease came from vital humors corresponding to elements of the human body. There were four—black bile (representing earth), yellow bile (fire), blood (air), and phlegm (water)—and all these bodily elements needed to be in balance and of good quality for a person to be healthy. This system was better than previous ideas that the gods or evil spirits were responsible for illness because it made tentative connections with real live body fluids such as blood and bile. It was at the University of Padua that the idea of the humors was first rejected and new ideas about disease took their place, ideas that have been confirmed over the centuries. At Padua, medicine was divided into theoretical and practical disciplines. The theory was based on Galen, Hippocrates, and renowned Arab physician Avicenna (Ibn Sina) who is considered the father of medicine. The practical application of that theory was based on both Avicenna and fellow Persian physician Rhazes (Muhammad ibn Zakariya al-Razi) and the analysis of previous cases.[41] These two approaches soon combined with the Paduan innovations of human dissection and bedside hands-on experience to mold medicine as it is practiced today.

THE BIRTH OF HUMAN ANATOMY

One of the first professors to break with the idea of humors was Andreas Vesalius, a Flemish-born anatomist and physician who started his career and made his mark on the future of medicine as a medical student and then as a professor at Padua. Now considered the father of human anatomy, Vesalius moved past the traditions of Hippocrates and Galen by regularly dissecting human corpses. Galen, he discovered, had only watched his assistant cut up Barbary macaque monkeys and therefore had never gotten his hands entangled with blood and guts, and had never really understood the human body. From Vesalius's dissections at Padua, where he also made students cut up bodies to see how they had functioned, he published *Tabulae anatomicae sex (Six Anatomical Tables)* in 1538. The book had clear and detailed illustrations and most of it was at odds with other treatises on anatomy. Soon after, Vesalius published what is considered to be the first anatomy textbook, *De humini corpus fabrica* (The structure of the human body). It was printed in Switzerland and appeared as seven volumes with finely drawn dissected bodies posed in ways that showed off the subject at hand, such as various muscles or skeletal parts. Vesalius was a master at dissection and his book was the very foundation of anatomy, physiology, and medicine, not only because it differed so radically from past ideas of vague humors but also because he got everything right. He corrected misconceptions perpetrated by Galen, such as the fact that the mandible is one piece, not two, and that men and women have the same number of ribs. He also set out the correct structure of the heart. In these drawings and explanations Versalius erased fourteen hundred years of wildly incorrect notions about human physiology. His work also showed that observation and hands-on practice were the only ways to conduct medical practice and make sense of health and illness. And it was through dissection, autopsies really, that anatomists realized that the cause of death sometimes came from the failure of organs.[42] Now we investigate living bodies—CT scans, X-rays, blood work, urine analyses, biopsies, the stethoscope, and blood pressure cuffs are among many instruments of investigation—but all that began with dissection and seeing how the body worked and what could go wrong. Vesalius spent the rest of his life living in other places as a court physician and such, but his influence is felt today every time a medical student opens a textbook or a child looks through a pop-up book of the human body.

Dissection as a method of study was possible for Vesalius because the university and the republic allowed him the perfect space and lots of bodies. Fifty years later, dissection had become so normal at Padua that in 1595 humanist and botanist Girolamo Fabrici d'Acquapendente designed the oldest permanent anatomy theater in the world with help from Father Paolo Sarpi, the believer in separation of church and state (see chapter 4).[43] The theater is on view today at Palazzo Bo, the main building of the University of Padua. The Venetian state had come to a compromise with the church about dissection and allowed four to five official autopsies a year. The bodies of dead criminals were ferried over from Venice, and the dissection took place at the bottom floor of the elliptically shaped and very intimate anatomy theater.[44] Students stood in the six tiers above the dissection table, leaning over the wooden balustrades to view the open body. The theater was was lit by torches and candles; without any windows the air became pungent. To make the experience more pleasant, a chamber orchestra played from the top tier giving the demonstration a classy and intellectual flair.[45] Unknown to the authorities, there were also many "illegal" dissections every year there. Medical students were sent out to rob graves of the recently dead and sneak a body in. If someone opened the theater door uninvited during the secret autopsy, the anatomy tables could be flipped and the body dropped into a box underneath where it remained until the corpse could be removed.

Several physicians and anatomists at Padua in the 16th century also made particular "discoveries" within the human body that today we think of as obvious body parts. Anatomist Realdo Colombo was the first person in Western culture to correctly describe pulmonary circulation, and d'Acquapendente identified bits of membrane inside veins and christened them "valves." In the late 1500s, Venetian priest Father Paolo Sarpi, a statesman and theorist, accurately described the contraction of the iris. But Gabrielle Fallopio was the anatomical star in the mid-1500s. He correctly described the complete workings of the inner ear,[46] noted the sternum was one bone (not the Galenic seven), gave us the term "fallopian tubes" for the connection between the uterus and ovaries, thus countering Galen's pronouncement that male and female reproductive organs were analogous. He also apparently discovered the clitoris, although he might not have been aware that it was an organ of female sexual pleasure.[47] Fallopio was also a practical man; he was the first person to describe a condom and then test it as a preventative against syphilis. After gathering some sexually active men who might have been at risk for syphilis, he asked some of them to wear a condom (a piece of linen soaked in some unnamed chemicals) and then checked for syphilis after sexual behavior. Those who wore condoms did not contract the disease, upholding Fallopio's hypothesis. This study is one of the first examples of a clinical trial.

THE BIRTH OF MODERN MEDICAL PRACTICE

There was also a growing cohort in the Padua medical school that was responsible for dramatically changing the actual practice of medicine, changes that have been little altered since that time. For example, Dr. Gian Battista da Monte made his students take their noses out of Greek texts and go to the bedsides of the sick, and in doing so he pioneered the practice of clinical medicine in 1539. It is difficult for us in modern times to think that doctors used to not even see their patients because medicine was about theory, not observation.

In general, the very idea of evidence-based medicine, something that is put forth today as if it were new, was invented at Padua in the 16th century. On March 29, 1561, Santorio Santorio (with other combinations of names including Sanctorio, Santorii, and Santorious) was born in Capodistria, in what is now called Slovenia but back then was part of the Venetian Republic. He became a physician, then a professor at the Medical School of the University of Padua, and then spent the latter part of his life in Venice as a Venetian citizen. Santorio was also friends with humanists, intellectuals, and geniuses like Galileo who also taught at Padua from 1592 to 1610.[48] Santorio and Galileo were in sync about one necessary change to medicine: everything could and should be measured, put into numbers of some sort, and analyzed, and understood. Yes, Galileo was the genius inventor who perfected the 32-degree telescope while he was at Padua, but above all Galileo was a mathematician, and he saw life in numbers. Although the exact story is not known, it seems that Galileo and Santorio worked together on an idea that others had suggested and produced an instrument that shows changes in ambient temperature. Called the thermoscope, this simple device was made up of a glass with liquid and a thin glass tube with a bulbous top end placed in that liquid. As the air temperature rose or fell, so too did the liquid. In other words, they perfected the thermometer. In essence, the thermoscope made real what the body felt—a hot day or a cold day. Santorio went on to add mathematical scale to the thermoscope, making it even more "scientific" and recognizable as today's thermometer.

As recounted in chapter 2, Santorio also invented the wind gauge and water current meter, both necessary for a life at sea but his other inventions were medical, and we know them well. The *pulsilogium* was a kind of pendulum made with a ruler, a silk cord, and a ball of lead. It worked by setting the pendulum swinging and then noting when the pendulum and someone's heart rate matched and looking at the corresponding number on the ruler.[49] The object of this exercise was to measure heart rate and therefore cardiac function; the *pulsilogium* was the first precision instrument of human wellness and first medical measuring device in history.

Santorio was also deeply interested in how the body functioned, what made it tick. A recently discovered book of Santorio's, found in the British Library by medical historian Fabrizio Bigotti, shows that around 1603 Santorio had rejected the common belief in the four humors, that is hot, cold, dry and wet, as the main actors of the human condition. Instead, Santorio suggested that there were infinitesimal "corpuscles," tiny things that we would now call cells, that were the basis of life. He had seen them in his urine and looked deeply into eyeballs and seen these fragments there as well. But how the body worked day to day was Santorio's intellectual or scientific obsession. He wanted real live evidence about the daily function of human physiology, and he knew that the only way to really study body function was to be his own subject. In particular, Santorio wanted to know exactly what happened to all that food and water he took in every day. In his mind, quantifying everything was the only path to knowledge, so he put together a giant scale and hung it from a beam. It looked sort of like a bed or a very big chair, as the small models at the University of Padua and various illustrations show.[50] Although weighing devices were routinely used back then for goods, Santorio intended this scale for himself. As he sat on this device, Santorio invented the first human weight scale, parent to the very same item now in medical offices and home bathrooms. When the physician's assistant tells you to step on the scale each year, or more often in the face of chronic illness, he or she is making a bow to Santorio and his idea that body weight is a significant indicator of human health and illness. But Santorio had something entirely different in mind when he sat on that scale every day for the next thirty years. He wanted to know what happened to food and liquid intake each day, how much was wasted and how much the body used to run itself. He weighed every single thing he ate and drank every day, and his corresponding daily output of excrement and urine. The result was for every eight pounds of food ingested, only three pounds of solids and liquids were excreted, suggesting that his body ran on the other five. We use this very concept today when we count calories, not weight, and then try to decrease that amount while upping our exercise to force the body to run on calories stored as fat. Following the math, Santorio also realized that "insensible" perspiration, or moisture leaving the body quietly through pores of the skin and through breathing, was how his body utilized most of his daily intake of food and liquid. This endlessly painstaking, and surely often annoying to his family, record-keeping was the first time anyone had made a case for what we now call metabolism. In essence, metabolism is the process by which a body takes in nutrients and water, expels some of it, and uses the rest to keep the body working. In 1614 Santorio published *Medicine Statistica*, the first study of basal metabolism and the first book that attempted to quantify the workings of the human body. Santorio's mathematically rigorous and long-term experiment on himself, and the publication of his book, are also the forefathers of the Fitbit, Apple Watch, or any self-tracker apps we use today to quantify caloric intake and output as a matter of good health. We owe to Santorio the idea that various aspects of physiology can be quantified and the basic theory that the body is an organic machine that is subject to quantification and analysis over time, no matter how the stars are aligned.

That self-conducted experiment by Santorio also formed what we now called "evidence-based medicine." Many of us believe that medicine has always been based on proof of care, that it is a science, but the truth is that much of medicine today is done on a try-it-and-see basis, which is why medicine is often called an art rather than a science. Even more common these days, physicians prescribe medications off-brand, essentially testing on the go the use of a drug that was never developed for that condition and was never officially evaluated for the safety or efficacy in that situation. They are just using us as human guinea pigs and trying it out. Presumably, this is why "evidence-based medicine," the idea that care should be based on studies that show actual efficacy seems so new, but it was born in Padua by revolutionary thinkers who dared to test previous centuries of canonical beliefs about how the human body operated.

That turn toward evidence-based medicine at Padua in the 1500s was also underscored by an active publishing and printing business in Venice that produced "handbooks" for physicians and surgeons. They also published larger and more detailed volumes we would call textbooks today. In 1472 Paolo Bagellardo published *De infantium aegritudinibus et remediis*, the first book on pediatrics and the first printed medical book. Physician and polymath Fortunio Liceti was teaching philosophy at Padua when he published *De monstrorum causis, natura et differentiis* (*Of the causes of monsters, nature and differences*) in 1616, where he described all the various genetic deformities that might appear in fetuses and newborns. The illustrated version appeared in 1634. Although Liceti was the first to suggest that diseases that occurred during the fetal stage could result in deformity, he took all this a bit far when he included mermaids and wolf children as real phenomena. His theory of headaches as mental lightning also showed he could go wildly off track. But still, Liceti correctly described the various stages of the human embryo and understood the placenta; he even suggested the then revolutionary idea that a woman contributed as much to a fetus as a man. But then he thought the female contribution came from all over the woman's body and that the actual embryo was present in those parts. One could take this as a genius idea of cells, or as a weird fantasy, like mermaids.

Decades later, in 1700 professor of medicine Bernardino Ramazzini wrote *De morbis artificum diatriba* (Diseases of workers) and established the discipline of occupational medicine, presaging the Industrial Revolution and what that kind of work could do to human bodies. Professor of anatomy Giovanni Battista Morgagni published *De sedibus et causis morborum per anatomen indagatis* (Of the seats and causes of diseases investigated through anatomy) in 1761 after he had been a professor at Padua for fifty-six years. He was eighty years old. That book, in five volumes, went beyond the only earlier pathology book printed sixty years prior and it outpaced that volume's methods and experience. Morgagni's book became a classic of pathology because of his attention to detail and his long experience dissecting bodies. He also noted death did not come calling through some ephemeral mist engulfing a body but from diseased organs and other body tissues. In that sense, he was surely the father of modern pathology. In the 19th and 20th centuries, Paduan professor Achille de Giovanni pioneered the idea that heredity could be the cause of diseases or abnormal conditions. At the time the field was called constitutional medicine, but today we might call it genetics. He also began the discipline of endocrinology.[51]

PHARMACIES, CURES, AND
HEALTH GUIDEBOOKS

The first botanical garden in the world was planted in the city of Pisa in 1544 and the second in Padua in 1545. The Padua garden is, however, the only medieval botanical garden that has stayed in the same place for close to five hundred years, and the site of the first conservatory ever built. Today the Orto Botanico di Padova is a famous town garden, a large circle surrounded by water and a wall, full of fountains and sundials, dotted with exotic plants brought in over the centuries. The aim of the garden now is research and preservation, but back during the medical revolution, it was an essential part of that reform, because the plants grown there and how they were combined addressed diseases of the newly described human machine. The Orto was first and foremost an "academic" garden planted to serve physicians and their patients, not really a place for Sunday excursions or to grow market vegetables. Before the Orto was built, medicinal plants in Europe were typically grown in monastery gardens. The monks were in charge of production and distribution along with their sacred duties. The monastic recipes for cures had long histories and were primarily based on Arab medicine. But the Orto was nondenominational and had been established specifically to grow "simples," the name for plants used in medical care, the pharmaceuticals of their day. We might now call them "medicinal plants," but back then they were medicine.[52] These plants were then dried, crushed, minced, vaporized, made into pastes and poultices in pharmacies all over the Venetian Republic. The fact of the garden and its close association with the medical department in Padua was unusually modern, but even more unique was the fact that these remedies were a regulated business. Three hundred years before the Paduan garden was producing plants for medicine, Venice had a law in place to regulate their distribution. From 1258 onwards, the Venetian Republic required that any plant treatment must have a prescription from a physician before it would be filled. In 1297 the Major Council of the government passed a law that regulated both the production of medicine and its sale.[53] According to that statute, pharmaceuticals had to be processed and sold at state-run shops, they were to be made with the highest-quality products, and formulas could only be made by licensed practitioners. By 1474 producers of simples could also apply for patents for their products, since Venice was the first place to have patent law as a regulator of commerce and inventions. Patent scholar Giulio Mandich quotes from the original Venetian patent statute concerning medication: "All medicines, syrups, and confections are to be made from the best materials available and to be sold only by those licensed." And, "If a physician makes a medicine based on his own secret, he, too, must make it only of the best materials; it all must be kept within the guild, and all guild members must swear not to prey on it...."[54] Obviously, pharmacists belonged to a guild and early patent law upheld their rights as a group.

Evidence of the pharmaceutical history of Venice is all over the city. The oldest pharmacy, Farmacia Santa Fosca, is still in operation on Strada Nova, one of the busiest Venetian shopping streets for locals. Go in and turn left to buy a Band-Aid or get your prescription filled. Turn right and you enter the preserved original pharmacy with its curlicued walnut cabinets, majolica containers lined up on shelves, and traditional Venetian lanterns hanging from the beamed ceiling. Or visit the island of San Servolo and see the apothecary there where they grew plants and used them to treat all their patients over generations. Here, too, the lovely ceramic jars are intact, the scales still work, and the mortars and pestles could be used right now to pulverize cinnamon or deer horn. This pharmacy has a particular history, since it produced medicines for the poor after the fall of the republic in 1797, medicines that were paid for by the remaining confraternities (see chapter 4) until they could no longer do that charity work after 1826.

Special pharmaceutical mention goes to the product Venice treacle, better known as theriac (*theriaca* in *veneziano*), a concoction that for centuries was a common cure-all around the world. Although theriac started out as an antidote to poisonous bites—Aristotle said it would also keep the plague away—it soon gained a reputation for curing just about everything. It held that reputation for centuries, perhaps because opium was also one of the sixty or so ingredients.[55] By law, the Venetian brand had to be mixed outside designated apothecary shops under the eyes of the public and the Venetian government.[56] Production followed set stages for adding ingredients and the instructions for the length of time for boiling it all down to a dark brown sludge and then allowing days of fermentation. Public visibility was meant to ensure quality, but when the potion did not work, bad ingredients or faulty production were blamed, not the medicine per se. It included things like snakeskin, spices such as cinnamon, and herbs like St. John's wort. Once these ingredients were smashed into powder and mixed together, honey was added to make the medicine go down.[57] Venetian theriac was manufactured from the 12th century until the 19th century, and it was considered the best-quality theriac and was exported all over Europe.[58]

As the confraternities provided health care for the poor, at least until the fall of the republic, and the government regulated everything to do with health, Venice made another contribution to the way we enact our own healthy living—the self-help book. Along with mass production of "handbooks" or textbooks for physicians, anatomists, surgeons, and pharmacists, there were also books for the common person on diet and lifestyle. That book list includes the first printed cookbook, *De honesta voluptate et valetudine* (On honest indulgence and good health), by Bartolomeo Sacchi came out in 1474. Sacchi's book was not just about cooking; it also provided ideas about healthy living. A hundred and fifty years later Giacomo Castelvetro published *A Brief Account of the Fruits, Herbs and Vegetables of Italy,* the first heathly cookbook. In 1558 Luigi Alvise Cornaro, an eighty-eight-year-old nobleman born in Venice but living in Padua, wrote *Della vita sobria*, the first treatise on how to live a long life, essentially kicking off the self-help book market. Cornaro also became the first person to write about the possibility of food allergies. Following in Cornaro's footsteps, Venetian alchemist Isabella Cortese published *The Secrets of Lady Isabella Cortese* in 1561, with all sorts of recipes for improving one's health and looks (see chapter 6). This particular self-help book was also the first book of cosmetology. Taken together, these books form the foundation for mountains of cookbooks and self-help books that now adorn our shelves as we try to better our health in some "natural" way. If only we had a bottle of Venetian theriac to make all that reading and cooking and improving a much less difficult route to good health and a happy, long life.

To these early men and women of science, the body was not so much sacred as it was an interesting subject for study. By yanking the human species down from the pinnacle of celestial glory, these men also paved the way for the very idea of evolution by natural selection for all organisms, including humankind. They had the support of a liberal government that believed in the individual's right to free thinking, and they were also unhampered by adherence to the dogma of the Catholic Church. The revolution in science and medicine at Padua paralleled the scientific revolution of the 16th and 17th centuries in other places, a transition from an organismic view of the world influenced by celestial bodies and the cosmos to a mechanistic vision based on mathematics. The idea was to measure everything and "make measurable" that which could not clearly be measured, to quote Galileo. That testable world was then subject to observation, hypotheses, and repeated experiments, forming what we now call the scientific method. And that method, we now know, can be applied to any study of nature, including the human body and its diseases.

From the scientific revolution that began in Padua we have natural laws that explain the world and an approach to the human body that paved the way for fighting disease. The anatomical and medical publications of the Padua faculty and all the years of hands-on experience that went into them demonstrate how essential the University of Padua was to the new face of medicine and the evolution of medicine today. The site of that revolution at Padua with the financial help of the Venetian Republic also underscores how freedom of thought and practice attracts great minds and those minds build on each other to create intellectual revolutions that change the world.

CHAPTER SIX

Venetians Invented Consumerism

Mastery of trade was God's gift to the Venetians.
> —Elisabeth Crouzet-Pavan,
> *Venice Triumphant*
> 1999

Perhaps more than any city, Venice was built on worldly goods.
> —Patricia Fortini Brown,
> *Venice Reconsidered*
> 2000

It was said of Venetian ships that they left such a profusion of pungent aromas in their wake at sea that they could be detected miles away given the right wind.
> —Andrea di Robilant, *Irresistible North*
> 2011

That's all you need in life, a little place for your stuff...That's what your house is, a place to keep your stuff while you go out and get...more stuff!
> —George Carlin, *A Place for My Stuff*
> 1981

THE WORLD'S BEST museum of stuff is in Oxford, England. Admittedly, there are lots of museums around the world that are repositories of human goods. Art, archaeology, and ethnographic museums display the things people have created, but they do so with care. Paintings in this place, sculptures on another floor, cultural artifact by ethnicity, bits and pieces of the past set out chronologically. But the museum in Oxford is quite different. Called the Pitt Rivers Museum after Augustus Pitt-Rivers, the man who first put it together in 1874, it is dedicated to a more garage sale approach to all the detritus of daily life.[1] You enter the Pitt Rivers by way of the Oxford University Museum of Natural History (the passage to the Pitt Rivers is tucked behind a fully articulated dinosaur) into what seems to be a three-story cross-cultural clutter of junk. Orientation to one section or another reveals a novel approach to material goods—objects are not displayed by country or culture but by type. Boats are over here, objects that mark off time in that drawer, combs in a glass case and shrunken heads from all over the world hanging together in their macabre glory. Most ethnographic museums that celebrate and educate the many ways that humans make and use objects are designed to explain cultural belief systems, to push the viewer out of his or her cultural envelope and see that there are many ways to mark a life stage with ritual or make a living by hunting, gathering, herding, or manufacturing. But the Pitt Rivers seems to be making a much more celebratory point—that humans all over the world make all sorts of stuff. But why?

It is a mark of human intelligence that humans can fix problems and we often do so by employing what anthropologists stiffly call "material goods." Things, in other words. This tendency to make and use objects is surely evolutionary. Some of our monkey and ape relatives with whom we share a genetic history are occasional tool users, but our closest primate relatives, chimpanzees, are more creative and adept. Chimpanzees break off twigs, modify the length just perfectly, and use them as fishing sticks to access insects in giant anthills. Chimps also crumple leaves into sponge wads to soak up and drink water, and they select appropriate stones to crack open tough nuts.[2] After the human lineage separated from the common ancestor with chimps about five million years ago, we made tools of our own; first evidence of human "tools" dates to about two to three million years ago. These tools are purposefully modified rocks presumably used for hunting and butchering. From that point, over millions of years and all around the world, only humans have invented, created, fashioned, made, manufactured, and become dependent on objects to aid our survival. In addition to the survival mode aspect of material goods, we humans are also apparently compelled to create and covet objects that have no practical value at all, except for their perceived beauty. It seems reasonable to suggest that the attraction to shine and sparkle came after the essential need for various objects, but in the end, humans have combined their aesthetic and practical sensibilities to become the masters of material goods.

Our unity as a species is underscored by this collective engagement with stuff as both a way to get through life and to have some fun. All those items are also used to construct and uphold our identities. We often cling to all these things, useful or not, because we have become attached to whatever meaning they might hold, what they meant in the past or how they might be used in the future.[3] These days, modern consumers feel defined by the things they own, and therefore they have to keep everything. American television programs such as *Hoarders*, the organizing advice of Marie Kondo and others, and the many stores selling storage designed specifically to organize all this stuff demonstrate the role of material goods in our modern world, including the consequence of all this owning. Our love and use of things in extremis have also resulted in mountains of discarded stuff we call landfills. Maybe our ancient ancestors were more practical about their stuff, but we are now a runaway train of consumption. Our very economic progress is fueled by this compelling urge.

We are now *Homo sapiens consumerensis*, and Venice played a major role in pushing or leading us along that path.

THE ART AND COMMERCE
OF VENETIAN TRADE

The most amazing, and one of the defining facts of Venice is that it took what should have been a disadvantage—a small, isolated place surrounded by water—and became a giant of commerce.

The Age of Exploration (or Age of Discovery) in Western culture spanned two centuries starting in the 1400s. International sea voyages required not just decent ships but also great navigators and maps (see chapter 3). Elementary school history books portray these voyages as adventures of great courage captained by heroes who braved great uncertainty and danger to see new lands. But the truth is that these voyages were nothing but business trips. They were not about discovering interesting new places or people; they were simply looking for new sources of goods that might appeal to the folks back home. Later, voyages such as Captain Cook's or Darwin aboard the *Beagle* were more set on cataloging the variety of people, animals, and plants in foreign places, but they came as afterthoughts to the global exploration for wealth. The great trading city-states and nations during the Age of Exploration (or perhaps better named the Age of Making Money) included Venice, Genoa, Pisa, Portugal, Spain, Flanders, the Netherlands, and Great Britain. They opened up the world to the West. The globalism we have today is founded on those voyages and on their goal of trading goods and making a profit. Complex networks of voyages across the oceans and seas pushed European economies to grow, made for pockets of extreme wealth, and financially uplifted many city-states and countries. Importing desirable goods gave birth to globalism and mass consumerism as the heart of modern economics. Historians mark this time for international trade as the Commercial Revolution (the seven hundred years between 1000 and 1700), a system that seems to have great sticking power—it still runs our lives today.[4]

Venice's history as a trading nation came about because Venetians were, unlike other traders, without any other options if they wanted to make money. Venice had no endemic resources, except fish and water, no place to grow crops for sale or fields for breeding animals such as pigs and cows. There were no forests to cut down and no mines to be dug up. By all rights, the early Venetians should have just starved to death or faded into the background of other city-states who had more to work with. But from the beginning, Venetians had that uplifting sense of a unified collective that served them well. The collective consciousness was as a people set apart from a hostile world, exiled and alone in a vast lagoon. That identity was part myth but it brought all the lagoon people together as a community, a governmental entity, and as an economy. If they wanted to remain as lagoon residents, they would have to take advantage of what they did have—the lagoon, easy passage into the Adriatic, and from there easy sailing (or rowing) into the Mediterranean. They harnessed that geographic gift and the sea became their métier. In that watery option, they soon found out that the best way to make money was by moving things around, buying and selling the goods that others produced or harvested.

And so, they became not just merchants and traders but also middlemen, a role they supported with their naval might, especially within the Adriatic. Given that just about everybody in town was involved in the merchant business—as sailors, stevedores, dock workers, shipbuilders, warehouse owners, shop and stall owners—the government of high-ranking, but also working, merchants readily chipped in with laws and regulations aimed at protecting Venice's mercantile income. The Venetian skepticism toward authority was also a great advantage for encouraging business. The pope was resisted, and the doge had no real power. Yes, the nobility after 1297 held the city in a tight economic grip. But still, everyone in the city benefitted as Venice, the government and the social entity, became more financially successful.

The government also made choices that furthered trade. In 1082 Venetian sailors supported their sort of Byzantine rulers in a war with the Normans and in return smartly asked for, and received, the right of tax-free shipping anywhere west of Constantinople—that is, the eastern Mediterranean. When the Crusades came a decade later, Venice was well placed to be the shipper and supplier for those Crusaders, and to make a huge profit for the effort in the bargain.[5] The Venetian Republic's manifest destiny could have been oriented toward conquering territory and expanding their power as other European city-states had done, but instead, Venetians quite famously made a collective decision to focus on money, not land or power.[6] When they did try a foray into empire building, it simply didn't work, and probably added to the city's eventual decline.

Venice's mercantile history began more on land than water. The merchants Marco Polo and his uncles, for example, set off from Venice by ship but were soon traveling to China across land. Other Venetian traders traveled long distances in caravans that took years to stock up with goods to sell back home (see chapter 3).[7] Those land journeys began and ended on water because there was no other way to reach home. Venice also had the geographic advantage of grand markets to the west and north. Early Venetian traders had at their disposal the complex river system of the Po Valley on the mainland and entry into markets across northern Italy. And once they crossed the Alps, all of Europe became reachable customers.

Only later, as ships got bigger and more sophisticated, did this mercantile nation take fully to the sea, going out into the Adriatic, taking over a necklace of convenient Adriatic ports where they could dock and sell. From there they entered the Mediterranean in a flotilla of merchant cargo ships. At the height of her trading excursions, about 3,300 Venetian merchant ships a year sailed out into the northern Adriatic (then called the Gulf of Venice), hitting Venetian-owned ports on the west and east as they moved south into the Mediterranean.[8] Then, ships routinely looped up through the Bosporus past Constantinople and into the Black Sea. They headed for their established and very busy trading ports, such as Soldaia and Tana, where goods from the Far East had landed. Ships also went to the Levant, the eastern Mediterranean, hitting Crete and Cyprus (which they owned) on the way, then down to Acre and Alexandria. "In the fifteenth century, say, a Venetian ship need put in at no foreign harbor all the way from its owner's quay to the warehouses of the Levant," writes historian Gerald Mandich.[9] Think of it like a traveling businessman's route, one covered over and over, a routine with established sellers and buyers. And because they were coming and going from many different cities, countries, and regions, they also had to navigate political strife, which could open or close a port overnight.[10] It's hard now to imagine the kind of sea traffic this involved, but old prints of Venice portray the Giudecca Canal, the main canal through which boats arrive in Venice then as now, and myriad types of boats. Some are passenger boats, but most are hauling goods, loading, and unloading. The city was also an open market. There were vendors in stalls all over but especially in the Piazza San Marco and other areas right on the water. The whole city was essentially a loading dock, since ships could anchor anywhere around its circumference. Many of them simply put down a plank and invited merchants and citizens aboard to choose their items.

It is from this activity that Venice gained the reputation as a prosperous and wealthy city, especially since all that commercialism was packed into a very small space.[11] As historian Elisabeth Crouzet-Pavan describes, "Shops and stalls tended to invade all available space."[12] That same commercial hustle and bustle saturates current tourist areas near Piazza San Marco today. Walking down streets from the Middle Ages, such as the Mercerie, a street that has always been the main shopping street for outsiders, is exactly like strolling through a shopping mall. One simply can't help but look at the glittering windows and be captivated by the goods. In that sense, Venice has remained, or perhaps is now experiencing a resurgence of, the kind of commercial capital it once was. The place is full of imported goods for sale (lots from China), alongside items made in Venice, places to eat and drink, and crowds spending their money.

During the trading centuries, the waters were packed with cogs and galleys, dozens of oars jutting out into the water, masts holding sails aloft and a general sense of chaos that was probably finely tuned for efficiency. Standing on the southern edge of the main city today, you'll see the same circus of boats pass by, everything from public water buses to traditionally rowed gondolas. Archival prints certainly make clear the endless comings and goings, the loading and unloading of Venetian trade. Because they were picking up goods that had been shipped or caravanned to their port cities, Venice was the broker that not only brought items into Venice but also pushed and pulled them up rivers to the rest of Europe.[13] The state also owned the ships flying the intimidating flag of the Republic of Venice, the lion of St. Mark, underscoring the Venetian sense of righteousness. There were, of course, competing cargo ships in the Mediterranean and Adriatic, and Venice "allowed" other participants in trade, but for a price. The Venetian military used its impressive naval power and skill to intimidate anyone who crossed the Adriatic.[14] They closely tracked all boat traffic with naval patrols, required non-Venetian ships to first unload in Venice no matter where they were headed, thereby wiping out any other middlemen, levied taxes on their loads, and outpriced competitors on goods, creating forced monopolies on many items. The naval patrols were a great safety net for Venetian traders but scary for everyone else. And Venice wanted it that way. They wanted the competition to cower.[15]

Venice was also happy to work with, or make a deal with, anybody. Although a decidedly Christian or Catholic country, they traded with Muslim countries, even when at war with some of them. They were also happy to house Crusaders as they gathered together to set out for Jerusalem; since Venice was considered the point of entry between Europe and the East, this made sense, and Venice took advantage of that fact. They operated hostels for Crusaders and sold them supplies for their journey. The best scam was the offer of indulgences (basically prayers) from God to the holy pilgrims to cleanse their souls of sin before embarking for the Holy Land.[16] Venetians used that very Catholic Fourth Crusade for Venetian economic gain, not the glory of God. When the Crusaders arrived in Venice they wanted to go to Egypt and from there into Muslim-held areas, but they were lacking ships to carry the devoted and supplies to keep them alive. So, they struck a deal with Venice that involved not only Venetian ships, rowers, and sailors but also food and other supplies. In other words, Venice funded the Fourth Crusade. In return, the Crusade detoured from the goal of entering the Muslim Middle East through Egypt to conquer the port of Zara on the east coast of the Adriatic and then on to ransack Constantinople, a city full of trade goods and traders. Venice gained both these cities as part of their empire and trading network. In other words, Venice strategically used a religious crusade to further her trade routes and wealth.[17] In essence, Venice was more about money than it was about amassing political power or owning land, except places like Cyprus or Crete that could produce agricultural products destined for the Venetian trading machine. Venetians were not a bunch of merchants selling trinkets here and there but voracious and talented businessmen backed by military might on the open sea. Lots of trading happened around the world before Venice sent out her ships, but no one did it more efficiently, effectively, and ruthlessly than Venice. As Frederic Lane has written, "Venice used her power to organize trading on her own terms."[18] Venice also had little conscience about its business deals. For example, although Venice did not use slaves for their rowed galleys, they were happy to move slaves from one port to another, making money in the process.[19] And they dropped slaves off in their colonies, such as Crete, to become indentured workers in the cotton and sugarcane plantations.[20] Some were brought back to Venice and put to work in shops and homes.

As international trade grew, so did the governmental bureaucracy that stuck its nose into every aspect of the business. As Venetian Alvise Zorzi writes, "In fact, the government knew only too well that the Republic's economic survival depended entirely on state control of the markets themselves, and the availability of vast sums of money to cover the staggering costs of maintaining her maritime supremacy."[21] Except for the part about wanting to maintain a maritime state, these very words could be written about first world economies today. The United State is in the middle of a trade war with China and other countries; Britain is creating havoc by pulling out of the European Union, and that havoc is mostly over commercial goods; and the health and well-being, let alone the economic privilege of being a first world citizen able to buy a lot of stuff, is now dependent on governmental regulation and economic diplomacy. And just like today, their capitalist spirit produced inequality between the rich and everyone else. At first, individuals owned the cargo ships and financed voyages, but eventually the oligarchy that ran the state after 1297 realized they could put regulations into play that would increase their personal coffers. The oligarchs decreed that the large round ships (cogs) that hauled mountains of cargo could still be built in private shipyards and financed by individuals, but the state took over the building of galleys at the Arsenal. Galleys are sleeker, faster boats, perfect for luxury items of trade, such as spices and fabric. Those particular goods didn't take up room and they were light weight, yet resulted in high profits. The bigger and slower cogs were more appropriate for mundane, and less profitable, items such as salt and wool.[22] Every year the state then auctioned off leases (mude) for convoys of galley trade ships, and the bidder had to ensure that lease by proving he had an income or estate that could cover the financial loss of the proposed load. As a consequence, the noble class of merchants then monopolized the import and export of goods that had the highest profit level.[23] Their overreaching governmental power also meant these noble merchants could coordinate the timing of trade expeditions and oversee the cycle of the entire trade market.[24] The rich just got richer, but they didn't manage all this on their own. Noble merchants needed the cooperation of non-nobles for the system to work. It went like this: The winner of a muda made a deal with a non-noble citizen to take a boatload of goods to other ports and pick up items that might sell in Venice and beyond.[25] And so they had to have cooperative and well-paid assistants at home and abroad as employees or partners in the endeavor. In the end, the collective of Venice, not just the noble class, underlay a system of successful trading that benefitted everyone, just some more than others.

The ruling oligarchy was so invested in making money from trade that they also put their governmental power to work enticing foreign merchants— their customers—to the city. They built housing for trade delegations and gave them lots of space to use as warehouses. There, these delegations could unload their goods and stock up on what they bought in Venice to take back home. Some of these traders' residences still exist and have been repurposed. The Fondaco dei Turchi (now the Natural History Museum) on the Grand Canal was for Ottoman traders and the Fondaco dei Tedeschi at the base of the Rialto bridge was built for German traders to both live in and to store their goods (fóntego means "warehouse" in veneziano). Making it easy for foreigners to use Venice as a trading base might seem like encouraging the competition, but Venice made lots of money off their deals within the city. By going to Venice, where so much had been imported and stored, these foreign traders could stock up without going farther afield. And the government made money when it levied taxes on goods coming and going. Today over two hundred emblems that denote trading entities survive on the balconies of the Fondaco dei Tedeschi. Venetian historian Alberto Toso Fei has called these inscriptions the oldest collection of medieval corporate insignias, or brands, in the world.[26] Over time, that building became the post office of Venice, and more recently it was remodeled into a luxury shopping mall, thus repurposing the building back to its original function. Although allowing residences and warehouses for foreign traders seems like a generous business perk, it is also possible that these accommodations were aimed at sequestering foreigners and keeping an eye on them.

The many layers of the ingrained and successful mercantile business of Venice have also been documented in various "merchant manuals" of the time. The oldest surviving such manual from any country or city is the *Zibaldone da Canal* written in Venice primarily in the 14th century and some of it in the 15th century. Historians have used the date 1320 as its origin and assume it belonged to a young noble male named da Canal who was preparing to take over the family business.[27] *Zibaldone* means "notebook," underscoring that this manuscript was not a literary track or textbook but a practical manual. It appears to be a multiyear journal of a young man learning the trade, although it looks like more than one person wrote parts of it—perhaps other members of the family. Keeping a running journal like this was a common practice at the time.[28] From a practical standpoint, the *Zibaldone* is full of arithmetic problems that might have confronted a merchant, and it always uses Venice and her shipping as examples. "If two ships sail from Venice in July...." The *Zibaldone* also has comparisons of weights and measures, advice on the condition and price of goods at various ports, and most importantly, currency exchange rates. At that time in the West, every city had its currency, so it was essential for a trader to know what money was worth no matter where he was. The manual also comments on what to expect at various ports, what the condition of goods might be in a certain port, and how to judge the quality of certain spices. For example, one entry translates as, "There are no state monopolies in the Tunisian village of Djdjelli and the people are nice."[29] It walks the reader through simple math problems then progresses to the complexity of international trade, just as a young man would be adding that information to his career. At the same time, the *Zilbadone* is also full of poems, stories, astrology, and whatever; it is a personal document with a practical application, much like the personal journals kept today that are full of reflections, memories, and goals, a document of someone's life but with their career at the center. Since there were no schools in Venice at the time, learning from others and this sort of manual was essential. Another Venetian merchant manual, *Tarifa zoè noticia dy pexi e mexure di luogi e tere che s'adovra marcadantia per el mondo*, came later in the 14th century; this document is preserved in the *Archivio di Stato* in Venice. Calling it a "tarifa," derived from the Arabic word that means "to inform" or "make known," decries its use. This small volume includes customs duties and the costs of export and import, and it is decidedly Venetian with Venice shipping as its center. Two hundred years later, in 1503 Bartolomeo Paxi added to this list of *tariffe* with *Tariffa de pexi e mesurei*, which tracks Venetian trade routes in detail, pinpointing the ports where various spices and such were picked up and dropped off. It gives a real insight into how the trade business was conducted and why it was so profitable.[30]

These manuals, combined with the global reach of trade, also initiated a body of laws spread across countries called the Law Merchant. These laws protected traders in foreign cities and made for standardization of the rights and duties of merchants. Locals and foreign traders could go to a Law Merchant court and complain and be given restitution. This overarching Law Merchant system was the foundation of what today we call commercial law or the Uniform Commercial Code.[31] Given the perils inherent in trade by ship—bad weather, shipwreck, pirates, illness among sailors, and a possible drop in price as a ship pulled into port—these standardized laws were essential to ensure that global trade functioned. Like any business regulations that might end up in court, the Law Merchant meant every ship kept very good records of their cargo. These surviving accounts are one reason we know so much about the Venetian trading network and how it worked.

THE BASIS OF VENETIAN TRADE— SALT, GRAIN, CLOTH, AND LUMBER

In international trade, there was always the hope, the search, for gold and gems, but it was also obvious that there was a broader market that could be enticed to buy more common consumer goods, such as food, clothing, and building materials. For Venice, it started with salt.[32] Just as people were beginning to populate the lagoon in the 6th century, they turned to the salt in the lagoon and sea as an obvious resource. Salt was important back then for seasoning, but even more important for preserving meat and fish at a time when there was no such thing as refrigeration in a climate that had a short cold season. Salt was also used in the production of cheese, so it was important throughout Europe. Early Venetians harvested salt from human-constructed drying pans on the various islands that make up Venice today and on islands farther out in the lagoon. From the 9th century, they were loading that salt on barges and porting it through the Po Valley for barter.[33] One resource claims that there were 119 saltworks around the lagoon in the late 1200s with the majority in the southern end.[34] Early Venetians had used landfill to build a peninsula on the edge of the lagoon, topped with the ports of Cervia owned by the archbishop of Ravenna, and Chioggia, which is still part of Venice. Both made excellent areas for salt production and a great rivalry. Today, the salt-works in Cervia established in 965 still produce a special salt that is sold around the world. Shallow areas were sectioned off, and the salt water flowed in. Over time, the sun dried up the water and a field of salt was left behind, a field that could be harvested by shovel and loaded into barrels for shipping. Eventually, the salt producers constructed a series of pools and moved evaporated and increasingly salty water from one section to another, making for a sort of factory line of salt production. The end of the line was the point where the salt formed crystals and sank to the bottom where it could be raked out.[35] In the late 1100s, the Venetian government realized they could make more money as brokers, rather than producers, of salt. As it was, they were shipping mountains of salt up the Po Valley, and people knew a moneymaker when they saw it. Venice started importing salt around 1240 from around the Adriatic, then putting it in warehouses and selling and shipping it to other cities.[36] Then they went even further. In 1281 the government legislated the salt trade when they instituted the *Ordo Salis*, a regulation that required all incoming Venetian ships to arrive with a load of salt as ballast; up to 33,000 tons a year were unloaded.[37] The city paid high prices for this imported salt, essentially an attractive subsidy, so no one complained. The salt was stored in twenty or so warehouses around the city; the most notable surviving cellars are at the Custom House, or *Dogana*, at the mouth of the Grand Canal. They now host art exhibits in their cavernous spaces; one is the home of the Bucintoro Rowing Society, the oldest rowing club in Venice. Merchants who brought in salt from other places in the Adriatic or beyond made so much money from the salt subsidy that they could then finance and risk a long trip to pick up more profitable, but riskier, goods such as spices and silks. Since Venetians paid well for salt, they squeezed out all the competition and controlled the market. In other words, the control of the salt trade pushed up the initial price of salt in the area, but that cost could be made up as the salt was moved into the rest of Europe by Venetian merchants.

Over the next centuries, Venice made exclusive contracts with various Italian cities and held the monopoly on salt in the Mediterranean and across Europe in its tight fist. All of the salt business was managed by the *Magistrato al Sale* and it kept track of every single grain. No matter that it meant sailing farther and farther away to pick up salt and transport it to Venice to fulfill the demands of the markets they ruled. The city also used its salt profits as loans to merchants for further trade and fed into a growing economy. As a civic duty, the city took some of the money made on salt and used it to maintain public buildings. They also used it to deal with the water damage that constantly plagued the city. "The grand and cherished look of Venice, many of its statues and ornaments, were financed by the salt administration," writes historian of salt Mark Kurlansky.[38] The government also used the money they made on salt to make sure even the poorest Venetians had something to eat. The state bought and stored grain in their giant grain warehouses.[39] On the northern arm of the Grand Canal sits a boxy building with not so much decoration called the Fondaco del Megio (*méjo* is the word for the grain millet in *veneziano*), built in the 13th century. Today this building is a school, but centuries ago it warehoused wheat and then the less digestible, less desirable yet hearty cereal millet which came in handy when the wheat supply diminished in Venice. In 1507 there was a major shortage of wheat, so the doge ordered local bread to be made out of millet, a grain that no one liked, especially those with little money and few alternatives. The doge was thereafter ridiculed as the "Millet Doge," underscoring Italians' long-standing refusal to eat bad food.[40] Another grain warehouse, the Fonteghetto de la Farina, operated from the 16th to the 18th centuries around the corner from Piazza San Marco. Today it stands in ruins and an excavation is in progress. At the moment, the dig is best seen from a bedroom window in the Napoleonic Wing (really the Napoleonic Palazzo) of the Correr Museum in Piazza San Marco. The sheer size of the foundation perimeter shows how much grain was regularly dumped there to feed the city and be made available for export to other places. That building was also the

Office of the Magistrate of Flour, the agency that tracked, taxed, and levied tariffs on wheat. Another grain warehouse, on the eastern edge of the city in the district of Castello, began as a salt warehouse in the 13th century, then turned into a grain warehouse in 1322.

Other staples such as corn, wool, cotton, and lumber supported the income from salt. Timber was felled on the mainland of the Po Valley and barged down to Venice and from there out to places around the Adriatic. But this natural resource soon gave out once the area was denuded. Given that Venice is built on a forest of tree trunks, they eventually had to scour the Mediterranean for lumber. There was a thriving business in making cloth from wool and cotton based on imported resources but manufactured in and around Venice.[41] By the 14th century, cotton, in particular, became a major trade good for Venice, second only to spices.[42] The cloth market, south of the Rialto bridge, filled huge buildings, where traders and citizens could stock up on the finest fabrics in the world. The guild system in Venice also helped protect the quality and sale of these products.

There might be a romantic notion that Venetian wealth and prosperity were based on the spice trade, or through buying and selling luxury items, but those presumptions are false.[43] The majority of trade was for more mundane essentials, and the luxury market was simply the cherry on top.

TRADED LUXURY

Trade in luxury goods in Venice was more profitable per load than the more everyday items, but that high trade carried enormous risk. Shakespeare's *Merchant of Venice*, written in the late 1500s, outlines that process and explains how the wealthy merchant Antonio is almost brought to ruin because he has financed four trade ships and they are all lost. He no longer has any capital and owes money to Shylock, a moneylender. Antonio had unwisely cosigned a loan with Shylock to help out his friend Bassanio, but the fortune he was counting on is now lost at sea. Shakespeare could write that twisted plot because the Venetian trade system was well known in Europe.[44]

The trade in exotic goods soon became synonymous with Venice's identity as an open bazaar between East and West. But it's not as if Venetians were sailing to the Spice Islands to get the goods. They were, instead, rowing or sailing along the eastern coast of the Adriatic where they had access to Grecian goods; to Alexandria, Egypt, for items that had been brought by others from the Far East; into the Black Sea to established ports where land traders from everywhere brought goods; sometimes up to Flanders for coveted European goods such as wine, paper, and refined silk. They brought this all to other ports on the way home or brought it back to Venice for sale. The trade networks for these more "luxury" goods went back and forth, and Venice was always the middleman, moving goods around and making deals. But those deals also required on-site loyal agents, people who lived in the target country or port, who gathered items from various sources and coordinated the offloading and unloading of goods. As such, the Venetian trade network was not just land and sea routes; it was also a complex hierarchy of personnel who had local knowledge and signed contracts of commitment. They, too, were interested in making money and they did so on commission by percentage.[45] To assure this arrangement, Venetians invented a new kind of economic contract called the *colleganza* (which translates today as "working relationship"), whereby merchants stayed at home but paid for the ships and everything necessary for a voyage, while their agents did the real work of sailing to foreign ports and interacting with other agents.[46] As economic historians Giovanni Tassini and Loic Sadoulet explain, "Companies today still exploit these mechanisms to succeed in underserved markets."[47]

Venetian historian Andrea di Robilant calls Venetian ships of that period "floating emporiums." Their cargos included all kinds of food, including candied fruits, sugar, wine, wool, herbs and spices for cooking, and cosmetics. Venice also exported and imported leather goods, furs, timber, fruit, soap, animals, and medicine.[48] Imagine a floating big box store with everything on it anyone might want or need. Now we have cargo planes that move consumer goods from place to place, but the cargo ships of today are mere echoes of the fleets of Venetian (and other) trade ships. A cargo ship today might be bigger and faster, but nothing could be more diverse in its cargo than a Venetian cog.

Spices were a specific success of Venetian trade. Spices arrived in Egypt and from Black Sea ports from South and Southeast Asia, both by ship and by land. The point was to market them in the Levant and Europe by way of Venice. The spice ships were not the good old cogs used for big items like wool and lumber, but sleek rowed galleys with armed guards.[49] The best part of spice as a trade item, unlike lumber, for example, is that the product is relatively small and lightweight, and it could be divided into small expensive packets for sale, upping its profit margin. And they were easily transportable in sacks and barrels, didn't rot on board ship, and were oh so coveted in Europe. The list of spices traded during the centuries of Venetian trading was extensive, and the most important was pepper.[50] This spicy spice was used in cooking, preservation of food, and as medicine. Venice ruled the spice trade from about the 9th century until Vasco Da Gama sailed around the Cape of Good Hope in the latter 1500s and opened up another way to get spices to Europe. Yet Venice had such a solid system for this particular commodity that it took another hundred years for their spice trade to dwindle in the face of competition.[51]

After experiencing silk from the East, a silk industry in Venice was initiated closer to home. The true center—where they grew mulberry trees, harvested cocoons, made silk thread and cloth—was the city of Vicenza on the mainland, about a thirty-minute train ride from Venice today.[52] In the way agriculture supports manufacturing (think wine) the area around Vicenza was uprooted and planted with vast forests of mulberry trees to feed the silkworms.[53] Making silk cloth was a labor-intensive process, from cocoons to looms, that required great skill, but it was also the cloth that was in high demand and had a high price point. Venetians made some silk on the mainland and they also imported silk thread from around the world.[54] Manufacturers employed cost-effective hydraulic silk mills along the rivers and canals of the mainland to replace much of the handwork performed by transient day laborers. Eventually, workshops opened in Venice and fabric was also made there. The Senate encouraged this manufacturing endeavor by granting several patents for new spinning machines that could be managed by one person—an example of early automation that replaces jobs and continues to plague the modern industrial workforce.[55] There are many other Venetian patents granted to inventors in the silk industry that protected the design of various internal parts of silk-making machines. But then came the usual heavy-handed control on the industry aimed to bring more money to Venice proper. In 1457, the Senate decreed that before it was boiled and dyed, silk had to be physically taken into Venice, inspected, and the manufacturer had to pay a customs duty on the lot. By 1505 the city put a tax on silk and divided the categories into raw silk and other silk that had been made into cloth or ribbon. Given the luxurious clothing of the 15th century onward, the tax on silk was a big moneymaker for Venice's coffers. Dresses, blouses, shirts, pants, hats, and jackets were made of silk. So were curtains, pillows, and bedcovers. Silk was draped about a house or palazzo, and there was silk ribbon and silk trim on everything. All these silk items were also sewn together with fine silk sewing thread also made by the same industry.[56] Compared to wool and cotton, silk was the luxury market, and in the mid-1500s Venice and its environs were producing thirty-five to forty tons of silk fabric a year.[57] The silk business was healthy into the late 18th century, when there were still over one thousand looms in operation in about five hundred workshops across the city.[58]

Other products were produced in Venice and on the mainland and then sent out on their trading routes (see specific examples below). Those goods succeeded because of the reputation for the high quality of Venetian-made products and the support of the government for manufacturers and inventors of new products that could add to the Venetian economy.[59] In other words, "Made in Venice" became a successful brand.

LACE

The contemporary Lace Guild of the United Kingdom claims that although no one knows who or where lace making was invented, the first city to be known for lace was Venice.[60] Lace making, most believe, is based on the practical need to tie things up. Specifically, men who made a living catching fish needed large nets to scoop through the water or trawl along the seabed floor, and they needed to make and repair those nets by hand. Fishing nets require tons of knots—all kinds of knots—and they require great patience for sitting quietly and squinting. In some sense, lace making is the female version of men making and repairing their nets. Burano is a famous fishing island in the Venetian lagoon where tying nets was part of daily life. So, it makes sense that Buranelli women might take to lace making. In fact, the individual stitch of Burano lace is called a *sagolà*, a word close to *sacolà* (fisherman's loop), says the Venetian-born ethnographer of Burano, Lidia Sciama.[61] But according to some Buranelli, lace making began when a mermaid enticed a fisherman with her song. She made him a lacy wedding veil and then other women saw that veil and attempted to copy it. Another legend claims that some sailor gave a girl on Burano a lacy-looking plant called "mermaid's lace" and she made a duplicate in thread.[62] A third fable also involves a mermaid, but in this version the origin of lace came when that mermaid slapped her tail on the side of a boat and produced a lovely pattern.

There are four kinds of handmade lace. Knitted lace is made with two sticks and can be either fine or chunky and is most often done in wool. Same with crocheted lace, but that type uses only one needle with a hook on the end. Bobbin lace is perhaps the best known; various strands of silk or cotton thread are wound around small wood spindles called *fuselli* in *veneziano*. They are passed over and under each other while the knots and lines that move the thread in various directions are pinned to a pillow in some delicate pattern. Bobbin lace was always made around the Venetian lagoon and it's possible to find old bobbins in antique stores in Venice. New hand-turned ones can be bought in one artisan woodshop. But Burano lace, called needle lace, is a very different kind of lace artistry. It requires only one needle, just a simple sewing needle, and there isn't a bobbin in sight. A pattern or design is first drawn on a piece of paper, usually brown wrapping paper, and then covered in a bit of waxed paper and sewn with normal stitches onto a bit of cloth. That layered contraption is then pinned onto a cylindrical pillow called a *tombola* (a tube of cloth full of soft batting and sometimes a utility drawer to hold needles and thread). A much smaller rod of wood is placed between the design and the pillow providing a hard surface for passing a needle down and up while it traverses the pattern and creates a picture. The lace maker must sit on a chair, hopefully with her feet resting on a step that lifts her knees to a 90-degree angle, and holds the *tombola* in her lap. Once the needle is anchored in the design, she passes it right or left, always returning to the edge of the design to anchor the turn in a previous stitch. Most designs have sections that are cross-hatched into a net of webbing, but the most amazing parts are curlicues of flowers and such highlighted with nodules of knots. Only in doing this work can one fully understand how it's done, and how very hard it is to do perfectly. The subspecialty of needle lace invented in Burano is called *punto in aria (ponto in aere)* meaning points floating in the air. Unlike knitted or woven fabric, and unlike bobbin lace, it clearly shows the network of threads that backs up the piece. This lace blossoms as an intricate web of twisted and knotted threads that are not anchored down to anything but each other. With no substrate, everything appears to be suspended in space. And it is.

The paper pattern is carefully removed once the piece is completed, and only close examination (a magnifying glass would be of use here) shows how the stitches hold it all together.

At first, much of the lace produced in Venice happened in cloistered convents where lace making was a noble craft approved by God. Lace making was also an occupation for those who had no other means of support or nothing better to do, or women with an uncertain place in society who needed to stay out of the public eye. Spinsters, for example, could take up what we now call a hobby but which was an acceptable occupation back then. Women sold handmade goods to cloth makers and clothing designers. Merchants also exported their work to other countries. It was, in the usual Venetian way, these middlemen who made the money from lace, not those who produced it. Lace makers were woefully underpaid, especially given the physical disabilities that came with sedentary work and intense visual scrutiny.[63] Apparently, that close work could make a person crazy. Writer Laura Morelli claims that of the thirty women housed at the insane asylum on San Clemente in 1880 (see chapter 5), twenty-four of them were lace makers from Burano.[64]

The lace business began as an individual artisan craft on Burano and other Venetian islands then expanded in the late 1400s when it was helped by the patronage of Dogeressa Giovanna Dandolo Malipero who was nicknamed the "queen of lace" for her enthusiasm and support. She also supported the nascent printing business in Venice and that trade helped make Burano lace known when the printed patterns were distributed across Europe.[65] Around 1595 another doge's wife, Morosina Morosoni Grimani, loved Burano lace so much that she established a lace-making workshop in Venice with 130 needlers, many of them children and female prisoners,[66] to make lace for her garments and to give away as presents.[67] At that point, what was once a home craft turned into a moneymaking business. During periods when Burano lace was in high demand, workshops operated in an assembly line sort of way. Those good at the netting did that, those better at flowers did that, and pattern drawers stuck to their part of the craft.

Venetian lace was the darling of the noble class and royalty, as evidenced by the many portraits of these fashionistas and consumers. Lace adorned their necks, wrists, hands, hair, and clothing. Lace was also put on curtains, bedding decoration, and even doorknobs. Queen Elizabeth sported those exaggerated lace collars, made in Venice, to complement her ghostly white skin paste also made in Venice. When Catherine de Medici became queen of France in 1547, she brought along Burano lace makers to accent her wardrobe. By the 17th century, the business was in full swing in organized workshops, mostly in Venice. There was also a busy workshop on the island of Giudecca run by nuns who were strict and punishing. They exploited the orphans, former prostitutes, and poor young women who were stuck in their care. Lidia Sciama points out that there has always been a connection between lace making in the Venetian lagoon and the church, a cultural attachment that deeply influenced the mind-set of workers; there was an overarching feeling that the guild system of the republic and the Roman Catholic Church would take care of the lace makers.[68]

The industry began to decline in the late 1600s as the Flemish and French needlers became more adept and began to imitate *punto in aria* with a French version named *point de France*, even though the French version was a cheap imitation of the real thing. The real death came with changes in clothing fashion. Revolutions in America and France in the 1700s were founded on the ideas of equality, so women were no longer necessarily interested in showing off their status. You could lose your head over that. Instead, clothing took on a less lacy air, and Burano took an economic hit. Into the mid-19th century, lace making was in decline, and there were no workshops at all. Women still made lace at home and sold it, but the income was sporadic at best. Then, after Italy became independent and Venice, and other parts of Italy, went into economic decline, there was a movement to encourage the women of Burano to find their way out of poverty through lace piece work. Andriana Zon Marcello, a Venetian countess, became the engine behind the opening of the Burano Lace School in 1872.[69] Marcello had money, influence, and friends in high places, such as Princess Margarita of Savoy, who were good customers. The school had a slow start but soon their work was gaining favor at craft shows and was selling. Royalty was buying again, as were British aristocrats; even the American department store Neiman Marcus stocked Venetian lace.[70] But the school was no school. It was, instead, a sweatshop that required "students" to work long hours for which they received no pay and were pounded with church doctrine. Even after eight years or so of instruction and working for free, when they should have finally received some pay, the young women were only allowed to work in a factory system such as those workshops of centuries before. And in the 1930s and 1940s, nuns were once again in charge of this lace workshop, as they had been four hundred years earlier in Venice, and they were just as stringently moral and oppressive. Ethnologist Sciama speculates that the production of lace in the home and workshops was bound up with the patriarchal need to control women, to contain their sexuality and stop them from having any sort of independence, especially financial.[71] The hope of lace lifting Burano out of poverty was never fulfilled. Well into the 19th century, and riding the waves of war that rocked Italy, the lace business waxed and waned. Buranelli made a wedding veil for Mussolini's daughter and a surplice for Pope Pius XI, but only financial help from the descendants of Countess Marcello staved off the steady decline of students, *maestre*, and foreign consumers who wanted to buy lace. The Lace School of Burano officially closed in 1972, after one hundred years of operation. Only the lace school that spun off from Burano, called Jesurum, is still in operation. And yet there is another minor revival of Burano lace happening right now. It's possible to take an individual lesson from a *maestra* at the lace store Marina Vidal on the island. You can buy lace-making supplies at a local stationery shop. And it's not hard to look past the lace made in China that is displayed in shops around Venice and find the real thing—the price point is the clue. But the most important revival is the Burano Lace Museum (Museo di Merletto), which recently received a face-lift. Drawers and drawers of unimaginable lace are laid out in what used to be the lace school. It is in those drawers that the ethereal artistry of Burano lace shines through.

THE VENETIAN PATENT SYSTEM

Trade was not the only way Venice laid the foundation for the Commercial Revolution. The city was also home to businesses that manufactured items put on ships and sent to foreign ports or carted to other Italian cities and into other European countries. At the time, Venice was a powerful city with a sparkling reputation that carried that brand. Today we are impressed by clothing labels like "Made in France" or "Designed in Italy" because clothing from countries with attention to fine fabrics and good tailoring fit better and last longer. In the same way, consumers during the Middle and late Middle Ages turned to Venice for that kind of quality knowing it was backed up by a well-honed business plan, an efficient way to ship items, and a government that favored making money.

Most of those Venetian-made items were products of trade themselves in that the resources for manufacture had to be first imported into Venice. But the city was a haven for manufacturing principally because it invented the first patent system in history. Attention to the rights of inventors occurred in 1297 with a law that regulated the fabrication and sale of medicine. In this case, the regulation was also aimed at establishing state shops as the sole proprietors.[72] Two hundred years later, in 1474, the state went further with a statute that was basically what we might call intellectual property rights.[73] But the Venetian patent statute was even more strict and protective than the ones we have today.[74]

The *Archivio di Stato* (Archives of the State) building in Venice houses the records of thousands of these documents, called "privileges," which we would call patents. Applicants had to explain exactly what they were proposing and why it would be good for the state and the economy. Note the caveat about "good for the state," a line that does not appear in modern patents. Inventors granted patents also had to share details of their plans, and the state then had the right to take it from there, if it chose.[75] Many inventors made sure their requests were vaguely written because they were afraid of idea theft.[76] If granted, a privilege allowed the manufacturer to work in Venice as the only producer of that product or the only one allowed to use a certain technique. Those privileges usually expired after ten years, much like the patents given out today in many countries. Privileges were essentially monopolies on a product or a manufacturing method or some type of tool to improve manufacturing.[77] They gave legal weight, and the chance to complain legally if an idea or product was imitated. Other European countries followed suit by simply adopting the exact same parameters.[78]

Although there is a current debate about the effect of patent law and intellectual property rights as an incentive for creativity, those laws made Venice a good place to be a manufacturer or inventor. And those laws were instrumental in encouraging foreigners—non-Venetians, that is—to move to the city. In 1434 the state even offered a 5 percent reward on any public invention, especially military invention, if producing it didn't harm the state.[79] The Venetian government even reached out to bring manufacturing and invention into the city. Someone outside Venice could apply for a privilege on a process or tool to gain a monopoly in the city and then export their goods. With patent laws in place, Venice was known as business friendly in the manufacturing sector.

Once a patent ran out, the holder would be integrated into the guild system, so they continued to be tracked, let's say monitored, by the state by way of the guild. For example, the knitting frame came to Venice from Lyon, France, and the manufacturer was granted a ten-year monopoly. Once the patent ran out, the Senate made a whole new guild to accommodate that sort of knitted cloth production. And they did more. As far back as 1272, the city offered foreign wool weavers a free house and ten years free of taxes if they moved to the city. That act was the first evidence of an economic tax inventive offered to a business.

PERFUME, SOAP, HAIR DYE, AND COSMETICS

Venice was well placed to make money from the grasses, herbs, spices, and animal products brought to town from the East and that's how it also became a hotbed of invention for the sake of beauty. Venice's history with Byzantium had brought with it both the knowledge and the mystique of working with fragrances.[80] Aromatic items in particular traveled from the Far East to the Levant, Egypt in North Africa, and other ports around the Mediterranean by way of the Incense Route, which had been around since 2000 B.C. That route was eventually overlaid with the Venetian trade network.

The list of imported plants and animal secretions used in recipes for fragrances, cosmetics, and medicines is extraordinarily long and includes musk, ambergris, civet, benzoin, agarwood, labdanum, camphor, storax, frankincense, astragalus, resin, clove, nutmeg, sandalwood, pepper, ginger, alum, mastic, myrrh, tragacanthin, carnation, roses, almond, mastic, aloe, mace, jasmine, spikenard, rosemary, citrus peel, and alum, among many others.[81] Musk and ambergris were prized items, and they were traded at high prices for the best quality. "Ambracan," which is a "bilious secretion" of the intestines of sperm whales, dolphins, and porpoises found floating in the oceans or washed up on shore, was also coveted. Manufacturers combined these bits in recipes for fragrant soaps, hair dyes, pomades, and cosmetics. Across Europe, items stamped with the brand "Venetian" denoted superior quality.

Soap has been manufactured since 3000 B.C., and its recipe has gone through various iterations. The first soaps were made with animal fat, a sort of industrial product mostly used to clean sheared wool. That soap was meant to be corrosive and stripping, and it was never used on human skin. For Venetian soap and perfume makers, the most important technical advance in this arena was the move from animal fats to alcohol as the basis for formulas.[82] When Venice entered the soap-making business, producers paid attention to the soap makers of Aleppo, Syria, who had already mastered the use of olive and laurel oil as the base rather than animal fat. That change in ingredient made the Aleppo soap much more delicate in its cleaning action and therefore easier on skin. The very same soap can be ordered today on Amazon, where it is touted as one of the few "natural" soaps because of its simple formula of oils without artificial ingredients. Venetians took the Aleppo recipe but instead of drying the slabs of newly poured soap on a factory floor, they dried it in the sun.[83] Venetians called their product *bianco di Venezia* (Venetian white) and it was one of the first soaps used exclusively for personal use and the first time scents were added.[84] This soap was famous and exported all over Europe, stamped with a seal of authenticity as early as 1293.[85] By 1489 the Senate moved to protect the soap business when it forbade the sale of any imported product.

In general, Venice dominated the soap business in Europe between the 14th and 17th centuries.[86] The extent of the soap business in Venice can be witnessed in the many *calli* in the city named for the *dei Saonàri* (or *dei Saoneri*), meaning street of the soap makers. In the 1500s there were about forty soap manufacturers in Venice and they produced about eight thousand tons of soap annually. The rise in oil prices was a death knell to the *saoneri* and as hard as the government worked to keep workers happy, the business declined dramatically in the 17th century.

The same sort of manufacturing shift occurred in the perfume business when Venice initiated the switch from essences diluted in oil or fat to *acqua vita*, or alcohol, as the base in the mid-1500s. That switch created *acqua mirabilis*. This substitution was monumental because alcohol preserves the infused essences and prolongs the life of any perfume. That meant Venice could export the product without worrying about a sell-by date. At the same time, the use of alcohol rather than fats to dilute the fragrances meant batches could be increased exponentially, which meant, of course, greater profits.[87]

Also, there had been a cultural shift in the latter half of the 15th century that put the spotlight on the importance of female beauty.[88] Now there was the compelling need to look good and smell pretty. The general thrust of that marketing ploy was that women could make themselves more desirable to men and the envy of other women not just with nice clothes and lots of fine jewelry, but also with the right cosmetics and fragrances to improve their appearance. Women of all classes bought into the new trend, a focus that haunts us today in magazines, on television, online, and in movies.

The perfume business in Venice began in the late 15th century, when the very first perfume-making shops opened. By the 17th century, their manufacturing technique and recipes had spread from Venice to the rest of Europe. A map in Anna Messina's book *The History of Perfume in Venice* shows that in 1568 there were at least twenty-four *muschieri*, establishments that made perfume, spread all over the city. Fragrances were used on the body, clothing, accessories such as gloves, and household items. Although we can assume nice fragrances helped with the unpleasant odor of a public that didn't do much body or clothes washing, the use of fragrance in Venice was wisely marketed as an extravagance. Venice was also instrumental in changing types of fragrance. In 1709 Giovanni Maria Farina (aka Feminis) created eau de cologne while living in Cologne, Germany. The desire to concoct perfume was surely passed down from his grandmother, famous Venetian perfumer Catarina Gennarri. Calling it eau de cologne rather than perfume suggested a difference. These days it usually means the bottle holds a slightly more watered-down fragrance and is often intended for men. Nonetheless, Farina's concoction was the darling of the heads of Europe; some thought it would ward off the plague if swallowed.

The strong connection between fragrance and Venice was recently underscored by the establishment of the first museum of perfume in the world in the civic museum of Palazzo Mocenigo in 2013. The display contains the original workshop of the Venetian soap and body product maker Angelo Vidal who began making industrial and home-washing products in 1900 and continued for decades. Eventually, Vidal made body soap, talc, and perfume. The company, now owned by Henkel, still sells basic Vidal body products. More interesting, a Venetian spin-off of that company headed by Vidal's grandson recently produced a new line of perfumes called The Merchant of Venice. This line is displayed in luxuriously decorated perfume shops around Venice and in the gift shops of the city's civic museums. The perfume business in Venice is not quite finished.

Venice was also ground zero for the most important cosmetic of the 16th century—ceruse, the white facial paste (let's call it foundation) used by women all over Europe, from Queen Elizabeth I of England to the least noble peasant on the planet. Romans had mixed up a similar product, called *bianca* (white), several centuries before and it was used by men and women. *Bianca*, and its offspring ceruse, is a white lead pigment, and everyone knew the dangers—they were warned of its evils by physicians and the church said women were punishing their vanity with that stuff. Pliny the Elder called *bianca* a deadly poison. But fast forward many centuries and Venetians nonetheless mixed up a special ceruse launched in 1521. They named it Venetian ceruse or the Spirit of Saturn. Venetian ceruse became the most fashionable and most expensive skin whitener available in the 16th century and women couldn't get enough of it and that was a tragedy. The recipe for basic ceruse is white lead powder and vinegar heated together in a furnace (in Venice that meant a glassmaker's furnace) for three to four days. The Venetian brand was the best because it contained the highest concentration of the whitest lead powder. The resulting concoction was then mixed with the ashes of burned green figs and made smooth with a little distilled vinegar. The resulting paste was opaque; when spread on the face it made a lovely satin finish that covered scars from the plague and other skin problems. Women never wiped it off but just added layer after layer. Over time, and with repeated applications, a woman's facial skin took on a grey cast, and then her skin dried up, wrinkled, and aged. Ceruse brought up undertones of yellow, green, and purple, and ended in rotten teeth, bad breath, eventual hair loss, and finally, death from lead poisoning.[89] Yet the consequences did not deter the fashionable.

And they loved the brand name "Venice." It had become synonymous with high quality and fashion. That branding continued for centuries. Elizabeth Arden, the mother of the oldest cosmetic company, initially launched her cosmetics line with forty-eight products called "Venetian" in 1916.[90] That launch was the beginning of the global cosmetic industry, an industry that has staying power.[91] Arden brought makeup into the mainstream with the help of a Venetian logo.

Venetian frippery also included doing things to hair, including changing the color and wearing a wig in the 17th and 18th centuries.[92] Hair bleaching was a long-standing tradition in Venice. Women sat on rooftop decks, called *altane*, and donned wide-brimmed straw hats (a *solana*) with the crown missing. Wrapped in shawls to protect their shoulders, women pulled their long hair through the top of the hat and doused it in some magical potion. The dye was called *bionda* (blonde) or *acqua gioventù* (water of youth) and it was applied with a sponge stuck on the end of a rod, sort of like a selfie stick.[93] As they sat there for hours, the formula worked with the sun to lighten hair and turn it blonde.[94] Once wigs arrived in the city around 1665, there was the need for *polvere di Cipo*, or Cyprus powder used to whiten wigs. This powder was manufactured in Venice in several factories across town; if a street name today includes the word *polvere* it means a Cyprus powder factory once stood there. This same powder, called *cipria* in Italian, is available today as a face powder.[95]

Underpinning the self-care industry in Venice were printed manuals or pamphlets on how exactly to work with homemade products and have a healthier life and more beautiful face and body. These works, many published in the 1500s, explained how to whiten one's teeth, clear up acne, banish freckles, and of course, dye hair blonde.[96] With the printing business expanding in Venice (see chapter 8), the city became the center of beauty advice books and medicinal cures.

The best seller of 1561, by Isabella Cortese, *I secreti della signora Isabella Cortese; ne' quali si contengono cose minerali, medicinali, profumi, belletti, artifitij, & alchimi* (*The Secrets of Lady Isabella Cortese*), was a book of recipes for cosmetics and hair dye and all sorts of household products. The recipes were based on formulas invented by spice sellers, alchemists, and pharmacists. Cortese's marketing ploy was to use the word "secret," to bring an allure to her manuscript. She explained how to cure various illnesses, make disappearing ink, use insecticides to get rid of bugs, make toothpaste, mouthwash, and lots of cosmetics that would improve one's looks and stave off aging. If you want soft hands, she wrote, rub in this mixture—the root of yarrow, wild cucumber, alum from wine sediments, burned white tartar, and broad beans. Grind all that in a mortar and then cycle the paste several times through a process of drying and mixing with egg white until it turns into a fine powder. Add a little water and wash your hands. Blonde hair? Take some soda water, put pennyroyal into it, bring it to a boil. Then wash your hair and go sit in the sun on your rooftop deck. For hours.

No one knows who Isabella Cortese really was; she might have been a man, some suggest. But her book brought chemistry and alchemy to the public. Maybe her directions for turning metal into gold didn't work, but she knew how to glue things together, polish household items, and make potions that would cure just about anything, including erectile dysfunction. The book was wildly popular and went through eleven editions. And she was a savvy author; Cortese's real secret to literary success was making her book a known secret. She cautioned the reader to hold her recipes close and burn the book after they had read it. Nonetheless, reprinted versions are currently available on Amazon.[97]

These recipe books also included medicines, some written by doctors. Of great import then was how to avoid the plague, and there was the misguided feeling that fragrance could someone protect a person. Alchemists and *muschieri* certainly tried, but no amount of flowery, spicy, or citrusy odor was going to stop bacteria. And yet we see that same type of hopeful alchemy at work today in the aromatherapy business that harkens back to the streets of Venice in the 1500s.

GLASS

There is a world of perfection in a Venetian bead. Ones made today sparkle with feather-light bits of gold buried in a swirl of glass, or they blossom with thin rods of glass cane fused into a miniature bouquet of colors. Other beads sport lines or dots of matt colors that crisscross with other colors; there is a tiny bit of circus in every Venetian bead. Most precious are the traditional chevron beads constructed of at least six layers of glass, added one by one as the blob is fired and blown. They become either round balls or more elongated, but always with the various layers spun out in rings of flat points that radiate back down the bead. Old Venetian beads have seen the world as they've traveled in the hulls of ships to be used for trade, and they are tainted with the stain of slavery; huge bags of beads were taken aboard Venetian ships going to Africa where they were exchanged for slaves to be taken against their will. Many of these beads have since returned to Venice and they end up in shops or on eBay.

Venetians did not invent glass. No one knows who did, but history shows that glassmaking was probably developed to mimic naturally occurring glass such as obsidian, a volcanic glass. Early glass pops up in Egypt, Syria, Greece, China, Mesopotamia, Bohemia, and on and on. Historians suggest that Venetian glass had a specific mixed heritage of Roman and Byzantine techniques and style.[98] When early Venetians ventured into the lagoon to live, they presumably brought their Roman skills, gained in the mainland city of Acquileia on the edge of the lagoon. The discovery of bits of glass on the island of Torcello and the remains of a glass furnace complex there shows inhabitants were making glass there from about 700 A.D.[99] The Byzantine influence is also clear in the mosaic glass tiles made in Torcello, the ones still on the walls in the Church of Santa Maria Assunta, dating from the 11th century. There is also a hint about the age of glassmaking in Venice from a party invitation. When the city dedicated a Benedictine monastery on the island of San Giorgio Maggiore in 982, a man named Domenicus Fiolarius, whose name indicated he was a maker of *fiole* (bottles), was a guest.[100] But then Venetians made the process all their own. Sand is the main ingredient of glass, but the lagoon provides little of that. Instead, early Venetians gathered quartz pebbles from the rivers of the Po Valley and ground them into powder. That powder was pure silica, whereas sand is less so. Sand also contains iron, which gives glass a green tint, so the pebbles made clearer glass. Also, Venetian glassmakers imported their soda ash, a product of biodegrading plants, from the Levant rather than Egypt. It was a better kind of ash, and as a result, they produced a superior product and no one could imitate them because the Venetians held a monopoly agreement with the Levant.[101] The Venetian recipe also added lime or calcium carbonate to hold everything together. This mixture was then subjected to intense high heat until it became liquid. Glass is such a marvel because it is easily made, molded into shapes, forever recycled, made into a million colors, and has myriad uses. Just look around. Windows and doors, drinking glasses and dinnerware,[102] eyeglasses, jewelry, lightbulbs, and oh so many kinds of containers. Yes, it breaks, but that's a small price to pay for such a useful human-made product. And Venetian glassmakers were very good at it.

The glassmaking industry of Venice was successful in part because they produced an amazing array of practical and luxury items.[103] They introduced the Renaissance idea of elegant dining with accouterments such as drinking glasses, goblets, vases, bowls, and candelabra. Paintings of this era illustrate these banquets in all their material sparkly glory. Clearly, such a table was an advertisement for refinement, consumerism, and envy. Conspicuous consumption of glass then led to the next stage of consumerism—collecting glass—and there, too, Venetian glassmakers were happy to feed the growing desire for things.[104] Murano glassmakers were also smart enough to produce luxury goods along side items for daily use.

Of course, once Venice became famous for glass, the state stepped in to foster the industry, control workers, and protect them. The most significant edict came in 1291, when the Senate ordered Venetian glassblowers to move to the island of Murano, just off the northern edge of Venice. Restricting glassblowers to Murano might have been considered banishment, but it made sense because there were so many glass furnaces in the city and housing was primarily wood. This decree was, in fact, the first forced exodus of a particular trade to protect a city.

Murano then became the center for glassmaking in Venice and it was well known throughout Europe. It maintains that reputation today, over seven hundred years later. To be a glassmaker on Murano has always been a position of status. Once the Senate solidified the various social classes of Venice, they elevated the status of Muranese glassmakers to the *cittadini originari* automatically making them and their families high-ranking non-nobles, not just laborers.[105] These Muranese glassmakers were immune to prosecution and could wear swords like noblemen. Murano was also granted a respected separateness from Venice. By the late 14th century, they had their own police force, their own small governing body and, of course, a particular dialect and accent of *veneziano*. By 1605 the island also had its own *Libro d'Oro* list of high-ranking members.[106] And when a Venetian noble daughter married a Muranese glassmaker, that was perfectly fine. The state was also a great supporter of their work. It helped out with large orders, which in turn encouraged standardization and mass marketing. For example, in 1295 the state supported the production and sale of a standard-size glass for windows and made sure they were installed in most of the taverns in Venice.[107] All that respect came with a cost, however. The Senate proclaimed in 1295 that glassblowers were not allowed to leave the republic and that sharing professional secrets with outsiders could be punished by death. Glassblowers did manage to leave and share their skills, but two were murdered in Paris by hired Venetian killers.[108]

The Venetian glass business doubled in size from the late 1500s well into the 1700s.[109] They had a strong guild that regulated apprentices, made sure recipes and techniques were closely held, incorporated ideas from other artisans, and fostered innovation. In the 1300s Venetians perfected flat glass, which is why we have windows today. In that century they also invented mirrors of glass backed by tin.[110] Writer Mark Pendergrast suggests in his book, *Mirror Mirror,* that the modern industry of mirrors was born there with the production of large and luxurious mirrors with gilded frames.[111] Perhaps the most important Murano glass invention was the discovery of *cristallo* glass, a crystal-clear, sparkling product invented by Angelo Barovier around 1455.[112] Barovier made this glass by first purifying his raw materials, sifting them through a sieve and passing them through filters, and sometimes re-crystalizing to get rid of impurities, especially any traces of iron.[113] The result was a crystal-clear glass of sparkling brilliance. The iconic glass bead of Venice was also invented by a Barovier, but it was Marietta Barovier, the daughter of Angelo, who first produced the rosette or chevron bead in 1480. That bead style is a concoction built of many layers of glass that are easily counted on the surface by the fluted layers at both ends of the bead. These beads made up the necklace that Cortez wore when he met Montezuma and exchanged it for the Aztec emperor's necklace dripping with gold.[114] In the 1400s the same Angelo Barovier developed a technique to make milk glass called *lattimo*, a glass infused with white lead intended to imitate Chinese porcelain.[115] He also concocted a recipe and technique for a glass that mimicked semiprecious stones. In the 1500s someone figured out how to suspend rods of colored glass in clear glass, allowing for stripped pieces. That technique was pushed forward when the glassblowers understood how to twist the color canes and sometimes have several intertwined spirals embedded in the glass, thus inventing the *filigrana*. These spiral lattices, usually white, commonly jazz up a Murano paperweight, vase, or perfume bottle. In the 17th century a great glass family, the Miotti, infused molten glass with shards of copper hung suspended in the blob of glass and produced pieces that glittered. In the late 1930s, glassmaker Flavio Poli perfected the trick of pushing a piece of colored glass into a larger chunk of clear glass, thus inventing the *sommerso* technique. This style is a great favorite today because it is easy to see that it takes a special skill to suspend a spot of color into a larger orb and leave no connection between the two parts.

It's hard to know if Venetian glassmakers or someone else invented eyeglasses. There is speculation that in the 13th century a glassworker on Murano invented some sort of magnifying lens and placed it in a handheld frame that was used for reading, perhaps the forerunner of spectacles. The only evidence of this is the street in the San Polo district called Ramo de l'Ochialier (street of eyeglass makers) and there is one mention of someone named Giustizia Vecchia, who in 1284 advised that *ochiali* (glasses) should be made from Murano glass and not the usual rock crystal. It is, however, certain that Venetians invented sunglasses. In the 17th century, they manufactured tinted lenses meant for entertainment or to hide behind.[116] These lenses, set in metal frames or a monocle, were multi-faceted and refracted multiple images, like the eye of a fly. Today you can buy the same sort of thing made of plastic, which fractures the field of vision in what is supposed to be a humorous way. In 2014 the Biblioteca Marciana in Venice hosted the exhibit *Spectacles Fit for a Doge; Sunglasses of the Eighteenth Century* curated from museums and private collections. During the 18th century, the exhibit documentation explained, Venetian opticians produced emerald-green lenses for commanders at sea and anyone else in a boat. The idea was to dampen the reflection of the sun on the water. They were called "Goldoni glasses" after the great Venetian playwright.[117] Some were framed like a handheld mirror and were just the thing for a lady to hold up to her eyes on a gondola ride. These glassmakers also tinted corrective glasses in blue or green tones to be worn indoors.

Chandeliers and lamps are, of course, made by glassblowers all over, but in 1724 Giuseppe Briati produced the first *ciocca* (bouquet of flowers) chandelier with polychrome glass flowers. Briati is also credited with reviving the Murano glass industry in 1740 after an economic downturn by bringing the Bohemian method of glassmaking to Venice.[118] Bohemian glass was, in fact, superior in quality to Murano glass at that moment in history, and while it might have rankled to admit that Venetians could learn something about glass from someone else, Briati was also Venetian enough to adapt the traditional process for purely economic reasons.[119] His *ciocca* became the model Venetian chandelier that hangs in palazzi, including the magnificent one in the Ca'Rezzonico palazzo, now a museum of the 18th century. These chandeliers burst with colorful or clear flowers and leaves, and once lit, they send out a planetarium full of light rays that add sparkle to any room. Although there are many contemporary kinds of Venetian chandeliers available on Murano, these beauties still sell well after three hundred years.[120]

When the republic fell in 1797, Napoleon closed all the guilds, including the glassmakers' guild, plunging that business into decline; the only product that came out of the island were beads. French troops also occupied Murano, a demoralizing moment for glassmakers in terms of the identity and the sovereignty that the island once enjoyed. Glassmakers across Europe had already begun to imitate the Venetian methodology (sometimes under the tutelage of expatriate Venetians), disingenuously calling their creations *Façon de Venise*, meaning "in the Venetian fashion," and the Venetian glass business went into decline in the absence of governmental support and in spite of the superior workmanship. Various families held on and several began making glass again in the latter half of the 1800s. Antonio Salviati played a major role in that revival when he figured out how to make glass mosaic tiles and won the contract to repair the murals in the Basilica of San Marco. One mosaic facility company is open in Venice today and a student can take a course to learn their techniques. By the late 1800s, Murano glass was again admired and the business was back on its feet. By the 1950s and 1960s, great quantities of Murano glass were exported. It was also available for tourists at every price point, so the glass factories entered the unglamorous but profitable industrial glass market. During World War II, Muranese made glass covers for navigation lights in airplanes and ships.[121]

Today there are hourly excursions from Venice to Murano to watch the glassblowers and to buy glass. All the main streets of Murano are devoted to shops selling glass, but interspersed with Muranese product are now tons of glass imported from China and other Asian countries. It would seem that the Murano glassmakers would have prevented this, and certainly, in Venice everyone talks about it, but perhaps it's just another adaptation for profit, like all Venetian history. Or they have no way to stop the artlessly made and cheap Chinese glass incursion because the Republic of Venice no longer exists to control the situation. One push back has been the specific MADE IN MURANO sticker that eligible shops display. This stamp of authenticity is affixed to shop doors and products to assure consumers they are getting the real thing, but still cheap imitations abound.[122] "When it's quickly turned out for a cheap profit among the tourist trade, frankly it can look hideous. When it's well done, it takes your breath away," writes Laura Morelli.[123] Choosing low price over good quality is one of the by-products of the Commercial Revolution and globalization that Venice helped initiate.

ART SUPPLIES

Underneath all the glorious art in Venice, much of which is still in place in churches and guild halls, is the moneymaking business of the production of art in which Venice excelled. Unlike artists in other cities of the Italian Renaissance, Venetian artists did not concentrate on fresco for the simple reason that lagoon water always seeps up into walls and frescoes fall off. So, artists in Venice mainly used canvas, and of course, with all the shipbuilding in Venice, canvas was readily available. According to art historians, Venetian sail canvas was not only easy to get but was also regarded as the best quality around.[124] The word *canvas* is derived from the Latin word for hemp, *cannabis,* the oldest fiber in the world to yield a fabric. But in Venice, the canvas was made from cotton, and that particular fabric soon spread across Europe.[125] Venetian canvas is a simple woven cloth with one weft crossing up and under one warp. It can be stitched together, as the great painter Tintoretto often did to produce a large canvas, and nothing is wasted. Canvas also has the advantage of being lightweight. This means a huge stretched and framed canvas can hang from a high ceiling without fear of its crashing down and killing people.

Besides the ubiquitous use of canvas, Venetian painters stood out because they quickly adopted oil paint first used by Flemish artists.[126] The advantages of oil paint over tempura, the other broadly used paint, is that oil is long lasting. At the same time, oil paint dries slowly and gives the artist a chance to manipulate stroke and color. Tempura, in contrast, is bounded by egg yolk and therefore dries quickly and cannot be manipulated much once it hits the canvas or wood panel. The Venetian painter Giovanni Bellini was the original master of oil paint, adding translucent, luminous layers and incorporating gold.[127] Using oil paint, he was able to capture fine details of objects and create accents of reflected light. This medium can also be easily glided against itself or pushed this way and that to give a flat surface a topography. With oils, shadows in garments and curtains take on very real elements of light and dark that give movement to a painting. Bellini's style was the foundation for all later Venetian painters such as Titian (Tiziano Vicelli), Tintoretto (Jacopo Comin), and Paolo Veronese.[128]

Oil paint is made by mixing powdered pigments with linseed oil. Before the late 1400s, pigments were purchased at an apothecary. But in 1493 Venetians created a special occupation called *vendecolori* (sellers of pigments); this was a profession unique to Venice and the city subsequently became a destination for out-of-town painters shopping for pigments. Venice was well placed to have all these colors because the raw materials that arrived from the East were pure and nicely ground for use.[129] The *venedcolori,* also exported these pigments to other countries and cities since such shops did not appear anywhere else in Europe for two hundred years.[130] Lists of *vendecolori* inventories from the state archives show that in the 16th century there was an ever-longer list of pigments, and art historians feel the *vendecolori* had influence on which colors were formulated, how they changed over time, and what painters had to work with. Also, the *vendecolori* shops were a nexus for color workers where they encouraged experimentation and shared information. Venetian painters were the first to "underpaint" in white, pink, or gray before adding layers of other colors.[131] They also often mixed colors to produce subtle gradients of hues and tones.[132] Venetian artists also developed "glazing," a mix of oil and transparent or translucent pigment. These layers shine through each other, like panes of colored glass set on top of each other. Together they create a new color with subtle undertones. Artists sometimes added sand or glass to oil paint to change its color, texture, and level of transparency, and they borrowed color from other mediums, such as the orange used commonly for majolica ceramics. Venetian painters were *colorito* in their approach, meaning they were all about color, while other Renaissance Italian artists were more *disegno*, meaning drawing oriented.[133]

The color palette for oil paint during the era of *vendecolori* was broad. In 1100 Venice first saw ground-up lapis lazuli arrive at its docks and that disembarkation was the beginning of the bright blue color ultramarine (from other shores).[134] The hard to find and manufacture blues were, in fact, the first reason painters started to come to Venice looking for supplies. That long history with blues also explains why the blues in Venetian paintings are blindingly bright and lush even after six hundred years.[135] Pigments also included materials such as lead, mercury, vitriol, and potassium nitrate among many others; *vendecolori* were not only colorists, they were chemists. Venice also specialized in reds, made cobalt blue, and they were known for vermillion and their love of orange. It's not particularly surprising that the first ever specialized art supply shop opened in Venice in 1534.

Art was business in Venice, and it was supported by the government and the community. Venetian artists belonged to a powerful and extensive painters guild, and the "figurers," such as Bellini and Titian, were respected but in the minority. Others used colors in different ways, to dye cloth, paint on ceramics, hand-color manuscripts with inks, and tint clear glass. Coloring was also important to the major part of the painters guild that made their living decorating playing cards, furniture, textiles, books, and leather, and those who laid the magnificent Venetian terrazzo floors or plastered the walls of palazzi.[136] In the Renaissance, nothing was painted without a paid commission in hand, and artists had to bow to the wishes of those who gave them the commission.

Venice, that tiny town, also harbored some of the greatest painters of all time, a fact that is often overshadowed by the reputation of Renaissance art from Florence, a city financed by powerful benefactors. In Venice, that support came more communally from the state, the church, and the guilds, and from wealthy families. During that century paintings became larger, topped by Tintoretto's gigantic *Paradiso*. Measuring seventy-four by thirty feet it was, until recently, the largest painting on canvas in the world. To stand in front of it in the Great Council Hall of the doge's palace is to experience the might of the Venetian Republic. As writer Gary Wills has explained, the state repeatedly expressed its will and power through art.[137] The surprising thing is that so much of that art is still in place, hanging in churches, guild halls, and palazzi, still underscoring how ubiquitous great arts were in the lives of every Venetian.

Two female Venetian artists stand out as innovators of the discipline. Painter Rosalba Carriera was the first woman artist to have a major impact on the development of art. She had started her art career by making lace and drawing patterns for her mother. During the 1700s, she was famous across Europe for her miniatures painted on ivory rather than velum. For paintings, she used pastels exclusively, while others had used pastels only for sketches. And Carriera is credited with initiating the change to the exuberant style of the rococo.[138] After traveling across Europe for commissions, Carriera returned to Venice. She had been unimpressed by royalty, and as a naturally introverted person, and as an artist who suffered from serious eyesight problems, she spent her years in quiet seclusion and died in 1757 in a house a stone's throw from Venice's magnificent art museum, The Accademia.[139]

Venetian painter Giulia Lama was also famous during that century. She was the first woman to stand in front of nude figures, both male and female, and draw them, and her bold practice broke with what was traditionally the exclusive right of a male artist. A cache of two hundred of her life drawings and an extant altarpiece in the Church of San Vitale prove her skill as an artist, but Lama was also an intellect, poet, philosopher, and mathematician, which meant she was a threat to male artists of her time.

Although Venice abounds with great art of the past,[140] much of it *in situ*, the city has also kept up with the artistic times. As posters about town proclaim, there are always respective shows of famous and not-so-famous artists and any number of changing exhibitions of modern international painters, sculptors, and photographers. And there are venues specifically oriented for current work, including the Casa Tre Ochi, which specializes in life-altering photography such as the work of Sicilian photojournalist Letizia Battaglia, the first woman to document mafia killings in Sicily as they happened.[141] Venice was also the first to hold an international art festival. The Venice Biennale of Art began in 1895 as a forum for international artists, and that festival continues every two years, and its format has been copied around the world. What makes the Venice show so unique is how it is housed not only in the ancient Arsenale but also across town in borrowed private palazzi because the exhibit deeply involves Venetians in the showcase.

THE CONSEQUENCE OF CONSUMERISM—SUMPTUARY LAWS

Although a lack of interest in material goods during the Middle Ages had generally morphed into a love of things by the Renaissance, Venetians became such high consumers that it scared the ruling class. What nobles feared most was social unrest—revolution by the have-nots against the haves. Although the economic health of the city required a solid, mutually respectful working relationship between the noble class and the *cittadini*, the nobles had undercut that connection when they'd made specific the rules of nobility and taken over the government and the highest-profit trade ships.[142] The result was disgruntled *cittadini*, the backbone of the trade business, and confused nobles. The lawmakers wanted wealth and exclusive status, but at the same time they also wanted the solidarity that the community provided for everyone.[143] After all, this was a hardworking city full of merchants, who were frugal people; even the so-called noble class held their place because, by and large, their families had earned their wealth and economic class without connection to any royals.[144] But oddly, they tried to correct the imbalance they had created by controlling their own. They focused on public displays of that wealth.

In an attempt to hide all that economic boasting, the Senate passed a series of laws, called sumptuary laws, to make rich people more circumspect.[145] In 1299, and again in 1334 and 1472, they enacted laws that limited the expenditure on dowries and wedding celebrations. And no more outdoor banquets that fed hundreds. The later versions restricted what food could be served and what dishes were used, and they forbade the use of gold forks if your family had them. And no gilded food, please.[146] Opulent palazzi and their decorations fell under the rubric of "what you do at home alone is your own business." Other sumptuary laws addressed dress and fashion. These laws targeted women, in particular, possibly to place a restraining hand on the many highly paid prostitutes and courtesans who walked about the city. In 1430 the height of women's shoes was restricted. There was an extremely high style, called *chopines*, that made walking almost impossible and the nobles wanted that attention grabber to stop. That law was followed by a restriction on cloth woven with gold and silver. When dressmakers slashed sleeves of more simple cloth and inserted silver, gold, or fur inside, the Senate went after that, too.

By 1500 the sumptuary laws began creeping into houses, telling nobles what they could or could not use for drapes and judging wedding outfits, again disapproving of women's clothing but this time even inside the house. And then they went after the gondolas. Rich people rode around in gondolas rowed by their private gondoliers while the common folk had less elegant private boats. So, in 1562—and reiterated in 1633—all gondolas had to be painted black. Previously, they had glided around Venice decked out in all sorts of bright colors, attracting attention and bringing a lot of envy with them. Painted black, the gondolas seemed more solemn, but that color change didn't alter the fact that gondolas were owned by the rich. These laws explain why today all gondolas are black and there is no movement to alter this long-standing tradition.

Sumptuary laws also restricted the wearing of pearls. At first women were restricted to one strand, but since they ignored this rule and the pearls crept back strand by strand, pearls were banished from Venice in 1497. Of course, women still wore them, and of course, the glassmakers of Venice figured out a way to make fake pearls from a mixture of powdered glass, egg white, and the slime of snails. These small balls of glass were dipped in coating that dried with the luster of pearls. They were, of course, not exactly like real pearls because they were smoother and of exact equal size.[147] Venice, after all, has been the European center of glittery jewelry ever since Marco Polo brought diamonds back from China in the 1300s. And none of those sumptuary laws stopped a gem cutter named Peruzzi, who, in the late 1600s when Venice was still the center of the diamond trade, invented a 58-facet cut diamond that introduced the concept of break and star facets. The Peruzzi diamond cut was essentially a triple cut, and it sparkled like no other. This configuration, with some variations, is essentially the round brilliant cut diamond that fuels the wedding ring business today.

The failure of sumptuary laws in Venice is a good example of the nature of *Homo consumerensis*. If the goods are there, we want them. And no call to civic duty, or desire for social peace, will stop us from buying and displaying what we have.

VENICE AND CONSUMERISM

Venice played a central role in what we now call the consumer culture.[148] Venetians not only participated in the Commercial Revolution; they also led the charge. The city was blessed with the perfect geography for a time when trading could be more efficiently done by sea than land, and Venetians were compelled to follow that specific economic path because they didn't have any other way to go. The thread that runs through Venice's commercial success is the fact that it was a small, socially interwoven community. Although the nobility got greedy and tried to separate themselves off from the rest of the citizens, they also seemed to retain a bit of social conscience. Of course they did. After all, it was pretty hard to ignore all the other classes while squeezing down tight *calli* and sharing the same *campi*. In other places, the poor and middle classes are often invisible, but in Venice, they were part of the essential fabric of daily life. And those of lesser means also had the guilds, the church, and their neighbors for protection. Venice was a kind of beehive, everyone working, everyone either becoming rich or getting by. And the identity of "being Venetian" as something special was the glue that tied everyone together.

As heralds of the consumer culture, they were savvy negotiators, traders, and sailors. They kept their eyes on the prize of money and not the expansion of empire. The phrase "Venice has always been about money" is often delivered as an insult, an explanation for bad tourist food and rude waiters. But that sentence is quite positive and could be admired. Modern Western culture runs on money, on buying stuff, an echo of Venice's history. Those who think Venice floated up in its current beauty ignore what built that grand architecture, glorious painting, and all that gilding. It was being unafraid of speculation and risk-taking, state control of the various avenues of business, knowing the supplies and the markets, and not just playing the game but inventing it. That history is why Venice is studied by economic historians and why it should be presented in business courses. Once students get through learning what happened to the American automobile industry, or what happened to General Electric, or why Nike is successful, they should look much more closely at the merchants of Venice. In the end, Venice created consumer demand and consumer culture.[149] And we, now *Homo consumerensis*, have to decide if we are thankful for that example or furious about it and need to stop buying all this stuff to save ourselves and the planet.

CHAPTER SEVEN

Venice Changes Money, Banks, Profit, and Brings Us Capitalism

Among the many cities men have made, Venice stands out as a symbol
of beauty, of wise government, and of communally controlled capitalism.
—Frederic Lane,
Venice; A Maritime Republic
1973

The moneys of our dominions are the sinews, nay even the soul, of this
republic (*Pecunie nostril dominie sunt nervi immo animo huius rei publice*).
—Venetian decree,
Reforms for Accounting Practices
1474

What is the source of this insatiable thirst for wealth that seizes men's
minds?
—Francesco Petrarca,
Italian poet and humanist
1304–1374

OUR ANCIENT ANCESTORS had no money, and probably no reason
to have any. For at least 90 percent of human history, people made their livelihood by hunting and gathering, and there wasn't a coin, paper bill, paycheck,
bank account, or credit card in sight. Transactions (let's call them payments)

were done with what was right there—hunted and butchered meat, gathered fruit and berries, maybe speared fish. Anthropologists assume that these early humans knew the value of their food and traded it among themselves, or traded the cooperation in the hunt for sharing knowledge about vegetable items that were part of the haul to come later.[1] That speculation is supported by the fact that chimpanzees and other primates do indeed conduct what might be considered financial business. African vervet monkeys exchange grooming for support in fights later in the day,[2] capuchin monkeys cooperatively hunt, share food, and seem to understand that food is the reward for helping out,[3] and chimpanzees clearly understand that a token can represent something else, like food.[4] They are, in their way, financiers with credits and debits and mental ledgers, keeping track of all these deals. And for those nonhuman primates, as well as ourselves, all that buying and selling, working and being paid, all that tit for tat, is enmeshed in interpersonal interaction, the social context of moving around things of value, be they work or goods. It's no surprise that the word "transactional" can mean a business deal such as buying and selling but it can also mean a social exchange between individuals. In financial exchanges in our current times, items or labor have a mutually agreed upon value and the transaction proceeds from there. But that transaction has a not so hidden profit. And so, each party pushes and pulls the value of whatever is being offered. It's one thing to have this happen between two people or two monkeys, and quite another when some sort of barter, trade, or transaction happens between populations. Then, the social context not only allows the transaction to occur, but it also colors every aspect of the moment of exchange.

Money is simply the representation of something else that has value.[5] Humans took to money early because what they were exchanging was hard to carry about and they needed something of like value but much more por-table, a symbol with an agreed-upon value. Money started as clay tables incised with words that stated what stood behind those tablets.[6] Their value was in the inscription of a promise, certainly not the clay itself. Other items have stood for money. Cowry shells, the mollusk species called *Monetaria moneta* (*moneta* means "coins" in Italian), were once widely used in Asia, Africa, and North America as money. They were gathered on beaches, strung together, then traded across continents. It wasn't their simple beauty or their easy portability, but the universally accepted idea that they represented value and could, there-fore, be accumulated and exchanged down the road for something of equal value.[7] Beads have also been money; they were appreciated for their decorative prettiness, but also because they were portable and could be passed along and traded for something else.

Money historian Niall Ferguson says that in general, money has to be available, durable, fungible, portable, and reliable.[8] If it is, whatever we use as money can be exchanged for something we believe is of equal value (faith in our assessment is involved here). Two hundred US dollars can be exchanged for a cow as long as someone thinks a cow is "worth" two hundred dollars. "Value," our evaluation of what something is worth in monetary terms, is, of course, the trickiest and most labile of concepts. When money enters the scene, the transaction is no longer an exchange of recognizable goods offered by two parties who can see, smell, and taste or whatever; it is but a moment of faith that what the money represents will hold its value over time. A lucky buyer also hopes the thing will not only hold its monetary value but will also rise even higher and be worth more later. "Money," Ferguson says, "is a matter of belief, even faith."[9] That would be faith that the person paying is offering something of value, and faith in the place or institution issuing the representational money.

We now think of money as coins and paper bills when in fact money is also virtual these days. Often, modern money appears as direct deposits in our bank accounts or automatic payments for our utility bills. We never even touch that kind of money. Also, paying with debit cards has, in many cases, even replaced carrying around real live money. Facebook creator Mark Zuckerberg recently testified in front of a Congressional committee defending his proposal for the creation of a cryptocurrency called Libra. "Sending money should be as easy and secure as sending a message," he said. It already is with cryptocurrency companies such as PayPal and Venmo, among others.

We can also move money around these days, in contrast with the Middle Ages, not only because of technology but also because money in developed nations pretty much holds a value recognized by others, until it doesn't and a loaf of bread costs ten dollars. It takes a stable government with an eye on the markets, domestic and foreign, to support its currency at home and abroad. But a peek at any currency converter app shows that even the almighty American dollar, once the global standard of value and still recognized as such in many places, fluctuates every day.

What money is and how we use it is the basis for an economy—that is, the financial construct of a country. If the economy is healthy, growing and stable, a country continues to be autonomous and functional. If not, that country devolves into chaos. A country's economy is the beating heart of government and everyday life for its citizens. In that sense, everything, absolutely everything, has a financial component, a cost or gain, a paying out and a receiving in, an accumulation or a deficiency; there is always an accounting. Although financial transactions seem, on the surface, to be void of emotion—it's just numbers, after all—we know money is as emotionally laden as love or hate. It can be the root of all evil and at the same time an act of generosity that changes someone's life. Money is often the cause of homicide and suicide, and yet it also buys engagement rings and dancing lessons. It leaves people homeless and builds mansions. Moments of pain and moments of joy all have their monetary implications, and that too is part of the value of money—its emotional power.

And it's not just about having or not having, it's about complexity and integration. Layered on top of coins and bills are institutions—banks, companies paying stocks, government bonds, investment houses, retirement accounts, pensions, import and export businesses, paying for health care, travel, housing, and food. In other words, everything is touched by money in some way. Just about everyone on earth now exists within a complex financial scaffolding that props up or lets down a populace.

And for many of those platforms and pipes of the scaffolding, we can thank, or blame, the Republic of Venice because Venice was the first European financial center.

VENICE AND TRADE BREED MONEY

It's no surprise that Venice, in particular, has had a significant effect on the evolution of money and finance. Having no land to speak of and choosing to make a home in an archipelago meant early Venetians had few resources to aid in their survival. They could eat fish and grow some crops on their dots of land, but eventually, they wanted more, and as the population grew, there were more mouths to feed. In the end, there were only two options—go to war and take over some else's land, or go into business and make money. They chose the second option. Much later, the Republic of Venice tried out the first option but that was a disaster. The better decision, to become businessmen, was a choice to become long-distance merchant traders. Trade had already been operating around the world as evidenced by what anthropologists call "cultural diffusion"—when cultural bits, especially material goods, end up far from home. But that exchange or appropriation happened mostly by land, and it was slow. Marco Polo's seventeen-year journey to China is a good example of such a journey. Also, the Silk Road has a long cross-cultural history of goods being taken from one place to another.[10] For Venice, as an economy, the real push was the salt business, and from there they expanded to become the most profitable mercantile business in the world. Venice held that title until other competitive nations moved out farther across the globe, seeking goods and bringing back even bigger loads for sale. But the basis of all that trade was not so much the direct exchange of goods through barter, although that certainly sometimes occurred. It was money.

Venice's turn to business occurred at a time when Western cites were expanding and many were turning away from the farming. Instead of growing or pasturing things to sell, they started creating a process we might call manufacturing in the primitive or individual sense. That kind of production paved the way for entrepreneurship and initiated the growth of commercialism, the need for a broader marketplace.[11] People began to support themselves not by their own farming and herding but by adding a layer of the production of goods sold or traded for money—which then bought food, drink, clothing, housing, and other necessities. Even large farms that had more agricultural and animal products than they needed began trading or selling their surplus. The population of Western culture was growing, and cities became nations with identities and expanding markets. Venice was nicely placed to join in the economic development of the Middle Ages and the Renaissance in Europe because Venetians had no feudal lords holding them back and a very strong sense of individual independence and entrepreneurial spirit. They also had a bracing self-confidence that they were specifically chosen to be Venetians, different from everyone else, and destined to be successful. Their governing bodies were made up of merchants who obviously wanted to make money and the Venetian government was supportive and watchful about business more than anything else. And, of course, Venice was not under the thumb of any pope who might disapprove of individual initiative or tithe money for his sacred works. There were endless opportunities for hardworking people to make money.

The Crusades pushed the drive for monetary success even further because a huge group of consumers would periodically descend on Venice and spend money. They were Venice's first tourists. Crusaders needed to be housed, fed, and clothed while hanging out in town figuring out how they were going to get to the Holy Land. They also needed supplies for their journey onward. As one historian put it, "Thus, the Venetian Republic had a firm hold on its own fate and course of progress."[12] It was a lucky synthesis of good geography that fostered shipbuilding, the position of the city relatively close to exotic markets, and business sense with governmental support.[13] These factors combined to turn Venice into the perfect moneymaking machine. But to be flourishing businessmen they also had to understand money, exchange rates, loans, financial risk, mathematics, and accounting.

At the base of all that business success was a deliberate and canny shift in the use of the very numbers that would track their success. Arabs were initially the best mathematicians, and they developed a form of Hindu numbers that worked easily for arithmetic and algebra.[14] Their mathematics and geometry came with them to Spain and were exported to other European countries from there. In 1202, the mathematician Leonardo Bonacci of Pisa, better known as Fibonacci, wrote a revolutionary treatise, *Liber abaci*, wherein he introduced Hindu-Arabic numbers to the West, and Venice was the first Western city to embrace fully the use of this style of number notation over the traditional Roman numerals.[15] They did so because it is very difficult to perform mathematics with the bulkier and less flexible Roman numerals. When using Roman numerals, a trader needed to stop whatever he was doing and pull out an abacus, or use some method of moving pebbles about to multiply, divide, subtract, and add. From the 1200s on, Venetians had the highly malleable Arabic numerals while the rest of Europe still clung to Roman numerals for centuries more.

Venice also had ways to pass this mathematical knowledge along. Chapter 6 describes the *tariffe* manuals, basically training books for traders. But there were also special schools in Venice for young men, called *abbaco* schools, that taught students about Arabic numbers and arithmetic. Students came from all over Europe to Venice to attend. Luca Pacioli, the father of accounting, both attended and taught at the most famous of these schools, the Scuola di Rilato (see below). The *abbaco* method used pen and paper rather than an abacus and it was much easier and more efficient. Students also practiced commercial math problems and were introduced to everything about money that was necessary for being a merchant. The *abbaco* school perspective on numbers was practical and nothing like what the general public did with numbers—predicting dates of holidays and nothing else. It was also not how theoretical university mathematicians, who were called astrologers, viewed numbers in a way divorced from real life. The *abbaco* method focused instead on the pressing need to count goods, settle on prices, and figure out if a profit was at hand. For that reason, it used Arabic numbers and arithmetic. Venice was instrumental in that leap because it was, from the 13th century on, a mercantile nation, and business gave birth to math as we know it today.[16] And that math was all about money.

MAKING MONEY, LITERALLY

During medieval times, every city or region in Europe had its own money. Before the introduction of the euro, figuring out how to spend one country's currency before passing through the border to another country used to be part of the journey. Today, the euro makes European travel so much easier. During medieval times, items designated as money were produced by those in power, usually feudal families or a powerful city-state. There was no paper money back then, but there were many types of metal coins representing individual monetary values. The weights, sizes, imprints, and metal content of these coins were highly variable and changed over time, certainly with every new ruler. This variability was a challenge for merchants, especially global merchants like Venetians because they had to keep up with exchange rates and how they translated into Venetian money.

Just like other city-states, Venice coined its own money. But unlike other city-states, the government was made up of merchants dependent upon money having a stable value and they needed the value of Venetian coins to hold fast. This translated into a promise of continued value in all trade. In other words, the Venetian reputation for reliable and honest trade was upheld by their standing behind their money, imprinted with whatever doge was in power, St. Mark, and Jesus Christ.

The first coins came from a mint in the Rialto, on the east side of the Rialto Bridge. In the 9th century, this area was part of the financial district of early Venice. The original name for the Rialto Bridge was Pons de la Moneta, meaning "the bridge of money" in *veneziano*.[17] In 1277 that mint moved around the corner from Piazza San Marco. The building remains today, facing the Giudecca Canal and attached to the Marciana Library. It is still called *zecca*, "the mint," in *veneziano*, and that word is probably based on the Urdu (a language spoken in India and Pakistan today) word *sikka,* which means "coins."[18] The republic was very protective of her coins, and only verified citizens were allowed to work in the mint. The fear was that others could steal their forms for minting coins or copy their process. Government oversight was by committee, not individuals, and even the mint master who was in charge of production served a short term.[19] The government, for their pains, also took some money off the top for supporting the production process, a common practice in other mints called the *Seigniorage*. The coin called the *tornesello* was a big moneymaker in that regard, for the city and collection of the *Seigniorage* helped the republic pay for wars and put off taxing citizens for a while.[20] The government also kept the best coins of quality for use within the republic and used less well-made coins for transactions with foreign countries. The republic protected the value of its money by forbidding coins from other places to enter the city.[21]

When the *zecca* was in operation, the coins were made inside by order of the current doge. They were handmade, so minting was a rather variable process. As a control, each coin was evaluated for metal content and weight. Over the centuries there was a very long list of Venetian coins that popped up, were used, then fell out of fashion or were so devalued they were worthless and replaced. Coins might be made of gold or silver, the most precious, or bronze, or a combination of various metals. The first Venetian coin was the *denaro*, a silver penny that circulated in the 9th century. In 1202 Venice produced a large silver coin called the *grosso* or *matapan*. That coin was pure silver, big and heavy, and respected outside of Venice.[22] Along came the *piccoli*, small coins worth less but very useful for insignificant household purchases. The first Venetian gold coin was made in 1284 and was called the ducat, a word derived from *duchy* and so an echo of the doge. This coin was made of 3.5 grams of pure gold. The Venetian ducat was the most reliable medium for exchange for long-distance trading at that time,[23] and it became the standard for trading nations in the known world, including Africa, the Middle East, all of Europe, and India, among other places.[24] The Venetian ducat was accepted everywhere because it held its gold value and it was backed by the Venetian Republic. The state was so behind this coin that hands were lopped off, eyes blinded, and noses removed if a person was caught fiddling with the ducat in any way.[25] In its day, the ducat was the American dollar, but heavier and made of gold. It had a purity of 0.986 and was finely minted, that is, the imprint was well done and the edge was strongly made to cut down on "clipping." Clipping was a major issue for coins back then because the higher values were made from precious and rare metals, so a clever person could shave off a bit of gold and no one would notice. There was also wear and tear, that skimmed off a little more metal. That's why many transactions that used these coins were calculated by weight, not counted.[26] The ducat was the coin of choice across the globe for five hundred years and only fell out of circulation after the fall of the republic in 1797. These coins have been found closed up in ancient Indian temples and drilled with holes so they could be worn as expensive jewelry.[27] There was only one major adjustment to the ducat in the mid-1400s, when gold rose in price. The republic changed the name from ducat to *zecchino* (named after the mint, or *zecca*) and based its value on a silver standard even though it was still made of gold. The *veneziano* word *zecchino* is where we get the word *sequin*[28]—a shiny, flat object—and this most stable of currencies was accepted until modern times. No other coin has lasted as long and held its value so well.

Venice also made other coins besides the *ducat*. In 1472, they minted a silver coin called the *lira*, meaning "scales" in Latin. That Venetian *lira* was the forerunner of the most recent Italian state money before the euro. The word for Italian money, then, came from Venice and that fact is yet another example of the infiltration of *veneziano* into common language. Today there is an Italian designation for gold, *oro zecchino,* meaning the most precious and pure type of gold, or "solid gold."[29]

Venice was also the only city-state that was a bullion market.[30] Since the city was the port of entry and exit of metals into Europe, Venetians had first-hand knowledge about where metals such as gold, silver, lead, copper, zinc, and others came from and what they were worth before minting with them. While making money, they had their hands and their business all over actual money.

VENETIAN BANKS

The first national bank in the world—one upheld by the state—opened in Venice in 1157. But it wasn't a normal bank as we think of banks today. In 1171 the Venetian state began a program that forced loans on citizens to pay for a war. This so-called bank was no more than a holding company for forced loans imposed on the aristocracy to give the state money to wage a war. These nobles didn't pay taxes, per se, so these forced loans were not all that bad, because the nobles received 4 percent interest per year on the amount. It was, in effect, a tax with the bonus of interest eventually paid to the holder in full over so many years.

Unfortunately, the value of these bonds fluctuated over time and a holder could lose his shirt after doing his patriotic duty to support a Venetian war. Such "loans" were often hard on less wealthy aristocrats who then had to cough up money to the state, not on currently generated income but on the value of their estate. In 1523 the aristocrat Andrea Arimondo died from what was called "a melancholia of the bank" because he could not pay off his loans.[31] At some point in the 1200s, the citizenry holding these notes began to trade them among themselves to pay off debts, treating them like a form of money that would pay the debtor an annual fixed interest. Contrary to church rules, Venetians had no trouble with giving or collecting interest and claimed it was not diabolical or usury, since the rate was reasonable.[32] And so, this money that was paid to the state wandered about town as a kind of state-backed stock that grew interest but who knew when it could be paid back in full.

At first the government kept a record of the loans in a ledger rather than issuing notes, but eventually, there were paper drafts of these contracts. Today we call this type of system—giving the government our hard-earned money for a secure promise of annual interest—government bonds, and Venetians were the first to allow or force citizens to own a piece of their government as a sort of investment. It was considered a patriotic act, but only a fool wanted to make these payments. Today these government bonds come in various forms: city, state, or treasury bonds. In Venice, this system walked a fine line, and in 1351 the Venetian government outlawed spreading rumors intended to lower the price of these government funds, thus instituting state control of fluctuating financial markets.

The operation of banking Venice was tied to the mercantile dealing of many Venetians and the ebb and flow of money among persons, traders, regions, and companies. Before there were banks, most people had little money—that is, money as we know it, or the coins and bills we have now. Venetians invented the first letter of credit in 1107; if you had one, you could buy things with it if it was honored by the recipient.[33] Once there was money, people needed a place to put it. More importantly, they needed someone to count out all the various forms of coin and tell the merchant what that haul was really worth.[34] They also needed someplace safer than under the mattress and someplace that would make sure the money could be taken out whenever. In ancient history, some establishments or organizations dealt with accumulation or transfer of wealth, but they were skeleton operations. Individual rich nobles then started opening their own private banks and offered them as a place to deposit money after a trade. There was no paper money in Europe yet, although there was in China, and the idea of paper, in and of itself worthless, had not yet caught on. But the appearance of private banks meant no one had to carry around heavy gold and silver coins that were of variable value. At the bank, the clerks would count the coins, weigh them, and produce a note reflecting the actual and verified total value of the deposit. For large deposits, this service was welcome, even essential. The banks, therefore, became a great place to deposit "specie," that is, heavy coins. These private banks were also an efficient way to pay off debts. Two merchants—the one giving and the one owed—just had to show up at the bank together and ask for a transfer of funds from one account to another.[35] Even back then, citizens thought of this transfer as imaginary, as we do today as we do with online transfers.[36] No one visualizes stacks of euros or piles of coins flying through cyberspace to someone else. In the early banks of Venice, some of the pile was left for future use. This kind of bank was called a "bank of deposit" because that's pretty much what it did. It wasn't a place of safety so much as a place for storage.[37] These banks were called *banco giro* (*banco del giro, banco del ziro*); *giro* in this case means "transfer."

The very word bank comes from the Italian and *veneziano* word *banco*, which means "bench," while the word *bankrupt* comes from the simple phrase *banco rotto* or "broken bench." Indeed, early bank transitions were conducted at tables outside, especially in the Campo San Giacomo di Rialto in the Rialto, the financial district of Venice. There, the clerks sat on benches in front of tables for a few hours to conduct their business with depositors and those they owed. A clerk would write down any requested transactions in two books and that was proof of transfer from one account to another. The merchant himself did not need to keep a record because the bank always had one. These banks worked with personal accounts but also provided maritime insurance.[38]

Premiums were calculated as 6–12 percent of the value of cargo depending on cargo size, time of year, route of travel, and type of vessel.

That easy exchange of money with a deposit bank was further enhanced when Venice invented the "bill of exchange" (*lettera di cambio*) as an international transaction.[39] If a merchant showed up in Constantinople, for example, carrying one of these bills, he could pay for a pile of silk with the document. That bill of exchange would be honored in Constantinople as if it had been real money. Venetian merchant traders began using these bills of exchange in the 13th century and explains the efficiency of their long-distance trade business and their necessary financial acumen.

The private bank system lasted for almost three hundred years in Venice, but it was not without problems, scandal, and failure. Banks were named after the families that owned them, as in Pisani Bank or Lippomano Bank. Part of the privacy was that whoever kept track of the money didn't make a distinction between the rich aristocrats' private assets and the bank's assets.[40] As historian Frederic Lane put it, "In sixteenth-century Venice banking was not a specialized business, it was part of the larger operation of merchant nobles."[41] These banks not only took deposits and transferred money from one account to another in-house, but they were also deeply immersed in loaning money or credit to the state to fund wars and other governmental needs. They also loaned money to friends and family.[42] Most of these loans were lines of credit and they eventually took down most of the private banks; they were loaning money they didn't have, and when depositors wanted their money back, there was none to be had. For example, in 1499 Lippomano Bank owed money to 1,248 depositors; 700 of them were nobles.[43] Nevertheless, private banks also took over Venice's governmental debt, floating it around, which made them "cashiers of the state," a risky move.[44] But the idea of transferring money between accounts was so efficient that it stuck. In 1619 those transfers could move beyond the confines of one bank and happen between two different establishments, creating the first bank to execute bank transfers (called *scritta di banco*).

In 1587 when most of the private banks had failed (there were 103 banks in Venice three years before this crash[45]), the state stepped in. For the three years after the banking crash, there were no banks at all in Venice, an inconvenience that led to financial chaos.[46] The state decided to come to the rescue. It opened a new public bank with a major reserve of five million gold ducats to reassure depositors that they could take their money out at any point. Other city-states had public banks by that time, but still, Venice was deeply ingrained in the evolution of the bank system since their first public bank opened in 1157.[47] The government also kept a close watch on banks, even when they were private. The long list of changing regulations enacted by the Senate demonstrates both the follies of bankers and the long arm of the government trying to stop such follies.[48] The state went so far as to require that depositors could have a look at the bank's written record of their accounts and that the look-see had to occur right on the table, in front of everyone. Those tables had to hold regular hours, marked by an hourglass and opened with the tinkle of a bell.[49] The traditional opening bells of current day stock exchanges come from these tables of Venetian bank clerks. The Venetian Republic, therefore, regulated the banks, initiated what we now call wire transfers, invented government bonds, and put into place many banking traditions that we now hold dear, even without an hourglass.

DOUBLE-ENTRY BOOKKEEPING

As all that money was being carried about Venice, as debts were settled at tables set up in outdoor squares, and as bills of exchange were ported to other countries, it became clear that there needed to be a way to organize business accounts. Venice had become a successful mercantile empire, and just about everyone in town was involved. This business was not only physically complex; it was also financially complex. It required some kind of account keeping and some kind of periodic reckoning to determine if a business was profitable, by how much, and what was working or not.[50] People had been keeping track of transactions for centuries, using tokens in envelopes and then later by noting the exchange on pieces of paper.[51] Sometimes they separated what was spent and what was brought in, but the system was rudimentary, more a diary than an account book. In 1494 a Franciscan monk and mathematician named Luca Pacioli, who was not Venetian but taught and wrote in Venice, published a book there on algebra and accounting that revolutionized the Venetian trading houses. The initial purpose was to keep track of cash flow in foreign markets and keep track of the agents who ran that end of the business.[52] But this system was so clever, so simple, and so powerful at understanding the real value of a business that it revolutionized how businesses around the world operated.

More remarkable, this system, known as double-entry bookkeeping, is used today in small businesses, corporations, conglomerates, cities, and nations.[53] Double-entry bookkeeping is the system that brought us modern finance and imprinted on the West the overarching concepts of credit and debit, a lens that now influences everything from relationship choices to job choices.[54]

Double entry was not invented by Pacioli. There are hints of some sort of elementary double-entry-type bookkeeping in Arab and Asian areas,[55] and it was fully developed and used by the merchants of Venice before Pacioli dissected the method and wrote it down for publication. This system was initially known as bookkeeping "*alla viniziana*." Pacioli was born in a small town in Tuscany in 1494, well after Venice had achieved its reputation for business. There is some indication that he might have learned about math and perspective from the artist Piero della Francesa, who also lived in Pacioli's hometown. Della Francesca was a mathematician and an accomplished painter. Pacioli left for Venice where he attended and taught at the *abbaco* schools. He also had a job giving private math tutoring to two sons of a successful merchant. After this first stint in Venice, Pacioli lived in various cities, took on the mantle of a Franciscan monk, and ended up as a great friend of Leonardo da Vinci; Pacioli returned to Venice and lived with Da Vinci in the same house for a while. They were known as great friends, both mathematically oriented and also humanists. In 1509 Da Vinci drew the illustrations for Pacioli's book *Divina Proporzione* (Divine proportion) where he sets out the math and uses of perspective for artists.[56]

Among other books, Pacioli wrote a fun one containing mathematical games, proverbs, magic tricks, and card tricks. And he wrote the first-ever book on chess, *De ludo sacchorum* (Of games of chess), which was only discovered in 2006. That book empowers the queen and makes a coalition of the queen, bishops, and pawns. One surviving painting of Pacioli shows him in his monk's robe with the hood up, a pointer on a chalkboard and his finger on a book. There is also a young man behind him who is probably the Duke of Urbino, a student of Pacioli. But more telling is the crystal object floating in space off to his right. This blob of faceted glass is many sided, large, and sort of creepy. But that makes sense because at that time mathematicians were also called magicians.

But Pacioli is best known for his mathematical encyclopedia, the first one printed in Europe, called *Summa de arithmetica, geometria, proportioni: e proportionalià* (Everything about arithmetic, geometry, proportion and proportionality). It was published under his daily oversight in 1494 in a Venetian printing house.[57] Printers had had trouble with symbols and numbers, although they printed one of the first math books in 1478, so presumably, Pacioli's manuscript was not much of a printing challenge. It was, however, an unusual math book. because Pacioli understood both practical bookkeeping and the theory under-pinning mathematics, algebra, and geometry, usually the purview of university mathematicians. Most notably, and the part that has given Pacioli continuing fame, is the twenty-seven-page section in the *Summa* on bookkeeping that sets out how to do the kind of double-entry bookkeeping that Venetians had developed and mastered.

Other parts of the *Summa* are also notable; it gives the first printed expla-nation of the math involved in perspective. Pacioli also insisted on writing it in Florentine vernacular, that is, today's Italian rather than Latin, so that it could have a wide audience.[58] And it did. Since publishing in Venice was all about making money, his book received wide distribution. The *Summa* was printed again in a second edition. That's saying a lot for a book that was gigantic. It is twenty-five by thirty centimeters and has 615 very thin pages full of tiny type and numbers. Eventually, the bookkeeping part was pulled out and reprinted and circulated by itself.

The section on double-entry bookkeeping is called *Particularis de computis et scripturis* (Particulars of reckoning and writings). The purpose of the method, Pacioli wrote, was to let a merchant know where his business stood at all times. He claimed it had been around in Venice since the 1200s; here he simply refined and codified the method.

Previous bookkeeping methods in history had normally used one column to enter both debits and credit in running order, but the double-entry method separates the two into opposing or adjoining columns, with debits on the left and credits on the right. In an open ledger, it would take up a right and left page, but it is better done on one page where the corresponding debits and credits are easy to compare. In that case, one simply draws a large T and makes two columns. Then—and this is key—every transaction must be recorded twice; if a credit is marked down in the right column, a corresponding debit must go on the left-hand column. In just a glance, a businessperson has everything laid out. For a periodic reckoning, the two columns can be added up for a calcula-tion of all debits and credits. In that organized and mirrored process, mistakes stand out and can be corrected—what we now call "closing the books." All the figures can also be used to produce profit and loss summaries. A plus or minus summary figure can be entered against the original capital account. There you have the real state of the business, what was done against the original capital. Pacioli recommended that these summary calculations should be done every year. That singular advice to businessmen and women has been followed for the past six hundred years all over the world. It's called the annual report. This simple and elegant bookkeeping system, Pacioli felt, was necessary to run any business; by having a running record of transactions in this organized way, an owner could, at a glance, rest assured that things were going well, or maybe not. And he could have a periodic sense of the real value of a profit or loss, and why it occurred.

Beyond that major structural and mathematical approach for credits and debits, Pacioli also recommended that business owners keep three books for accounting. Before starting the books, however, the business owner should take an inventory of everything he has—money, bank accounts, stock, jewelry, and such. That inventory should also include outstanding debts owed to the owner, and debts the owner owes to others. With this inventory as a starting place, bookkeeping begins in the first book, called the *memoriale,* or memorandum. That book was basically a diary and history of double-entry bookkeeping. Jane Gleeson-White says it was called the "waste book" in which the owner records, one after another, all transactions during the day, every single minute.[59] The memorandum book was an ongoing record of the days and nights and how the business was navigated, a running narrative of the actions of a business. Pacioli said those transactions should then be transferred to a second book, the *giornale,* or journal. In this book, the owner or accountant should make those quick diary entries in a more systematic fashion. They should include the precise direction of the transaction and include the words *per* meaning "from" and *a* meaning "to" with the double slanted "//" in between. The *per* is a debit and the *a* is a credit. The third Pacioli book was the *quaderno,* or ledger. First the inventory is entered in the *quaderno,* then the bookkeeper should draw two columns, one for debit and one for the corresponding credit—thus the double entry. That method can be expanded to various accounts, such as assets, liabilities, and equity, and keep track of virtually everything. This system explains the use of the word *books* when referring to accounting, but today bookkeepers do not use the memorandum volume or require others, such as cashiers, to narrate their transactions minute by minute in a personal diary. On the other hand, running receipts and credit card charges, from the owner's point of view, are indeed a running record.

Pacioli also explained how this system could be adapted to deal with banks, how it would work with partnerships, and how it could prevent embezzlement. He was constantly cautioning businesses to stop writing things in a haphazard way that only they understood and to organize how they calculated what things cost to acquire and what they sold for. "He who does business without knowing all about it sees his money go like flies," wrote Pacioli.[60] He insisted that the books should be opened with a message to God, because according to him, good bookkeeping was a moral obligation. Double entry, with its comprehensive and orderly style, certainly had the feel of the kind of honorable stance that an honest businessperson might aspire to. And in that orientation, making money would not, in the Renaissance, be considered something bad, some usury that the church forbade. Instead, in Pacioli's orderly, mathematical world, double-entry bookkeeping made for a kind of sinless profit that was surely acceptable. The positive moral philosophy underpinning double-entry bookkeeping also drew bookkeepers and accountants to the forefront of business, giving them moral authority based on the beauty of numbers. In that, double-entry bookkeeping also established the profession of accounting and gave accountants the reputation of being honest, stalwart, and to some, boring, because they dealt with quantification, and numbers don't lie.

The effect of the codification and implementation of double-entry bookkeeping went beyond tracking businesses. According to Jane Gleeson-White, Pacioli's book and the idea of quantifying and working with numbers initiated a scientific revolution: ". . . without a knowledge of mathematical sciences no good work is possible," Pacioloi wrote.[61] We might look to great art and books as defining the Renaissance, but there were also earth-shattering changes happening in medicine, astronomy, botany, pharmacy, philosophy, and others. These disciplines were moving toward qualification, hypothesis, and proof, and they did so mathematically. None of that could have happened if there hadn't been double-entry bookkeeping and a focus on profit then used to pay for all those intellectual and artistic endeavors of the Renaissance.[62]

But the most interesting observation about the effect of Pacioli's double-entry bookkeeping points to its universality and staying power.[63] It certainly helped that the invention of printing and publishing made more copies of *Summa* available, and they were much cheaper than hand-scribed books.

According to Gleeson-White, 150 books describing double entry bookkeeping were published in Europe in various languages between 1500 and 1800.[64] That book also helped financially organize the Industrial Revolution and the stock exchange. It is ubiquitous in businesses, including utility companies, banks, transportation companies such as railroads, the health care industry, and manufacturing. Double-entry bookkeeping is also used today in small businesses run by the self-employed, giant corporations and partnerships, and complex businesses with shared capital. More forbidding, Gleeson-White also notes that we now are accounting for our planet in this way, the value and cost of cleaning up the air or allowing for the extinction of species and disastrous weather conditions.

Today we have computer programs that make a bookkeeper's life much easier, even apps on phones that are useful for household accounts, but someone still has to figure out who is the creditor and who is the debtor, and someone has to enter in all the data, line by line, as if it were still the 15th century. It has to be done in business, but it wouldn't hurt to have our personal finances organized this way, because, as Pacioli warned, "…if you are not a good bookkeeper in your business, you will go on groping like a blind man and meet great losses."[65]

CAPITALISM

Double-entry bookkeeping had an end game, and that end game was figuring out a profit.[66] Before this method took hold, merchants and manufacturers did not focus on profit as a thing to strive for.[67] But that changed with double-entry bookkeeping and Pacioli's admonition to business owners: "The end of objective of every business is to make a lawful and satisfactory profit, so that he may remain in business."[68] Of course, any business owner wanted to stay solvent and make a decent living, but there was much less push for making sure a business had a growing profit margin. With double entry, the whole point was to calculate a profit (or loss) and that is the paper trail of accumulating money, which is what capitalism is all about. The history of the Venetian mercantile business is the first real example of the correlation between double-entry bookkeeping and capitalism.[69] It is also about consumerism, because having items to sell is the way that profit happens.[70] "It is simply impossible to imagine capitalism without double-entry bookkeeping; they are like form and content," wrote economist Werner Sombart in his groundbreaking book *Der Moderne Kapitalismus*, the book that defined capitalism when it was published in 1902.[71] Capitalism, as Sombart explained it, is an economic system based on trade. It includes the owners of the means of production, workers who produce goods, and consumers who shell out the money to buy things.[72] He also wrote that double-entry bookkeeping might have "activated" the forces for capitalism or gave "rise to capitalism out of its own spirit."[73] The evidence for double-entry bookkeeping as the instigator of capitalism is the fact that this particular book-keeping method is aimed at figuring out a business's capital assets, that is, the total worth of a business owned by a person. Individual wealth, in other words.[74] The daily process of entering transactions with a focus on profit is a mind-set that fosters acquisition. With that attitude, capitalism and consumerism are born. Both are acts of the individual, now underscored by corporations that serve individual stockholders who are also looking for acquiring individual wealth or capital.

Capitalism is based on a drive for acquisition, fueled by competition in the marketplace, and it has taken on the idea that such an economy is rational and not personal, although we all know it's very personal.[75] French historian Fernand Braudel pointed to Venice from the 13th to 15th centuries as the oldest example of a capitalist-oriented economy in Europe and the first industrial center in Europe.[76] He based this designation on the development of the wool industry there, not the ship building at the *Arsenale*. Add long-distance trade for individual profit, and the result is capital owned disproportionally by some to the exclusion of the many.[77] Surely, we all recognize this system because it seeps into every minute of every day in Western culture.

Karl Marx famously warned about the evils of capitalism because of the social consequences when one person owns everything and the rest have to work for him.[78] According to Marx, capitalism contains within it the "seeds of its own destruction" as the rich overreach, consolidate their power and grip, and choke off their own economic growth. Venice was, in fact, a prime example of Marx's philosophy. When the nobles codified their class, then ousted those not in their *Libro d'Oro* from government, it probably initiated Venice's eventual economic decline, even if it took centuries.[79] Also, the con-tracts between nobles and *cittadini* for trade expeditions, with each "owning" shares in the eventual profit (in this case money from the noble and work and responsibility for the *cittadino*), are much like corporations today as they pool their capital to make a profit. Historian Eve Chiapello claims that Marx's philosophy came right out of a knowledge of double-entry bookkeeping and its goals—profit and accumulation by one business or person.[80] It seems that Friedrich Engels's family owned an English cotton mill. There is a record of Marx asking Engels for an example of "Italian bookkeeping," and presumably, Engels had an example at hand.

In general, Venice was the dominant financial center in Europe for centuries. "Venice, in the fifteenth and well into the sixteenth century, was the center of the Renaissance world economy," writes historian William McCray.[81] It had become the richest city in Europe by the end of the 14th century.[82] Jane Gleeson-White charmingly, and with accuracy, calls Venice the New York City of the times.[83] All that hustle and bustle on the docks was the outward manifestation of what can only be considered an accumulation of money, goods, status, and rampant capitalism. In Venice, one could join the citizens who were dedicated to making money in a governmental atmosphere that allowed individual freedoms, especially the freedom to get rich.

Venice was set to give birth to capitalism and pass along the financial genes oriented toward the God of consumerism; that path has become the dominant financial system in Western culture, for good or ill.[84]

CHAPTER EIGHT

The Written Word

I do hope that, if there should be people of such spirit that they are against the sharing of literature as a common good, they may either burst of envy, become worn out in wretchedness, or hang themselves.

—Aldus Manutius, Venetian Publisher
1496

Inventions introduce innovation into our society, but revolutions happen when innovation spreads and dramatically changes our everyday life.

—*Printing Evolution*
Exhibit, Correr Museum, Venice
2018

Industrialization, globalization, and marketing: Renaissance Venice already had it all. We may be talking about an enterprise dying back half a millennium, but the productive and commercial capacities expressed by the book capital of the world in the first half of the sixteenth century easily rival those of today's tycoons of the information age.

—Alessandro Marzo Magno,
Bound in Venice; The Serene Republic and the Dawn of the Book
2013

A TALL BUT slender ledger rests alone in a glass case, an ordinary but ancient account book set by itself. There's nothing fancy about this ledger. It's not showy or decorative in any way, just a bunch of yellowed pages with a brown leather cover and a simple flap for closure. Inside are lines and lines of old writing, with numbers by each entry. This logbook could be any accounting book before the age of computers and spreadsheets, a record that would be of interest only to the business owner keeping track of sales and receipts. But this book, called *Il Zornale* (The Journal, or "Day Book" in *veneziano*) is no ordinary record of accounts. It is the five-hundred-year-old Rosetta Stone that explains the birth of the book publishing business in Venice. And so, it sits in a glass case as if it were a precious jewel.

Written in the late 15th century, *Il Zornale* was tossed aside, left to rot like most business accounting records after a business goes bust. It suddenly appeared again in 1810 as workers were throwing out old papers that had been left in the attic of the Palazzo Ducale and the Basilica of San Marco, both buildings in the Piazza San Marco, Venice.[1] The title, however, still shows clearly on the cover and the writing on its 160 leaves is easily legible. *Il Zornale* also bears the mark of one Francesco de Madiis, a Venetian bookshop owner, and inside is the running record of all books his shop sold over three years (1484-1487) and his overall inventory for one year (1488). Listed there are 12,934 book sales, to whom and for how much. They succinctly explain 6,950 transactions of 25,000 copies of books, some of which are for gifts or barter. The ledger also follows, in detail, the journey of the books that went through Signore di Maddis's bookshop, if they were exported or sold locally. From this ledger of business accounting, we know the market value of books in the late 1400s in Venice, who was buying what and where they lived. We also know the titles of best sellers and duds during that period.[2] Through the precious lens of *il Zornale*, scholars and bibliophiles have been able to gain a dynamic and authentic picture of how a bookshop worked back then and why Venice, in particular, was so successful in both printing books and selling the printed word.

Within its fragile pages, and along its handwritten entries, *Il Zornale* is evidence of the human passion for reading books that went viral in the late 15th century because the printing press had been invented and because Venice produced not only multiple copies, but also knew how to distribute them all over Europe. Venetians might not have invented printing or the printed book, but they certainly invented the book business, initiated the singular act of reading books for pleasure, made reading more popular, gave us the type fonts and punctuation marks we use today, and established the very look, form, and feel of books we now purchase and read.

THE EVOLUTION OF BOOKS

The first book-like items were invented in ancient Egypt. They were lengths of papyrus or animal skins that were rolled up for convenience. In Central America, such long missives were folded, however long, into accordion shapes that could be easily carried, like their roly-poly cousins. Scrolls were written by hand, one by one, and often illustrated with fanciful decorations and illustrative scenes. The first flat-form books, the type we are used to today, first appeared in Rome in A.D. 100, but they didn't catch on in the East, where scrolls continued to be the preferred media format for centuries. Called codices, these portable flat tomes full of writing were formed as a stack of singular pages made from paper, papyrus, or, most often, velum (well-worked animal skin). An owner would cut a scroll into equal pieces of reasonable size, flatten each one, then place one on top of another in sequence to produce a codex. Although not the unwieldy length of a scroll, codices were bulky in a different way. The stacks of pages were inches high, and the leaves quite large. After the pages were sewn and glued together along one edge, the paper stack was attached to a hard spine. Then as now, the spine was designed to hold everything together while allowing a reader to turn the individual pages, one by one. A codex also came with a protective hard case made from wood, from which we get the term *hardcover*. Still, codices were considered an improvement over scrolls because they were more convenient to read and certainly took up less room to store. Codices were, of course, like their cousins, handwritten scrolls adorned or "illuminated" with gloriously colorful borders and capital letters. People who handwrote these manuscripts were called scribes, and their job was to draw letters into readable and beautiful perfection. Think calligraphy. Reading material changed again with the invention of incunabula, printed books that were stamped a full page at a time. The printer or artist would carve a complete page of text into a block of wood or cast that page of text in metal, then those blocks were inked and stamped onto paper with a press. The word *incunabulum* (singular of *incunabula*) refers to any of this type of printed book that appeared before 1501, but the date is arbitrary. Incunabula were considered a significant change for readers because these works were printed, not written by hand. And yet, the process was very slow, because each page had to be composed, or carved, one page at a time. Also, these books were produced on order for a particular customer, and they were manufactured and sold locally. Incunabula had limited circulation; they were considered precious objects and weren't lent or borrowed.

The real revolution in reading material came in 1449 when German goldsmith Johannes Guttenberg perfected a way to mass-produce pages through the use of movable type, oil-based inks, and a press modeled on the device used to press grapes. Movable type meant that letters and words were individually cast and could be moved about to make sentences. It had been used previously in China with porcelain molds, but Guttenberg made his molds out of long-lasting and sturdy metal. Those molds were versatile, and they gave a certain durability to printing multiple copies swiftly, and with less individual labor. That efficient process meant the cost of a book was much cheaper than either handwritten codices or incunabula. The ability to move metal letters around and have them in various sizes and styles also initiated the design of various fonts that gave a whole new flavor to printed matter.

The Guttenberg flexible printing technology spread quickly from Germany across Europe. It initiated an explosion, a revolution, in the art of printing in the 15th and 16th centuries. The populace was, at first, surprised by these new mass-printed books. Unlike handwritten manuscripts, those very precious objects, these comparatively hastily made volumes were made from inferior inks and paper, and all that smelly glue that held books together.[3] But the increased availability of relatively inexpensive printed books meant that reading was no longer the sole provenance of priests, scholars, doctors, and lawyers. It was now available to anyone who read. There was little change from that printing revolution until, perhaps, the current digital age, about five hundred years later. But even digital letters, words, and ebooks are simply electronic variants of the usual visual format of books made on a printing press.

Once movable type appeared and printing became faster and more economical in terms of materials and labor, a new business was born. That business required a printer who had a press and the skills to use it and financial backers to bankroll production. The new job of compositor—the person carefully translating written manuscripts—involved placing movable type in the correct orientation for printing. It required workers with skill and experience. Printed books would not exist without the typographers who invented fonts and made molds for crisp letters or the men and women who operated the presses and stamped the paper (called pullers). After printing, binders had the new job of putting printed sheets in order, sewing and gluing them together for posterity and easy page-turning. Also, artistic illustration might not have been a new technology when the printing press arrived, but when it did, artists had to use wood blocks or engravings to include them in printed books. Artists had to adjust to the fact that their prints would be embedded within the text and in black and white, not color.[4] And most important for a moneymaking business came the very new idea of distribution. For the first time, books became less precious because they could be produced more cheaply and faster, and that meant they could be exported and sold far and wide.

THE PRINTING PRESS COMES TO VENICE

Printing was introduced into Italy in 1467 into the town of Subiaco, near Rome, where a group of German monks lived in a monastery. Those monks had brought some of Guttenberg's movable type and a printing press with them and began turning out printed matter. Two years later, in 1469, the new technology arrived in Venice in the suitcases of German printer Johannes di Spira (John of Speyer).[5] Di Spira's first printed book was *Epistolae Familiares* by Cicero, which he cranked out the same year; that book is presumably the first printed book in Venice. On September 18, 1469, the Venetian Senate gave Johannes di Spira the first legal privilege for printing. That decree would have given di Speyer a clear monopoly on printing with movable type for five years, but he died a year later and the privilege was forfeited. Frenchman Nicholas Jenson moved to Venice in 1470 and set up a print shop, filling the void left by di Spira.[6] Jenson had a background working at the French mint, and he designed and cut exquisite type. His first printed book was a short-paged set of rules, called *Decor Puellarum,* intended to advise young Catholic girls how to behave. It was probably written by a Venetian monk intending to keep Venetian girls in line.[7] Jenson went on to be a very successful printer who was in high demand.[8]

Following Jenson, printers from all over Europe poured into the city and set up shop.[9] Why did would-be printers go to Venice and not some other city? Because Venice was known for its liberty and freedom of thought, and that translated into freedom of the press. Before the Roman Inquisition seeped into Venice, printers never worried that the government or the church would interfere. There was also no censoring of any printed matter in those early years. Up-front financing—books had to be printed and sold before they gen-erated income from sales—was also readily available in the city. It was here that "publisher" was first used to denote the network of investors, printers, paper manufacturers, salesmen, authors, and editors needed to make and sell a book.[10] On the consumer side, there were also many humanists in town devoted to reading and writing, and these clients were both authors and consumers. Also, Venice's level of literacy was quite high across all the social classes and there were plenty of consumers of the written word even inside the city.

Venice also had geography going for it. Paper, one of the most expensive requirements of mass printing,[11] was readily available from the forests near the Alps, especially from the area around Lake Garda; those trees were easily transported to Venice down the rivers of the Po Valley. But most important to the printing revolution was the very economic history of Venice as a mercantile trading empire. Venetian merchants already had a tried-and-true distribution network across Europe, and even farther, and this efficient method for distributing books was Venice's major contribution to the printing revo-lution.[12] At the same time, those trade networks and in-place agents provided printing houses with on-the-ground information about what would likely sell in other places.[13] In that, Venice also invented the active book market that targets consumers.

Because Venice treated book printing not as an individual intellectual exercise for the privileged but as a business opportunity, this city essen-tially invented the book business. By 1500, there were about two hundred printers in Venice cranking out twenty times the volumes the citizenry could absorb.[14] By the mid-1500s, Venice produced and sold more books than any other city in Europe.[15] One historian says that 16th-century Venice produced over thirty-five million books, and they surely made a good profit on those volumes.[16] That competition set the stage for technological advances and demanded a certain competitive quality for work produced.[17] As a result, Venice soon gained a reputation as the place to go if you wanted something printed well. The Venetian work was, by and large, artistic and often beau-tiful. The books were also nicely designed to make reading more inviting. Venetian printers tended to include things like a table of contents and an index, and in the late 1400s, Venetian printers were the first to apply page numbers to books.[18] More importantly, their movable type was generally crisp and readable.[19] Also, presses were not just printing books. They were also making manuals on everything from how to draw clear letters to curing various illnesses, and there were pattern books on sewing, embroidery, lace, and ideas for costumes.[20] These were texts aimed at the general public. The exploding Venetian book business was also unusual for the times because women in Venice not only wrote and published books; they also worked as editors, illustrators, and proofreaders. They even set type and operated presses, and some had their own book businesses.[21] For example, Antonia Pulci was the first writer of religious plays for street performances, and she was the first person to publish in Venetian vernacular.

Besides printing books and maps, Venice was also busy printing music.[22] In 1498, the state granted the first patent for using a double-impression technique that allowed for printing polyphonic music, in this case, voice, organ, and lute on separate lines. This patent was granted to Ottaviano Petrucci, and he held it for twenty-one years.[23] His initial production in 1501 was the first printed volume that was all music with no text.[24] Petrucci's designs used movable type and contained many lines of tiny notes, which made for a difficult printing process and varied quality, especially since each page passed through the press two or three times and everything had to line up perfectly.[25] These particular pages took skilled labor and time and they were expensive to produce. The technology of music printing improved when in 1520 Andrea Antico, working for Girolamo Scotto, the great publisher of music and books, combined woodblocks carved with music set next to text composed with movable type.[26] Venice published more music than any other press city in the 16th century.[27] In the late 1530s, a single-pass printing process for polyphonic music arrived from London, and it was utilized by the Scotto family and the press of Antonio Gardano, another great music publisher at that time in Venice. The Scotto family also published books and other printed matter, but Gardano specialized solely in music. As a result, Gardano produced a higher number of printed music pages. But Scotto was a master businessman with an efficient distribution system. As a result, he was responsible for the spread of printed music all over the world.[28] Scotto also formed partnerships with other printers, and his company seems to have rented out type to other presses. The music printing business in Venice also had a particularly inbred quality to it. Presses that attempted music printing—which often printed books as well—were owned by extended families and marriages across presses often secured a business.[29] This subset of the publishing business, therefore, had its own social structure, alliances, cooperative projects, and of course, interpersonal conflicts and competition. But it meant many kinds of printed music from Venice found their way across Europe.[30] The availability of all that printed music surely must have inspired Paduan Bartolomeo di Francesco Christofori when he invented the piano in 1709.

On the other hand, all that printing eventually brought censorship. In fact, the first-ever literary censor was Venetian Andrea Navagero, a printer himself, appointed by the state in 1516 by the Council of Ten, the secret police trying to oversee everything.[31] Venice formed its own brand of the Inquisition in 1540. Instead of following the pope, the Senate appointed their own Venetian band of six inquisitors, three clerics, and three nobles to look for the bad morals and acts of heresy in behavior or works of Venetian citizens. They made all sorts of accusations and some prosecutions, but their effect was minimal because the booksellers objected on financial grounds. And the sheer volume of Venetian printing during the last half of the 16th century made it impossible for the Inquisition to keep up. But sometimes books were confiscated and their publishers made to stand trial.[32] There was one point when the inquisitors held their censorship ground and many books were not allowed to be printed in Venice, but only for a short time.[33] And they regularly burned piles of books in Piazza San Marco or in front of the Church of San Domenico. They even followed the rest of Italy, under the direction of Rome, in burning the Talmud in 1553, an act that halted Jewish publishing in the city for ten years. Several times, the panel made an index or list of forbidden books and then had to step it back because printers and others made a stink. They also went after art and artists and this censoring went on forever until Napoleon disbanded the members of the Venetian Inquisition. However, the overall effect on their power was diluted by the Venetian sense of liberty and lack of respect for the pope.[34] Mostly, the inquisitors just caused havoc and injected a lot of paranoia into the book business. They also fueled antipapal attitudes and the ever-present Venetian dislike of Rome.

No matter, because there were so many bookshops and stationery stores selling books that the city was, as historian of Venetian books Alessandro Marzo Magno writes, "...a sort of year-round book fair."[35] Printers from Venice were also by far the biggest and best-known publishers at the two book fairs of that time, held in Lyon and Frankfurt.[36] Suddenly there was a tornado of printed matter rising out of Venice and spreading all over Europe and beyond.[37]

HOW VENETIAN PUBLISHING
CHANGED THE WORLD

Recently, the importance of Venice in the printing revolution of the 15th century was analyzed and confirmed by scholars from Oxford University with collaborators from other universities and the Marciana Library in Venice.[38] Their ongoing project, called "15cBookTrade,"[39] shared preliminary results in the fall of 2018 with a lively and engaging exhibit at the civic museum in Venice, Museo Correr.[40] The exhibit ended in the Grand Reading Room of the attached Marciana Library, the place that houses so many of these ancient books, including *Il Zornale*.[41] As the exhibit explained, "It was via physical books that ides and knowledge spread around the world."[42]

From the 15c project, we also know how and why the printed book business was successful, especially with Venice as its engine. Based on a collection of half a million books from that era (now held in four thousand libraries), the project has been able to document the fact that natural history, theology, mathematics, rhetoric, arithmetic, astronomy, and medicine were popular subjects. The cheapest volumes were religious books, astrology, rhetoric, and grammar. Apparently, there were loads of books published for children but they were tossed, so few have been unearthed.

The project approached the list of surviving books as if it represented a historical DNA analysis of a population long gone. Making use of printers' marks, ledgers like *Il Zornale*, and receipts and such, the 15c team was able to track and map where and when books were printed and where they ended up, to gain a real scope of the importance of reading in the 15th century. Of course, this analysis is not perfect; some possible best sellers were often missing from their cache, just as Tom Clancy or Stephen King books are now dropped into recycling bins. Interestingly, the Bible was neither the most popular printed book nor a best seller. Moneymakers were textbooks aimed at the four universities close to Venice—Padua, Bologna, Pavia, Ferrara—mostly in the fields of law and medicine.[43] Some of these early books have annotations in the margins and between the lines, just as people mark up books today. Based on prices in *Il Zornale* and other receipts, a Venetian wage earner could afford books. Ongoing competition between major printers and publishers also held prices relatively low.

The Venetian book publishing business also left a permanent mark on how we conceive of writing today.[44] In 1486, the first-ever book copyright was granted to Marco Antonio Sebellico for his book *Decades Rerum Venetarum*, a history of Venice. Note that this privilege was not given to the printer but to the author. That specific book copyright decree was followed in 1545 with another decree that writers had an "artistic personality" and therefore they were entitled to compensation. Along with patents, started in 1474 in Venice (see chapter 6) and privileges (monopolies), in place even earlier, these laws protected the printing trade because they covered typefaces, print technology, and then content, and they are the first evidence of intellectual property rights anywhere in the world.[45] The move to protect an individual's work came about because printing multiple copies introduced the possibility of hijacking. Instead, the states rights put the ownership of words at center stage.[46] In Venice, rights on non-tangible skills had first been given to guilds as a way to keep their tradecraft secret, then it was given to authors.[47] Because of the rise in consumerism in Venice at that time, and the extent and financial success of the book business, it's no surprise that the government was handing out protections to encourage the print businesses and keep these workshops active in the city.[48]

The printing business flourished in Venice for about three hundred years. Even after the end of the republic in 1797, when every business in Venice was hit hard or destroyed, publishers stayed afloat and even tried to restore Venetian printing to its previous glory. But eventually that effort failed, and Venice lost its title as the greatest and longest-lasting master of the publishing business. Today there are still several publishing houses in Venice and in cities close by on the mainland, but the flurry and success of the current book business in Venice is best seen in the many fine independent bookshops that dot the city.

ALDUS MANUTIUS

The most famous publisher of Venice was Aldo Manuzio, better known by his Latinized name Aldus Manutius (at times is Latinized further to only the first name Aldus).[49] He was born in the small village of Bassiano near Rome sometime between 1449 and 1450. As a young man Manutius studied Greek and Latin, which was unusual for that time. Then he was both a student and a teacher in various Italian cities. Manutius came to Venice from the town of Carpi in 1490 when he was forty, and he began a new life. There is no evidence of what Manutius did his first four years in Venice, but scholars assume he was learning the printing trade. Evidence for that is a few books printed with established and successful printer Andrea Torresani. Torresani had been trained by Nicholas Jenson, so Aldus was trained within Venice's best lineage.[50] Eventually Aldus married Torresani's daughter, who was thirty years his junior, and they had five children.[51] In 1494, Manutius set up his own print shop, calling it Aldine Press. The building still stands close to Campo Sant'Agostin in the area of Santa Croce, a tall yellow house bearing a small stone plaque noting that Aldus Manutius was "the light of Greek Letters returned to shine upon civilized peoples."[52] When business became slow, he moved into Torresani's workshop on the other side of the Grand Canal. This printer, working for twenty-one years in Venice, developed the highest standard of printing, and in doing so he changed the look and quality of printed books. He also invented many features of books that we think of as "normal" today, not realizing they were created in Venice five hundred years ago.[53]

Above all, Aldus Manutius was an educated and dedicated humanist. His goal was to use the printing press to revive disappearing Latin and Greek classics so they could be read once again. Manutius began with a grand project of printing the complete works of Aristotle, in Greek. He also printed Greek classics by Thucydides, Herodotus, and Sophocles, along with Latin texts by Virgil, Catullus, and Cicero. Early on, the Manutius shop also printed short pieces by members of the Roman Catholic Church. Some historians feel that Manutius, alone, was responsible for the renewed interest in the classics during the Renaissance.[54] Printing the Greek canon—before this only three books had been printed in Greek in Venice—meant that Manutius and his typographer, Francesco Griffo (see below), needed to figure out how to add diacritic marks necessary for Greek letters. Other printers had solved that problem by using a separate line of type for the marks. But Griffo supplied the accent to individual letters as they were set, a much faster and economical way to solve the issue.

Books, Manutius believed, provided an antidote to barbarous times and should not be hoarded by a few. He was a man on a mission to make reading available to everyone. To that end, he aimed to make printed books visually so close to handwritten ones that people would be comfortable switching from manuscripts to the printed word. Manutius also formatted his products to make that switch easier. He was the first printer to set type in double columns on a page, like a handwritten book. And he worked with goldsmith Francesco Griffo to invent various fonts that have survived today, including italics, designed to mimic handwriting (see below). Manutius is also credited with inventing the paperback book (known as *libelliu portatiles*, or "handy book") in about 1501 as a way to make reading a more portable endeavor.[55] Manutius went into production with this series of "undersized" books with red covers, starting with Virgil's *Georgicsi* printed entirely in italics to save space.[56] Clerics had used smaller octavo-sized versions (sixteen pages of type printed on one sheet, then folded into eight leaves) of religious works to carry with them on journeys, but Manutius adapted this form for classical texts and secular works. Unlike large and heavy codices, needing to rest on a pedestal or desk because they were so heavy, these small books could be held in the hand.[57] Since his list was secular and less expensive than larger printed books, his works were read outside the traditional reading sectors of universities, monasteries, and private collections.[58] These little red books also became portable status symbols; sometimes they show up in Renaissance portraits of those hoping to show off their intelligence and class.

In focusing on small books for everyone, Manutius introduced the idea of reading as a pleasurable pastime, and reading alone, with no need for desk or lectern.[59] Reading small volumes that can be carried is so completely a part of life now that it's hard to imagine being without a stack of books checked out of the library, bought secondhand, borrowed from friends, or purchased from the walls of bookstores. After five hundred years, Manutius's invention has morphed into the even more portable e-book—probably something he would not approve of given that e-books are available only to those who can afford the device. Manutius printed and sold so many books, and such a variety of titles, that it's also reasonable to suggest he created the "best seller." He sold about 100,000 various editions of *Canzoniere*, by the famous Italian poet Petrarch, for example. Francesco Petrarca had spent time in Venice, but he was long dead when Manutius printed his work. In 1528, the inheritors of the Aldine Press also printed one of the most famous books of the time, *Il Cortigiano* (The book of the courtier), written by Baldassare Castiglione, a famous statesman and diplomat and all-round gentleman. *Il Cortigiano* is a book of instruction for men who want to get somewhere in society. It tells them how to behave and what to know, and it became a best seller throughout Europe. Manutius also printed in Hebrew and tried out Arabic. To sum up all his works, and presumably to increase sales, in 1498 Manutius printed a list of all the books available from the Aldine Press, inventing the idea of a catalog, a handy item that is still used today to show customers available volumes for sale.

Manutius was a careful editor, always wanting the most correct versions of a book. He might have even compared several manuscripts of an intended volume before setting type to figure out the most reliable text.[60] And he had an eye for beauty on the page; he knew the correct proportion of text to border that made the open page pleasing to the eye.[61] As an aside, Manutius inadvertently started one bookmaking tradition by mistake. He had left a blank page at the top of a stack of pages. After realizing the blunder, Manutius wrote a note to the binders who were putting the books together to take it out.[62] But they didn't. From then on, blank first pages have been the norm in book publishing. Aldus also changed the book world by imitating the binding of ancient Greek texts. Aldine Press books were bound in olive green leather embossed with gold flowers, figures, and lines. This type of "classy" binding today is seen in private and public libraries, and that leather binding always signals a classic.[63]

Manutius was greatly concerned about how the Aldine Press printing style and format were soon copied all over Europe. In response, he was the first to use a colophon, a publisher's mark, to systematically trademark books from his press.[64] That emblem was an anchor with a skinny dolphin wrapped around its shaft and the accompanying Latin motto *Festina lente,* meaning "work quickly but slowly" (or better translated as "quickly but accurately"). The first time he used the motto was in an early book, *Hypnerotomachia Poliphili* (see below); the protagonist Poliphilo's motto, "*Festina lente,*" sent him on a path of discovery through a dreamscape.[65] The dolphin presumably represented the quickness and the anchor was the dependable and stable thing that holds that quickness to a standard of quality.[66] In 1503 Manutius became so angry about the Aldine fakes produced in other cities that he warned customers to watch out for certain telltale mistakes, similar to an altered trademark on an imitation luxury bag sold on the street today. He even printed lists of the mistakes he found in the fake Aldines. Like modern-day hackers, the counterfeiters just took those sheets of corrections and fixed up their books and put them back on the market. Aldus also told his buyers to look out for textual errors and to check the quality of paper since copies were always made with low-quality paper. Some fakes even had a bad odor and what Manutius called a kind of "Frenchiness" in their typography, whatever that meant.[67]

In his lifetime, Aldus Manutius was famous. Many bibliophiles collected his books, and many writers hoped he would print their words. In 1502 he was named the "Official Printer of the Venetian Republic." When he died in 1515, of the plague, his coffin, dramatically surrounded by copies of the books he had printed, laid in state in his home church of San Paternian. The business was then taken over by his father-in-law, Andrea Torresani, but it languished, even though the Aldine Press published the first Greek Bible in 1518. Aldus's grandson, often called Aldus the Younger, took over in 1574, but no one was able to meet the standard of the first Aldus and the business failed. As the Aldine Press started to go under, in an attempt to make at least a little money, the grandson added a list of back stock to each book sold. The Aldine Press of Venice closed officially in 1597, but the name Aldus Manutius is carried on today by those who collect original copies of his books, and by the honorifically named Aldus Corporation which developed the computer program PageMaker, released in 1985 to produce integrated graphic documents such as newspapers and books on a computer, replacing handset type.[68]

TYPOGRAPHY AND PUNCTUATION

There would have been no print revolution in the late 1400s without the invention of durable movable type. That Guttenberg, a goldsmith, cut type using metal is no surprise. Goldsmiths work metal into fine details that include making perfect lines and spaces. Carving or "cutting" sets of typographic letters, one by one, is called "punch cutting." At first, printers in Europe used only Gothic typeface, as did Guttenberg. But soon, a whole new discipline grew out of the very idea of artistic letter styles. We owe the long column of possible "fonts" in word processing platforms to typographers of the 15th century and their expansive imagination. Setting type was also an art. To make a readable page, the typesetter, now called a compositor, had to place the words right to left—that is, backward and upside down. They did this by picking letters from an array and lining them up in a chase, a metal holder that could accommodate any number of lines, with thin horizontal pieces of flat metal used to make spaces between the lines. Line by line a metal "chase" can hold a passage, ready to be inked and stamped on paper as the weight of the printing press pressures the letters onto paper and ink is transferred in whatever shape the type defines. Some works back in the 14th and 15th centuries combined metal type with woodblock illustrations, but later those illustrations were added using engraving or lithograph plates.[69] The process of typesetting stayed the same until the invention of the typewriter in the later 19th century; the typewriter made us all typesetters. The recent invention of computers and digitization makes this process even easier and faster, and potentially mistake-free because of cut, paste, and spellcheck.

During Venice's peak as the largest publishing place in the world, each house had its own set of typefaces, usually commissioned for that singular press. The big challenge for early printers and publishers was creating metal letters that were easy on the eye and that could seduce people away from handwritten works.[70] In 1470, Nicholas Jenson, working in Venice, invented the first Roman type based on measured typographic principles rather than more flowing calligraphy. Still, Jenson based his Roman-style letters on well-known handwritten manuscripts. Roman type, not exactly Jenson's but close to it, is a font option on Microsoft Word.

Designers of early typefaces had to be loyal to a printing house. That contract was one issue between Aldus Manutius and the revolutionary typeface designer Francesco Griffo who lived in Bologna and was the first person to make varied type for mechanical presses. Griffo was basically contracted to Manutius although he was eventually unhappy with his mandated exclusivity to the Aldine Press. Griffo, also a former goldsmith, was the designer and cutter of various forms of Roman type, all based on the original Roman type of Jenson. His many variations included a modification called Bembo Roman, or just Bembo, named after the great Venetian humanist Pietro Bembo. Griffo experimented with several variants of Bembo while Aldine was setting up in 1497 to print Bembo's travel book *Da Aetna*, about the author's hike up Mt. Aetna in Sicily.[71] Some think these experimental variants were aimed at mimicking the variation in handwriting, or maybe Griffo was just playing with his art.[72] Overall, Griffo's work was unmatched, perfect, and artistically formed, with the kind of subtle sweeps that calligraphers make with their special pens.

Surprisingly, those same incremental sweeps make for easy readability. In all, Griffo's fonts set the standard for all modern type designs. His forms and shapes are familiar to all of us because the more readable type today is based on his work. Griffo is also famous for including the horizontal line that now makes up the standard printed *e*. And he was the first to design and cut Greek and Hebrew alphabets with their diacritics, in addition to Latin letters, all for the Aldine Press.[73]

Griffo also famously worked with Aldus Manutius to create the first italic font (*corsivo* in Italian). They tested it in five words when Aldine Press printed some letters of St. Catherine. They also used italics for the explanatory notes to figures and diagrams in a 1501 two-volume set that was full of math and geometry. That usage prefaced the way we use italics today, to emphasize or set off text.[74] The italic style was based on the handwriting of Florentine humanist Niccolò de' Niccoli, who, on purpose, wrote in a slanted and curly way that was christened cursive. Manutius thought people would be more willing to read in italics because those letters mimicked handwriting more than the block letters of Gothic type. The fact that italic print took up less space and therefore used less paper, the costliest material for printing, was an economic bonus. With that new italic typeface in hand, the Aldine Press printed the first-ever book in italic type, *Le Cose Volgari,* by Petrarch, in 1501.

Griffo and his typefaces are part of the current world of digital fonts. For example, a Griffo Classico Roman and Griffo Italico are available online and attributed to a collaboration between the very dead Francesco Griffo and Swedish type designer Franko Luin, who died more recently in 2005.[75] Bembo type, also an honorific for Griffo and humanist Pietro Bembo, was designed in 1929 by Monotype Design Studio in Great Britain, based on Griffo's Roman Bembo.[76] It is a standard of the modern book publishing world because it's so clear and elegant. And italic fonts in printed Spanish are known as *letra grifa*.[77]

The use of movable type required, unexpectedly, a whole new world of punctuation that was mostly initiated by Aldus Manutius. Older manuscripts were written in a continuous flow. They were meant to be read out loud with the speaker determining what to accent and where to pause. That oral tradition of reading was specially established among the humanists who believed that knowledge should proceed from "oral discourse," be memorized, and then shared.[78] When reading became a solitary pursuit, Manutius realized it would also be good to organize words in a way that provided the accents and pauses. The period existed as a full stop, but Manutius invented a new kind of stop called the colon. It was one period over another and could be used to end a sentence firmly or as a link between independent but related sentences. His books are full of colons where we might today use periods. Manutius also altered the shape of the *virgule*, a slash that sometimes showed up in ancient texts. He added a curvy tail to the slash, creating the first comma, a mark of the short pause. Manutius also introduced the semicolon, a combination of the colon and the comma to denote a short pause that connects two related sentences.[79] And he started using apostrophes and all those accent marks that help us pronounce various languages, including Italian and *veneziano*. The Manutius influence on punctuation was furthered by his grandson, so-called Aldus the Younger, who wrote and printed the first book on the principles of punctuation, *Orthographiae Ratio*, in 1561. "He [Manutius] and his grandson generally created our modern conventional signs," says Lynne Truss, author of the best-selling book on punctuation, *Eats, Shoots & Leaves*.[80] In her analysis of the beginning of official punctuation, Truss says the idea was to standardize syntax. Marks such as colons, commas, and semicolons are either "separators or terminators," according to her, but their main job is to organize words into "sensible groups" and make them stay put.[81] A glance at old manuscripts shows a continuous line of words that goes on forever. When reading older books, it often seems like a different language, due to differences between modern punctuation and punctuation from the 1500s. Those sentences take up full pages and include endless clauses, commas, and semicolons.[82] It's not so much the language, per se, of these old books that seems odd, but the punctuation, which changes in form and evolves over time. With punctuation, the reader is shown visually where to pause and where to stop and take a full breath, as if someone were reading aloud. But after they were invented, punctuation marks were often overused.[83] It took *The King's English*, written in 1906, to clean this up, at least for English.

THE FIRSTS IN BOOKS PUBLISHED IN VENICE

Because Venice was such a hot spot for publishing, it attracted authors from all over, and many of the books that came out of Venetian printing presses were influential experiments in publishing or books that defined a new discipline. As a result, there is a long list of "Venetian firsts" within many book genres.[84]

The best place to begin is with the publication of *Hypnerotomachia Poliphili* (The dream of Polipholus), printed and published in 1499 by Aldus Manutius and Aldine Press.[85] The first word of the title combines the concepts of dream, life, and strife, and all that endeavor is done by the protagonist, Polifilo, who yearns for the damsel Polia.[86] This book of 234 leaves contains 171 detailed woodblock illustrations (including thirty-nine fancy capital letters that begin each chapter) set among type designed by Francesca Griffo. The artistry in this book is further enhanced by how the movable type was set. Sentences are not always right and left justified, as expected for a book. Instead, the lines of words might narrow to a bottom point, forming a visual upside-down pyramid, or morph into the shape of a chalice.[87] Pages are also proportionally enclosed in broad white borders, setting each page off as an individual artistic masterpiece.[88] The combination of illustrations and visually set type makes *Hypnerotomachia Poliphili* the first picture book in the world, and one of the most beautiful books ever printed.[89] It was also a very controversial publication. Written by a Venetian monk, Francesco Colonna, *Hypnerotomachia* is a love story, a dream of a man or boy for a woman and his searching through a dreamscape for her. It is also full of eroticism and philosophy, written in a mix of several languages, including Tuscan Italian, *veneziano*, Greek, Arabic, and Hebrew. *Hypnerotomachia* is both a romantic tale and a dream fantasy with allegorical meaning.[90] It provided the first general guide for fancy woodcut capital letters to begin chapters. These capitals also turn out to be a kind of mnemonic, because set next to each other, they spell the author's name.

Hypnerotomachia was privately funded by Venetian aristocrat Leonardo Croass. Between the text and the woodcuts, it must have cost a bundle.[91] Making this extravagant book required a team of artists to design the illustrations, three engravers to make the woodcut block illustrations (also called xylography), and they sometimes carried the story beyond the text, and composition designers for the unusual text layout. Altogether, it's such an unusual book. The reader's eye is forced to move back and forth from text to illustration rather than referring to illustrations that are bunched together. Historian of Venice Gary Wills claims the illustrations are essential to the book because they "kick-start the story and over and over."[92] *Hypnerotomachia* also has an all-woman cast except for the hero, Poliphilo; there are goddesses, nymphs, and real human women. No one knows why a priest wrote this book, or why Aldus Manutius agreed to print it.[93] Surely its erotic content was a risky move.[94] But the beauty is unsurpassed and the layout unique, even today. As historian of *Hypnerotomachia* Helen Barolini writes of its impact, "The Guttenberg Bible is somberly and sternly German, Gothic, Christian and medieval; the *Hypnerotomachia* is radiantly and graciously Italian, classic, pagan and renascent."[95] This book also set the standard of what printers could do, showing that they could do just about anything.

While *Hypnerotomachia Poliphili* set a high bar for complex text and magnificent illustrations, other first books made in Venice had significant social, religious, or literary effects.[96] How to be a mannered person was a favorite theme. Before the printing press in 1402, Pietro Paolo Vergerio "published" an incunabulum of instructions for young men to become educated persons. This self-help book, called *On the Conduct of Honorable Men*, is cited as harboring the beginnings of humanism at the University of Padua. Venice was also a ground zero for the beginning of many genres and pastimes. In 1551, Zoan (Giovanni) Francesco Straparola published the very first book of fairy tales, twenty-five in all. He followed that book with a volume of forty-eight more stories, published in 1553. Straparola's books were tales full of magic, some based on traditional oral folktales and some made up. He also seems to have invented the "rise tale," a story arc of a fairy tale that leads children into a dangerous dark wood and then rescues them into the light. In Straparola's Venetian version, those woods are the narrow, dark, and labyrinthinely scary *calli* of Venice. He also initiated the rags-to-riches story arc, wherein characters in poverty end up with a better life. A northern Italian working in Venice, not the Brothers Grimm of Germany, invented these traditional fairy-tale formats.

Venice, ever the cosmopolitan city based on international trade, also extended the various languages seen in printed books, and this happened often with religious volumes. For example, in 1452 Venetian Giovanni Leone translated Leon Albert Battista's *De Pictura* from Latin to English, thereby giving the English-speaking world the mathematics underlying art and science. Most notably, this book, which had only appeared in Latin, brought to the wider world the math that artists needed to compose their works with perspective. As stated previously, the inheritors of the Aldine Press printed the first Greek bible in 1518 with all the correct diacritics of the Greek language. In 1537–38, Venetian printers Pagiano Paganini and his son Alessandro published the first printed version of the Koran and did so in Arabic with its required diacritics.

That Koran was lost for centuries. It was found again in 1987 by library science scholar Angela Nuovo, when she followed its path to the small hidden collection of manuscripts in the Franciscan monastery on the cemetery island of San Michele, off the northern shore of Venice.[97] Daniel Bomberg, working in Venice, was the first printer of Hebrew books, and the first non-Jewish printer of Hebrew books. He designed and printed the Babylonian Talmud and it remains the standard page layout and pagination of the current printed Talmud.

Other printers affected how the country navigated among their base language of Latin, the vernacular of *veneziano*, and the eventual national language of Italian. Nicholas Jenson printed the first book written in Italian rather than Latin, *Decor Puellarum*, in 1461. Fra Nicolò Malerbi, along with collaborators, was the first person to translate the Bible from Latin into Italian, in 1471. That version is known as the Malerbi Bible and it is now owned by Oxford University. Centuries later, in 1798, Venetian Giustina Renier Michiel was the first to translate Shakespeare from English to Italian. Perhaps even more significant, the well-respected humanist, aristocrat, and international statesman Pietro Bembo was the first Venetian to write in Tuscan when he published *Prose della Volgar Lingua* (writing in the vernacular tongue) in 1520. That book helped established Tuscan Italian as the literary language of Venice and the national language of Italy once it became its own country in 1861. Other Venetian books changed the act of writing. In 1558, Aldus Manutius's grandson published *Eleganze della Lingua Toscana e Latina*, the first modern thesaurus. And Vespasiano Amphiareo, a master calligrapher, published the first and best-known manual on calligraphy in 1518, a book so popular that it went through nineteen editions.[98]

There were also long-winded Venetians who changed the face of history with their books. Marco Polo's four-volume series called *Travels of Marco Polo* (also known as *Il Milione*, or Book of marvels of the world) appeared in 1300 and it was probably written by his cellmate Rustichello da Pisa. When the two men shared a jail cell in Genoa after being captured during a sea battle, they had nothing better to do than talk. Either Polo talked a lot and Rustichello took notes, or Rustichello made it all up, but it's hard to imagine that the experienced romance novelist could imagine all those foreign adventures by himself. Polo never refuted the book and made lots of money from it, so presumably he told the story and was happy to have a scribe. Polo's book was, of course, a written manuscript, because there was no printing press in 1300. Nonetheless, the book was translated into many languages across Europe and copied for sale so many times that it can only be called a best seller. Other traders and explorers wrote about their travels (see chapter 3), but the fame of Marco Polo's book, even today, certainly makes him the first and most widely read travel writer in history.

Another internationally known Venetian, Giacomo Casanova, was even wordier than Marco Polo. His twelve-volume memoir was written in French, the language of intellectuals at that time, and was finished in 1792. Here it might be said that Casanova invented the memoir, at the very least the very long memoir. His endless tales of a passionate love life and his adoration of women must have initiated the idea of a Latin lover; Casanova's stories certainly helped men better understand female sexuality. Before Casanova, for wordiness there was Venetian aristocrat Marino Sanudo who sat in the Venetian Senate every working day from 1496 to1533 and took detailed notes. His fifty-eight-volume diary gives a clear picture of Venetian life and government in action.

On the more regular side of life, we owe Venetian printers and authors other debts. In 1555 Giovanventura Rosetti published his two volume *Secreti nobilissimi dell'arte profumatorio* in Venice which was the first book in Western culture with a recipe for dying fiber. We pay attention to time because Venetian Giovanni de' Dondi wrote *Tractus Astarri*, the first description of clockworks in 1364. Venetians certainly knew something about the marking of time, since their clock tower (Torre dell' Orologio) built at the end of the 15th century in Piazza San Marco, is all about announcing the time to everyone in town (see also chapter 4).[99] We also owe the genre of cookbooks to Venice, which means we owe to Venice the possibility of making different kinds of meals at home. The first printed cookbook, *De honesta voluptate et valetudine* (On honest indulgence and good health), by Bartolomeo Sacchi (also known as Platina), appeared in Venice in 1474. It came out in Latin and combined recipes by Martino with ideas of healthy living.[100] A little less than two hundred years later, in 1614, Giacomo Castelvetro published the first book that connected eating and health, *A Brief Account of the Fruits, Herbs and Vegetables of Italy*. More important to many was a book printed in Venice in 1565 that contained the first Western description of the cocoa plant and how exactly to make chocolate.[101]

The discipline of architecture was already well established by the time the printing press came along, but it was completely changed, molded, and rocketed forward by books made in Venice. Those changes have given us a long list of buildings and styles that reflect both the history and culture of a place. The first true book on architecture, *De architerctura*, was written in ten volumes by a Roman named Vitruvius, probably in the 1st century B.C. It was, of course, a handwritten treatise with maybe a few illustrations, all soon lost. In any case, those drawings would have been awkward, since, as Gary Wills puts it, "Scribes were trained to form letters, not to be draftsmen or artists whose impression could be trusted."[102] They also didn't yet have the mathematical ability to chart scale or elevation or make plans as we know them now, but budding architects worked with models, not plans. There was nothing else written about architecture for the next five hundred years, until 1486, when Leon Battista Alberti, of Genoa, published *De re aedificatoria* (The art of architecture). Alberti's book used only words and no illustrations to describe architectural forms, and it was written in Latin and aimed at the already educated. The book was redone in 1511 by Venetian Fra Giovanni Giocondo. That edition included original woodcuts and they showed visually what Vitruvius was saying. The Giocondo edition sold so well that it was eventually printed in a smaller format.[103] This book changed the very face of architecture because it combined illustrations with written architectural theory. Because of that combination, the language of architecture was systematized, and the profession morphed into what historians Mario Carpo and Sarah Benson call the "typographical architect."[104] "After centuries of the primacy of the word, the architectural discourse could, at last, put its trust in images and make use of images that faithfully reproduced and transmitted the appearance of original archetypes," they explain.[105] Now the printing press could use finely carved woodcut blocks for "mechanical reproduction of images" of classical buildings, and these illustrations informed a textbook that became available across countries.[106]

The visual and textual underpinning of architecture changed again in 1537 when Sebastiano Serlio, who lived and worked in Venice for thirteen years, published the first architectural treatise on the mathematical relationships between columns and bases. It also plotted the relationship of the long lintels that include the architrave, frieze, and cornices which are supported by a row of columns. Those relationships are the foundation for Renaissance architecture and architectural scholarship.[107] That book was called *Fourth Book*, but it was published first in a series of five out-of-sequence volumes.[108] *Fourth Book* was also the first architectural book published in vernacular, not Latin, and the first to have original illustrations. There were 126 drawings, almost half of them taking up a full page.[109]

But the most profound effect on architecture, on what we see in Western cities today, was made by a lifelong citizen of the Venetian Republic, Andrea di Pietro della Gondola, or Palladio. His four-volume work came out in 1570.[110] "The whole neoclassical movement would bear a Venetian stamp because of this man who was born and worked in the city's terraferma," writes Gary Wills.[111] Palladio also designed churches in Venice, civic buildings, a theater in Vicenza, his native city, and any number of villas in the area around Venice. We also see Palladio's vision echoed in American Federal-style buildings, Greek Revival homes across America, and any number of European buildings. A Palladian building is, by its very mathematical proportions and symmetry, pleasing to the human eye and brain. The immediate reaction is one of peace, comfort, and a sort of internal understanding that this building is "just right." A Palladian building, if we know the name or not, is now part of our soul, given to us by a Venetian with a good eye for perfection.[112]

VENICE, A PLACE FOR READERS?

With all these word firsts, it would seem that Venice was and is a city dedicated to reading and writing, but this is not completely true. Information about other cultures and business opportunities flowed into the city, but that information was more about making money than learning. The city was certainly brimming with humanists and intellectuals; the city of Padua, with its revolutionarily free-thinking university, was part of the Venetian Republic. But there were deeper issues with the written word that placed Venice outside the realm of more intellectual and learned places. The Venetian attitude toward books as primarily money-making opportunities rather than knowledge sources set Venice apart. Also, continued use of *veneziano*, rather than Tuscan Italian, made the city more dependent on local discourse than the words of intellectuals and writers. To Venetians, Italian was apparently as foreign as Latin, and they had real trouble writing poetry in Italian. The singsong, elided nature of *veneziano* always expressed itself more lyrically in song than on paper.[113]

Also, long ago Venetians were not really readers, even though Venetian glassblowers invented "reading stones" in the 13th century. Reading stones were handheld magnifiers set in a frame that made reading much easier, especially for those past age forty. But Venetians of the 15th century didn't revere books as might be expected for a city so steeped in printed matter. In 1366, the famed poet Francesco Petrarca gave his library to the city. Rather than cherish that gift, city officials put the volumes somewhere and allowed them to disintegrate. And this was a 14th-century collection of books from a famous intellectual. Along the same lines, in 1468, Cardinal Bessarion, an outsider, left his extensive and precious library to the city rather than Rome because he admired the lack of censorship and literary freedom. That collection included 752 books, 482 of them in Greek. The bequest lay unused and ignored for almost fifty years. Aristocrat and official historian of Venice Pietro Bembo was supposed to take care of these books, and he eventually called on the architect Jacopo Sansovino to design and build the Biblioteca Marciana in the Piazza San Marco. Once that library was established, a 1603 law required every printer in Venice to deposit a copy of every book they ever printed at the Marciana. That regulation speaks to a newly found governmental respect for books and reading.[114] Today the Marciana is filled with quiet scholars who often find historical treasures hiding in the stacks.

In present-day Venice, an interest in books, newspapers, and the written word is all over town. Great bookstores abound, and two universities also make for an intellectual level not often found in such a small place; Venice is a college town now. The Republic of Venice's contribution to the use of information, words, and reading in modern life is perhaps, today, dramatically demonstrated by two Veneto scientists who would have been citizens of the Venetian Republic if they had only been working about 175 years ago. Federico Faggin, born in Vicenza and a graduate of the University of Padua, led the team that invented the first single-chip processor in 1971. Twenty-six years later, in 1997, Massimo Marchiori, professor of computer science at the University of Padua, invented the HyperSearch, which ranks the relationship among several web pages. His work paved the way for Google and other search engines, changing the life of every reader on earth.

THE BEGINNING OF NEWSPAPERS

Venice also gave birth to newspapers. In 1563 the government began to distribute handwritten paper documents called *avvisi* (which means "notices" in *veneziano* and Italian) that came out every week.[115] These notices consisted of several small pages, but the writing was oriented so that the document could be folded in four and then read, like a pamphlet. By the late 1600s, the *avvisi* were printed on a press.[116] The point was to broadcast news, and control it, but also to make the news interesting. The *avvisi* were full of information about the workings of government, updates on military operations, diplomacy, and especially commercial news about the trade business.[117] The audience included government workers, nobles, and merchants—that is, the more elite members of society. These news sheets were not free. The price of an *avvisi* was two of a particular coin called a *gaxeta* (*x* is the *z* of *veneziano*), so these notices were also known as *gaxeta* (or *gazeta*). As a result, *gazette* is the word often used for newspapers around the world. Historically that includes the *Boston Gazette*, the *Providence Gazette* of Rhode Island, the *Phoenix Gazette*, the *Pittsburg Post-Gazette*, among many others. Today the word *gazette* still tops the front page of newspapers such as the *Charleston Gazette*, the *Delaware Gazette*, the *Gazette of Montreal*, the *Daily Gazette of Schenectady*, the *Saudi Gazette*, the *Bulgaria Gazette*, *Gazeta* of Poland, and so on. Like the *veneziano* word *ciao*, *gaxeta* is a Venetian word that has traveled the world. The Venetian *avvissi* were also filled with news about other countries, so they were the first international newspapers, and they have been found as far away as London.[118] The style of journalism they employed—short sets of news items, forwarded from a particular city, written under the name of that city and the date on which they were sent—was adapted for most of the early printed newspapers.[119] Today Venice carries on the tradition of *avvisi* with the local daily paper *Il Gazzettino* (little newspaper) began in 1887. It is currently owned by a wealthy consortium and has sections on all the various cities and towns in the Veneto mainland and for other islands in the lagoon.[120] They use, as all newspapers do, the traditional newspaper layout mapped by Gasparo Gozzi who launched the newspaper *Gazzetta Veneta* in 1760. Gozzi ironically also held the post as censor of the very press he published for some years.

The impact of newspapers was also deeply shaped by Venetian Felice Beato. He was the first photojournalist to document a military campaign as it unfolded during the Second Opium War in 1856–1860. The courage of Beato and those who followed him brought us photos of World Wars I and II, and the photographs and films of the Vietnam War, visuals that were instrumental in turning the American public against that war. Today we are barraged by images of war zones taken by a long list of photojournalists who have shown the disaster of combat, the difficult job of being a soldier, and the effect of war on local citizens with an impact that mere words cannot describe.

The tradition of public announcements in writing is also carried on today in Venice with bulletin boards at water bus stops that bear the notice AVVISI. They announce changes in stops and rerouting. An even more public, personal, eclectic, and transitory kind of written announcement also has a long history in Venice. Ever since the 1500s, citizens of Venice have been pasting up notes full of their disapproval or praise, usually anonymously, for everyone to see.[121] Today, there is an ever-changing flurry of posters glued to walls, buildings, community bulletin boards, and construction site plywood enclosures.[122] They announce lectures, art shows, concerts, and upcoming protest marches. And they are effective. In June 2019, after a cruise ship ran into a river cruise boat and an embankment on the southern edge of Venice, a call for a mass protest the following week went up in three days and was plastered all over town. Police figures noted that between eight to ten thousand people showed up at the march (about 25 percent of the Venetian population). More remarkable, the day after the protest march the call-to-arms posters were pasted over with thank-you notes to citizens from the organizers, surely the first-ever written thank-you note in protesting history. There are other kinds of temporary notices with words that come and go daily in Venice as well. There are taped-up pictures of loved ones announcing funerals, colorful graffiti usually aimed at some notion spray-painted on walls, and well-executed protest stencils with slogans such as "We insist on seeing the horizon." In the summer of 2019, an official Banksy showed up on the side of an old house, right at the waterline, and only accessible by boat but easily viewed from a small bridge. The design is a spray-painted portrait of a child holding a torch emitting a pink cloud, and it was aimed at countering the contrasting pomposity of the Venice Biennale art exhibit. Nowhere are written public announcements more visible and more up to date than in Venice.

CHAPTER NINE

The Venetian Way of Life:
Leisure Activities

Venetian opera was in every way larger than life...Venetians loved special effects—magic spells, shipwrecks, storms, and anything else that went boom.

—Thomas Madden,
Venice: A New History
2012

[Lottery] is a new method of commerce, giving hostage to fortune: it began in a small way. First anyone who wished to adventure had to give 20 soldi, then it grew to 3 lire, then to a ducat. And the prices (prizes) were carpets and other things; now there are money prizes, 200 ducats, and a piece of cloth of gold has been offered.

—Marino Sanudo, Diarist of Venice
1496–1533

The mask spared ceremony while preserving respect. It furnished a common footing without denying status and saved face when one's dignity was at stake. It was a token of privacy instead of the real thing, a manufactured buffer that licensed genuine aloofness and unaccustomed closeness. Its ritualized "anonymity" could be acted on or ignored at will. The mask honored liberty in the Venetian sense, which meant a measure of autonomy within jealously guarded limits.

—James H. Johnson, *Venice Incognito*
2011

Why, this Satan's drink [coffee] is so delicious that it would be a pity to
let the Infidels have exclusive use of it. We shall fool Satan by baptizing
it and making it a truly Christian beverage.

—Pope Clement VIII

Evaluating the evils of coffee

for the Venetian clergy

The way to enjoy Venice is to follow the example of these people and
make the most of simple pleasures.

—Henry James

VENETIANS DIDN'T INVENT fun, but they certainly know how to
have it.

Every month, overlapping entertainments collide in Venice as many
posters and invitations pop up overnight on walls and buildings. One pasted
over another, these public advertisements encourage everyone to come and
see. Fall opera schedule at La Fenice here, small art gallery opening there, the
usual Vivaldi concerts are ongoing, of course, but here's a film schedule that
just came out and a notice from the university about a new book just coming
out—the author is speaking at 8:00 P.M. Come as you are, come if you want,
don't worry about the right outfit, just join us for something interesting. As
a small, yet culturally sophisticated and cosmopolitan city, Venice is unusual.
Throughout history, and certainly today, there has always been an ever-changing
roster of things to do beyond the workday. Paintings from centuries ago show
Venetians watching plays in *campi*, gambling all over town, wearing masks to
hide bad behavior, and dancing and singing during festivals. Today, posters
bring in tourists, but audiences are also full of resident Venetians who like an
art opening or new theater production as much as the next person.

Both historical and current goings-on in Venice paint a picture of a place
where everyone, tourists and locals alike, take their leisure time seriously. The
beauty of Venice might be found in its art and architecture, but the soul and
genuine identity of the city are displayed when the citizens gather to celebrate
and have fun and even share those events with outsiders.[1]

More surprising, in their pursuit of pleasure Venetians have been responsible for some common entertainment firsts, inventions which then spread throughout Europe and became part of the Western entertainment canon.

GAMBLING AND VENICE

Gambling—that is, risking money on a chance of a payout of even more money—appears to be pretty universal. There also seems to be variation in the desire to gamble, which can be an underlying seduction during the game. Some think gambling is fun, while those who don't still see the allure. Some are addicted and can't stop; these players are also attracted to any kind of risk game, from betting on cards to football scores, like any addiction that continues in the face of ruin. Gambling is a curious case of human behavior, because it is not a substance that one ingests, but a behavior that one does. The best comparison might be addiction to sex or shopping, but gambling still differs from those addictions because it has such a seemingly low rate of pleasure return. Even those who find gambling "fun" and are attracted to the possibility of a big win know that overall, they will come up short and that there is a great risk of financial and personal ruin if they don't stop. Why, then, do people gamble?

There seems to be some link between dopamine release and gambling, putting gambling in the same category as other behavioral addictions that risk great harm because they are fed by chemical brain changes. But the link between brain chemistry and gambling is complex. Surprisingly, it's not about the money but the broader, and even more insidious, pull of possibility. Researchers call this the "attractiveness of uncertain rewards."[2] It works like this—the craving is most intense not with big wins but with "near misses." Gambling hooks people because we are, many of us, drawn to a sense of not knowing what a reward might be or how much will arrive. This attraction is also found in other animals, including birds, so the willingness to risk something simply for the chance of reward must have deep evolutionary roots.[3] And that makes sense. In an uncertain environment, brains that were designed to take on risk might do better than those that were unwilling to take on risk. Resources such as food and water are often unpredictable in both time and space, and the cues we and other animals use to find those resources are not very good. The color of fruit might be a sign of ripeness and high sugar content, or it can spell rot, for example. And so being attracted to unpredictability, rather than avoiding it, might lie in the very need to stay alive. Our minds were presumably designed to judge unpredictably, jump at a chance for a reward even based on little information and a fuzzy idea of the size of the reward. Evolution, then, would favor those who persist even in the face of loss, but that persistence only makes sense if there is the possibility of some reward, even a tiny one. Researchers also suggest, as a caveat, that this cycle is highly motivating when the actor is dealing with some sort of psychological or physiological deprivation in their past or present.[4] Uncertainty is the quality that pushes repeated behavior, especially when someone is primed by some sort of trauma, and there we have gambling.[5]

Venetians have a long history of gambling, and they brought gambling into the forefront of society. Gambling is also wrapped up in various attitudes about Venice as a city that suffered from (or was enriched by) a reputation for decadence in the 17th and 18th centuries. Gambling was both loved and frowned upon, and it was and is often the subject of moral judgment from both the church and society. Yet gambling is all around, and it's still in Venice. Venetian historian Alberto Toso Fei says gambling was not only rampant in Venice—it was a Venetian obsession. He claims even the threat of losing an eye for excessive gambling didn't stop anybody from rolling dice or picking up cards.[6]

Gambling became legal in public early on, but only in a restricted way. Three stately columns, so the story goes, had arrived by ship a hundred years previously as a gift from the Byzantine emperor. They were to stand at the entrance of Piazza San Marco as a reminder to all those arriving in the city of the might of the Venetian Republic. But one column had landed in the water and had yet to be recovered, while the other two had been unloaded and left like beached whales because they'd been so heavy, unwieldy, and tall that no one had been able to figure out how to raise them. Engineer and architect Nicolò Barattieri, the man who had designed the first wooden Rialto Bridge in 1181, and who had figured out how to use cranes and counterweights to erect the first *campanile* in the piazza, made a deal with the city of Venice.[7] In 1172 Barattieri had been asked if he could somehow raise the huge marble columns lying in wait in the Piazzetta, the space between the Palazzo Ducale and the Marciana Library in Piazza San Marco.[8] Barattieri attached a wet rope around the top of the columns and anchored the other end to the ground near the base. Once the wet rope dried and constricted, the columns rose just enough to allow supports to be shoved underneath at the top end. This process was repeated again and again until the columns were vertical, as they remain today. The government was so happy with the feat that they offered Barattier a reward. He asked for the right to gamble in the piazza, and the Council of Ten granted his wish, with the caveat that his privilege only applied to the area between the two columns. Since gambling in public was banned at that point, this concession marks a change of heart by the Venetian government. And afterward gamblers in Venice were called *barattieri*.[9]

Gambling was eventually so public and widespread in early Venice that in 1254 the state took a stand and banned any games in or in front of the basilica. The next year they forbade gambling from the large central courtyard of the Palazzo Ducale or anywhere near the *Sala del Maggior Consiglio*, the giant state council meeting room. Surely some councilors were gambling while conducting governmental business. In 1266 gambling was finally banned from anywhere inside the Palazzo Ducale, the seat of government. By 1292 the only games allowed were chess and board games. Three hundred years later, still unable to stop ruinous gambling, the Council of Ten restricted the games again, but all bets were off (or on rather) during Carnevale.

In general, gambling was considered amoral in Venice, but there were times when it took on the whiff of goodness. As historian Jonathan Walker points out, between 1500 and 1700 gambling in Venice went public.[10] Seen as an amoral activity in the early 1500s and condemned by the church, when gambling went more public it took on a positive aura that included notions of the "self-control" and "magnanimity" required of a player.[11] At the same time, with all the gambling and prostitution, outsiders viewed Venice as a debauched city in the 16th and 17th centuries.

Venice is, of course, most famous for its February Carnival, and that month of partying is now part of the iconic imagine of Venice for many tourists. Partygoers don't realize that today's Carnival is a kind of invented fake. The first Carnival of Venice was held in 1268 as a pre-Lent celebration. By the 17th century, Carnival was an essential part of Venetian life, the time when hardworking people let loose. It had also been growing steadily and attracting visitors from all over Europe; by the early 1700s, over fifty thousand revelers typically crowded into the city, adding at least one third to the population. Today, Carnival lasts over a month, and there are party venues all over town, including nighttime festivities in Piazza San Marco.[12] Debauchery extended to a wild Carnival full of people hidden behind masks, everyone throwing money at male and female prostitutes and gambling away their life savings. As with all the myths of Venice, this portrait has some truth to it, but it is also exaggerated. Nevertheless, the idea of a free-for-all Venice brought in lots of foreigners and tourist dollars pouring into the gambling houses and brothels and to merchants and hoteliers. And gambling was an integral part of Venetian life.[13] Venetians from all classes even loved to bet on election outcomes.[14]

A lottery is the king of common gambling because the stakes are so low, while the payoff is often large. Lotteries are also usually in the public domain, and many people can play at the same time, even those with few resources. Lotteries, especially now in America, are government-sponsored, and the money goes into the state or city purse to fund community projects such as schools. Thus, lotteries have always been a more acceptable—that is, more moral—form of gambling because they cross socioeconomic classes and often fund public good. Venice was probably the first place in Italy where lotteries appeared in 1522.[15] Scribe of the city Marino Sanudo wrote that the first lottery was conducted by one Geronimo Bambarara, who was a used clothing merchant at Rialto.[16] After that initial lottery offering by Bambarara, lotteries took off in the city. This first wave of lotteries was both private—run by a merchant—and public. The prize might be some mercantile good, such as a carpet, or something needed for the city. For example, governmental lotteries of that time were used for building the Rialto Bridge, paying a ransom to pirates for captured Venetian citizens, and donating to the city's poorhouses.

There were also many other types of in-house games in Venice, such as backgammon and other board games.[17] But most of all Venetians loved cards and dice, and they played them all over town.[18] Playing cards were dealt into Europe from the Islamic world in about 1360, but a decade later they were a common sight because of extensive trade between Egypt and Venice.[19] The new import was a fifty-two-card pack lacking the joker, an American introduction into the deck in the late 1800s. Cards, with their exotic figures, quickly spread across Europe.[20] New illustrations were invented to overcome

the Eastern origin of the drawings on cards, because they were misunder-
stood by Europeans who were wary of any sort of mysticism coming from
the Orient. A small book called *The Cards Speak* was published in Venice in
1545, and it contained explanations for the designs on fortune-telling cards;
swords were a sign of a mad gambler who lost his fortune, batons and clubs
represented cheating, coins were the nourishment of gaming, and glasses
of wine warned of drowning in drink.[21] Artists drew these illustrations for
cards in Venice and then printers made the decks. Today boxes of cards stand
in the windows of tobacco shops, and they look only slightly like the decks
most Westerners are used to. The Venetian ones are slimmer, and the suits are
different than the usual hearts, clubs, diamonds, and spades. Instead, there
are scepters, swords, medallions, and what look like hand weights for games
such as *trevigiane*, *triestine*, and *bergamasche*. As with card decks everywhere
now, some Venetian cards are artistic canvases that also advertise the city.
Venetian artist Giorgio Ghidòli sells his deck from his toy shop in Castello.
It includes drawings of the doge as the king, the dogaressa as the queen, and
the *capitano di mare* (sea captain) as the jack.[22] Even today it is normal to see
people playing cards in public in Venice; on a visit to one of the local wine
distributors, where a liter bottle can be filled at the tap, the owner behind
the counter is often engaged in a card game with an unrecognizable deck.

There are four kinds of card games. Tarot cards started as a game but
morphed into fortune-telling in the late 1700s.[23] In 1550 Francesco Marcolini
published in Venice *Le Ingeniose Sorti* with pages and pages of card face illustrations and what they meant in their various combinations.[24] The first official
printed explanation of how to actually use tarot cards also appeared in Venice
in 1575.[25] Another use of cards is focused on gambling with randomness as the
spark, since cards are drawn from a pack with no manipulation, and there are
no strategic decisions to be made after the draw. For example, *basetta* is a card
game attributed to Venetian Pietro Cellini, who invented it in 1593. Later, the
game was introduced into France by the Venetian ambassador, and the name
took on a French twist as "basset."[26] Venetian *basetta* is played by turning over
cards in a pack.[27] If a certain card ends up in the dealer's pile, he wins; if the
card shows up in a player's stack, he or she wins. In that sense, *basetta* is pure
gambling—it takes no skill and is won or lost solely on chance.[28] The third
kind of card game takes skill as a player decides to add or discard cards and
determines the value of his or her hand for making a bet against others. The

fourth kind of card game involves winning cards from others by trapping them in a card interaction. These "trick-taking" games include *trappola*, the first real trick-taking game invented in Venice in the early 1500s.[29] Along with *trappola*, a new kind of playing card was designed in Venice that had mirror images of the figures on the cards and why we now have the figures of king, queen, jacks, and jokers twice on our modern cards.[30] There is some indication that *trappola* has a Greek connection. Τράπουλα, the transliteration of the word *trappola*, is the Greek word for "playing card." That connection might have happened on the Greek Ionian Islands occupied by Venice in the 16th century.[31] In honor of that ancient game, there is a phone app called *trappola* and it provides a deck of cards on the go, but no instructions on how to play *trappola*.

THE VERY FIRST CASINO

Close to Piazza San Marco and sitting at the mouth of the Grand Canal is a complex of hotels and restaurants called, interestingly, Hotel Monaco & Grand Canal, which was renovated in 2002–2004 by the Benneton Group. The Monaco is a fabulous place to sit in a waterside café, have a Venetian spritz cocktail, watch the traffic on the Grand Canal, and track the sun as it slides into the lagoon and says goodnight. This building complex also has a hidden treasure tucked inside that is even more iconic for Venice. To the right of the front reception desk is a wide stone staircase that leads up one floor to the oldest public casino in the world.[32] Called Ridotto, this first-ever state-owned gambling house opened in Venice in 1638 with four floors of gaming rooms and other salons for conversation, rest, food, and drink.[33] *Ridotto* is an Italian word that means, among other things, "private room" or "retreat" and is connected to the verb *ridurre* ("to close off," "to make private"). In the case of Venetian gambling houses, *ridotto* also meant "foyer" or "lobby," which in the 17th century was a room at the back of a theater where people donned masks and hung about gambling or doing other frowned upon acts. Since Venice had an out-of-control gambling problem and state laws made no difference, the government decided to cash in on this vice, and thereby control it. Patrician Marco Dondolo offered up his palazzo just outside the Piazza San Marco for this purpose. It became the Ridotto. Although this first gambling house was open to the public, only the rich had the means to play, so in reality, only aristocrats with money and rich merchant-citizens could attend. The state—that is, a bunch of nobles—also decided that since it was their gambling house, aristocrats who had fallen on hard times and needed state charity could earn their keep as croupiers. Although all the gamblers at the Ridotto wore masks, these sorry aristocratic croupiers were forbidden to don masks; instead they were required to wear a uniform of toga and curly wig. In return, these fallen nobles were allowed to live in state-owned places, and in poverty, around Campo San Barnaba.[34] By working at the casinò they were able to pay back the state for their charity, but they couldn't hide their status from their gambling fellow wealthy aristocrats and foreigners who were masked, so their humiliation was public. Profits from The *Ridotto* went straight into the city coffers, much like public lotteries today. But this first casino made for other public problems, like gangs of thieves who waited in the neighborhood to rob unsuspecting and probably drunk customers. Eventually, in 1774 the state closed the doors on the Ridotto because it seemed to have caused more trouble than it was worth; the Great Council was also afraid that too many nobles were going broke.[35] And the state casino had created a shadow economy with money flowing from aristocrats to the lower classes, and they didn't like that.[36]

But the end of the big public casino simply forced the creation of many little casinos. Citizens of Venice moved their gambling into small private playhouses called *casini*, miniature palazzi intended for secret fun, places where one could act out without bruising the social order. In 1755 there were seventy-three of these tiny *casini* in the area of the Ridotto, but they were also scattered across town. Some still stand in their finery today and are used for concerts; they can also be toured.[37] Then, in 1768 the state opened another gambling house at San Cassian Theater. It featured the usual card games, including *biribisso* or *biribi*, a game like bingo with numbers drawn from a bag and placed on a large pictorial card.[38] Today there is one official casino in Venice, the Ca' Vendramin Calergi, and it used to be Richard Wagner's home. It opened in the 1950s and includes the usual card games, roulette, and slot machines, and you can play with any foreign currency. Another casino opened on the mainland near Venice's Marco Polo airport in 1999. Billed as the first "American style casino," this place is aimed at catching players right after they get off the plane.

THEATRICS AND ENTERTAINMENT

There was a thriving outdoor theater tradition with acts and puppeteers setting up stages in the *campi*, so Venetians easily took to staged performances indoors, where they were an enthusiastic audience. Another common outdoor entertainment was the *Forze d'Ercole*, or human pyramid, a test of physical strength and skill. Men from opposing teams climbed upon each other's shoulders, then a small child clambered to the top, making eight full tiers. The participants grasped each other in configurations that had names like Bella Venezia or the Colossus of Rhodes.[39] Recall that Giovanni Battista Belzoni, the great explorer of Egypt, made a living as part of a human pyramid in Venice before he found fame as an Egyptologist (see chapter 3). Entertainment was also provided by staged outdoor fights between groups of men called *guerre bastoni*. These guys met up on various bridges and tugged and pulled at each other while the crowd looked on from either side of the canal and windows of nearby palazzi. These fights were socially sanctioned and upheld by the state. The idea was to move what might have been gang warfare into a contest of physical prowess but without violence.[40] Ponte di Pugni (Bridge of Fists), the small bridge that links Campo Santa Margarita to Campo San Barnaba, still bears the outline of feet that mark the starting point where boys from two sides of Venice met for tug-of-war and a little fist fighting.

Venice was, therefore, the perfect place for staged entertainment because it had all the ingredients for success: the right buildings to accommodate an audience, investors and backup guarantors called *pieggio*, various performers and conductors, and that spice of Venice required to succeed—at least one aristocrat as a patron.[41] In 1545 an all-male acting troupe formed in Padua on the Venetian mainland. This company was the first professional acting company ever, and it legitimized acting as a paid profession.[42] Eventually, their style was called *Commedia dell'arte,* and the typical improvised play was designed to make fun of familiar personality types.[43] The actors donned masks and costumes that became standard figures of clowns, buffoons, manipulators, idiots, lovers, heroes, and heroines whom we all recognize in our daily lives. The performance was always comedic and driven by the same characters dressed in their usual costumes that appeared over and over in various simple plots. In repetition, the actors gave a ritual feel to their performances and provided an easy path for audience recognition and connection. Like a well-known and memorized puppet show story or a repeated fairy tale or fable, or even a TV sitcom that has lasting characters, they became "knowns" and part of the wider culture. The actors' skill lay in developing improvised plots and bringing a well-known character to life. *Commedia dell'arte* developed into a particular style of theater that had staying power and brought us many familiar characters and names. There are the *zani*, or minor characters of the *Commedia,* who drive the fun and bring on the laughs with their vulgarity and slapstick moves.[44] *Zani* is *veneziano* for John, but more significantly from those characters, we get the modern word *zany*.[45] The *zani* also included the servant Arlecchino, or Harlequin, that trickster dressed up in a costume of brightly colored diamond shapes who still shows up at Halloween or at masked balls. The character of Pantalone, the merchant, was identified by his ankle-length tight pants, bent posture, and big nose. The pants were especially funny; they were quite unlike the short breeches of nobles and just like pants worn by peasants in France. And so, in Venice, they were christened *pantaloni* after the character. Thus, *Commedia dell'arte* gave us the idea of long pants and the word *pants* as we now know them and wear them. The Venetian playwright Carlo Goldoni decided to clean up *Commedia dell'arte* in 1738 by writing real scripts with plots, but these more staid events were not as successful or entertaining as the bawdier free-form productions.[46] Goldoni was the first to use the term *Commedia dell'arte* for this type of play, but he meant it as an insult.[47] Goldoni was also instrumental in its death in Venice. Nonetheless, *Commedia dell'arte* was so popular that it spread across Europe and lasted for over two hundred years, well into the 18th century. It greatly influenced the comedy theater that came after. Even today we recognize these character types when they appear.

In the mid-1600s, actress Isabella Andreini was considered the first stage diva. She was born in Padua of a Venetian family. She was a main player in the famous professional Italian acting troupe *I Gelosi*, a traveling troop with which she was frequently on stage in Venice. Historian Rosalind Kerr has compared Andreini's fame to that of Lady Gaga, including the fact that both women switch their personas between male and female with ease.[48] Andreini had to keep her personal life anchored in the expected purity of marriage and children to keep her fan base, but she was also clearly a feminist. Andreini famously rewrote a play and changed the plot to give women, not men, power. Her many poems also sought to empower the usually quiet women of the day. Her written works included a kind of gender fluidity rarely, if ever, seen in those times, especially from women.

Opera arose from the spoken theater as a much more expensive production that required elaborate staging, costumes, musicians, and dancers. At first, opera began in the early 1600s as private productions. These small court operas probably started in Florence or Mantua, according to Venetian historian Thomas Madden, and it wasn't until opera came to Venice and became public theater that it grew into a popular art form for everyone.[49] Venetians didn't invent opera, but they certainly made it a thriving business and entertainment for the populace. Operas also ran the gamut from comedy to tragedy, so they expanded the experience of theatergoers in Venice.

Eventually, operas required many investors, so Venetians invented *carati*, or shares, so both profit and loss could be shared by the many who had invested in a production.[50] That arrangement is markedly different from the donations from opera buffs today that help finance productions. Of course, the Council of Ten was horrified by the vulgar nature of many theatrical performances, and in 1508 they proclaimed that all scripts had to be approved by the council. This rule was, of course, ignored; its power was, at best, weak, given that many young aristocrats acted in the very plays that the council might ban.[51]

The first public opera house in the world was Teatro San Cassiano opened in 1637 with a performance of the opera *Andromeda* by Francesco Manelli. Teatro San Cassiano was significant because it was the first time the general public, not just nobles, could go to the opera. This theater was in operation continuously until the beginning of the 19th century—that is, for two hundred years. Venetians also invented box seats (*palchi*) costing more and these private sets separated the nobility from the lower classes.[52] These boxes were first built in around 1581 in Venetian theaters showing *Commedia dell'arte*.[53] Since part of the point of going to the opera was to be seen, box seats were little stages that showed off the rich people in their finery, but they could also be used for assignations, as the owners used them like "motel rooms."[54] Boxes were often sold as a theater was being built or reconstructed to help with financing.

Venice was known as a place where women could sing in operas. In other opera houses, castratti sang the female parts, but women proudly took the stage in Venice.[55] In the Venetian desire to present operas with spectacle, Venetian works included mythological and spectacular figures such as women warriors or Amazons. Between 1650 and 1730 there were about 120 productions in Venice starring women warriors.[56] Their presence was an acknowledgement of women's relative freedom and power, for which Venice was famous.[57] The works in Venice depicting Amazons also created opportunities for professional female opera singers to shine and those stars paved the way for opera divas, a moniker only for temperamental and highly talented women. Roman opera diva Anna Renzi moved to Venice in 1640, where she was instrumental in making opera an established entertainment. Renzi was a great actress and a singer; roles were written specifically for her, and she was a master performer, repeating a performance over and over, never once slipping out of character or giving a less than perfect performance vocally or physically. She also single-handedly created the role of a madwoman in opera by pushing the standards for the "madwoman" performance now common in opera.[58] From Andreini the actress and Renzi the opera star we get the phrase *prima donna*, which translates literally as "first women." It underscores how female stage stars were both admired and so important that they were treated with kid gloves.

Theaters were businesses in Venice, especially for the aristocracy. Of the ten theaters operating in the 17th century, seven were owned and operated by noble families, and they were all moneymaking ventures.[59] By the 18th century, there were seventeen active opera houses in the city, and the opera season was firmly attached to Carnival and it took advantage of the foreign tourist wallet.[60] As opera spread to the rest of Europe and foreign attention began to wane, the theaters in Venice switched back to plays. Today, several theaters, large and small, put on plays and other performances that are viewed by both tourists and residents. The opera house of Venice, La Fenice (the Phoenix), opened in 1792, even as opera was becoming less popular. It was owned by the people who collectively by contract had bought the box seats. The auditorium burned to the ground three times, most spectacularly in 1996, but it was famously rebuilt according to the old Venetian adage, *Com'era, dov'era* (As it was, where it was).[61] Performances and guided tours of the opera house show that La Fenice retains its 18th-century glory and acoustics and it is still a coveted venue for operatic performances for the public if they can afford it.

The love of performance by Venetians is also kept alive today by the annual Venice Film Festival which began in 1932. It was the first international film festival in the world, a format that has been copied in many countries. Each year movie stars, directors, and production staff flood into Venice and over to Lido, the barrier island where the festival is held. Winning a prize at this festival is an honor and professional achievement, and it also predicts box office success.

COFFEE, ALCOHOL, AND OTHER PLEASURES

Nothing spells leisure more than food and drink, and Venice has its fair share of contributions in both categories. Although about sixty different plants contain caffeine, humans have globally and historically settled on the roasted coffee bean as their stimulant of choice.[62] Some authors have suggested that humans are attached to coffee because the stimulation helps in advancing reproductive success, as jacked-up people not only wake up and go to work but also look for mates.[63] Today, coffee culture is certainly global. In the West, coffee venues litter the landscape as places where patrons sit and talk (or work on their computers), while coffeehouses and bars are common in other parts of the world as well. The role of a coffee house or bar or café highlights how coffee has become the instigator of much of modern human sociality.[64]

Making an elixir out of coffee beans was probably invented in Ethiopia, where the plant is endemic. From there, beans, still in their skins, mixed with animal fat, were traded to Yemen, right across the Red Sea from Ethiopia.[65] Today coffee is the second most globally traded commodity, after petroleum. That trade amounts to more than seventy billion dollars a year, which shows the reach of coffee and human dependence on this bean.[66] Though Marco Polo never went to Yemen, he somehow brought coffee beans back to Venice. These beans came from the city of Mokha, Yemen. They were of the arabica type; perhaps they had a touch of chocolate flavor, although this has not been confirmed. But if so, it would explain why today we consider mocha a chocolate-flavored coffee drink because the cocoa beans that make chocolate never grew in Yemen.

Although coffee drinking was common in the East, where Polo surely sampled it, no one paid much attention when he brought the beans back to Venice. Two hundred years later, in 1591, Paduan botanist Prospero Alpini published in Venice the first description and drawing of the coffee plant. He went to Egypt and claimed coffee as medicine in his *De Medicina Aegyptiorum*. But again, no one paid much attention. But following Alpini's book, coffee plants were sold as medicine by 1600. In 1615 Venetian traders finally introduced coffee to Europe in a big way, and the first coffeehouse in Western culture opened in Venice in 1645.

Given Venice's connection to Constantinople, it's no surprise that the city was the first place where coffee entered the Western lexicon and Western culture. Venetian historian Alberto Toso Fei claims that the first Italian to mention *caffè* was Venetian, Gianfrancesco Morosini.[67] He had been bailiff to Constantinople and spoke of the Turkish habit of drinking in public inside street shops. The Turkish brew was made with seeds they called *kahvé* and was served hot, and the Turks claimed that coffee kept people awake. In 1720 the coffeehouse Alla Venezia Trionfante (known as Café Florian today) opened in Piazza San Marco and was the first coffeehouse in Italy to allow women.[68] Florian's is also one of the two longest-running coffeehouses in the world (the other is in Paris). Piazza San Marco eventually had thirty coffee cafés.[69] Today there are two in the piazza, but they are always busy. These cafés were venues for discussions, literary places where books were reviewed, and also the site of many political arguments. In Venice, coffee-houses also distributed the local newspaper, the *Gazzetta*. By 1759 this tiny city had 206 coffee shops. Italy, in general, has a well-established reputation for great coffee, and this is interesting given that no coffee plants at all grow in Italy. The Italian skill seems to be in buying (that is, trading for) the right beans in countries where they do grow and then roasting them in Italy. The city of Trieste, north of Venice, is ground zero of the coffee brand Illy sold around the world.[70]

The Venetian love of coffee was also famously adapted to dessert when in the 1960s or 1970s a cook at a restaurant in Treviso slapped together the first tiramisu, a sweet composed of hard cookies, coffee, brandy, and lots of whipped cream and mascarpone cheese. This dessert is served in fancy restaurants all over as a special treat that came from a city that used to be part of the Venetian Republic.

Italy is also famous for its various liquors and wines. Prosecco, the star of the Veneto region, can even be bought on tap in small shops called *vino sfuso*, where locals go to fill up their empty plastic liter bottles. In 1919 the Barbieri brothers of Padua invented aperol, the sweet and bitter amaro that is combined with Prosecco and soda water to make a spritz, currently the fastest growing cocktail in America and Europe.

Venetians back then and today devote much of their leisure time to life on the water. *Regatta* is, in fact, a Venetian word that probably derives from the *veneziano* word *riga*, the name of the line behind which boats assemble before the start of a race. Or it might be related to the *veneziano* words *regatàr* (to compete) or *remàr* (rowing).[71] In 1645 the first official regatta ever was held in Venice among gondoliers. And we all can be grateful for that orientation toward the water, especially in the summer, because Venetians constructed the first outdoor swimming pool in the early 1900s on the barrier island of the Lido. From that, the British now call any outdoor swimming area, or a beach for swimming, a "lido." Perhaps Brits had been swimming in Venice while there on vacation and brought the word *lido* back as a souvenir. It's certainly not a surprise that Venice, the city of water, brought the world a large human-made body of water simply for the pleasure of cooling off.

Tourism, High Water, and the Future of Venice

Nothing daunts foreigners. Nothing frightens them. Nothing stops them.
>—Luigi Barzini, *The Italians*
>1964

Though there are some disagreeable things in Venice there is nothing so disagreeable as the visitors.
>—Henry James

It's not possible that a city like this has become so cheesy. We are the sinners; we're responsible for [the deterioration of culture] that has transpired.
>—Paolo Olbi, Venetian Master Printer[1]
>2019

If it's tourist season why can't we shoot them?
Tourists go home. Immigrants welcome.
Tourists are Terrorists.
Fuck tourists.

>—Graffiti all over Venice
>2019

Being a foreigner. It's the lowest life form.
>—David Sedaris, *This American Life*

AS I WALK the glossy, grey streets holding my umbrella high and hoping the exercise class will still be held in the face of this deluge, the siren goes off. It's a loud, mournful sound from the center of the city; when the water level is rising to critical levels, the siren emits an extended wail that floats over the city. In any other place, this sound would send the populace running for cover, wondering if the end was near. But here in Venice, the siren is a touchstone as much as a warning. It announces that *acqua alta* (high water) is coming and soon the rising tide will breech the canals, seep up the drains, and spread across the city.

Venetians know this drill well. They put on their well-worn, unfashionable rubber boots and go on their way. Some stores move to put inserts of solid metal or wood into the bottom of entry doors, trying to keep the water out. City workers erect elevated sidewalks in areas of greatest flooding so foot traffic never halts, even for a second. And no one panics, because Venetians know that the water goes down when the tide goes out; they also know it will come again on the midnight tide.

Seasonal *acqua alta* is iconic for Venice, but it can also be destructive. One terrible flood made news in 1966 when both Florence and Venice were covered in water and there was extensive damage to the buildings and great works of art. At that time, volunteers from all over the world came to help save precious objects. There weren't many tourists going to Venice during the 1960s, but after the news coverage and an influx of government money, Venice cleaned itself up and the tourists began to come in earnest. Those numbers increased each year because the city gained a reputation for fragility, and the rumor began that Venice was "sinking." Now what threatens Venice is both the frequent flooding of water and the frequent flooding of tourists.

The story of what Venice is dealing with today and what happens to it in the future is part of Venetian identity, culture, and history. It's true that Venice has not created any original inventions to deal with high water or mass tourism, but because these threats are now such an integral part of life for visitors and residents, this book needed to address the current pressures that have engulfed the city—and what the potential for an inventive solution might be.

FROM THE GRAND TOUR TO MASS TOURISM

Luigi Barzini, a great Italian journalist and observer of his people, pointed out that during the 1950s, between 8 and 12 million tourists were coming into Italy each year for a visit. When Barzini wrote about that increase in his 1964 book *The Italians* he said, "They have now passed the twenty million mark, a proportion of more than one tourist to every two and a half Italians, and the total is still growing." At that rate of tourist growth, Barzini pointed out, the number of tourists per year would soon be higher than the total number of Italians. Today there are 6.5 million people in Italy, and exactly that number of tourists arrive each year; Barzini's prediction is on the brink of coming true.

Venice, in particular, has surpassed Barzini's prophecy. Today there are forty to fifty thousand full-time residents in Venice, and about twenty-two million tourists arrive each year, a ratio of 440 tourists to each Venetian. Compare those numbers with other popular tourist cities. New York City has a population of about eight million and attracts thirty-seven million tourists—that's 6.5 tourists per resident. Venice is also very small compared with the crowds that spread across the city every day. Venice is about 7 square miles (18 square kilometers), and New York City, in contrast, covers a hundred times that area, at 302 square miles (784 square kilometers). Amsterdam, another overcrowded tourist spot, gets about the same twenty million tourists a year as Venice, but the city is much larger than Venice, with 84 square miles. Amsterdam also has a much bigger resident population of 870,000, which steadies the onslaught of tourism. Barcelona, another European city under siege, has about 1.6 million people and covers 100 square miles but has only about nine million tourists each year.

Of all the tourist-crowded cities, Venice comes off the worst. It is almost entirely an ancient city, and because it is set in water and has no way for cars to get about, it is uniquely isolated and at risk. Absolutely everything—for tourists and residents alike—must be imported by boat and that type of transport also sets Venice off from the more typical European tourist cities. The water location also complicates the infrastructure necessary for dealing with hordes of visitors, including getting rid of their garbage and human waste. How did Venice become the epicenter of the tourist invasion?

With the invention of safe travel by train in the 17th and 18th centuries came the "Grand Tour," whereby upper-class British citizens, men and women, traveled leisurely for months across Europe soaking up art and culture.[2] This

adventure was designed as an educational experience and one that the sophisticated had to do. A much bigger bump-up in the number of tourists to Venice
in recent times has been facilitated by airline flights that bring people from
far away and cruise boats that ferry people into various foreign ports. Buses
and trains add visitors to the mix and allow even those with little money to
experience other countries. The most recent contributor to mass tourism is
the availability of very cheap flights crisscrossing Europe, making Venice an
economical adventure for just about anybody. The city is now trampled under-
foot every day, and the numbers have been growing exponentially over the
past five years. In 2015, the *Independent* said that twenty million tourists came
to Venice each year.[3] That yearly number averages out to 50,000 people per
day, although authorities suggest the number is 100,000 during peak holidays
and summer. One study conducted a decade ago, when there were more permanent residents of Venice, claimed that the maximum number of tourists
the city could reasonably manage per year was 86,000, and that a minimum
of 31,000 was needed to keep the tourism economy going. Both values have
been breached.[4]

Because of the overwhelming number of tourists in Venice, there have
been very real structural changes in the topography of the city and alterations
in the way it functions as a social space. In eleven years, from 2008 to 2019,
the rate of the number of rooms to rent in Venice rose 497 percent, and the
number of restaurants rose by 16 percent. Also key to these numbers is the fact
that during the same period the resident Venetian population decreased by 13
percent.[5] Keep in mind that tourist rentals do not come from new construction
but from renovating local residences into rental apartments. In one *sestiere*, San
Polo, the arrival of the Airbnb system increased the number of available beds
for rent by 1,635 percent over that decade.[6]

Theorists of tourism cite the word "over-tourism" to describe a place
that has gone past its "tourist carrying capacity," which loosely describes
the point where the tourist population overwhelms the number of residents.
Under that transition, residents have a hard time getting through their daily
life because of the mobs.[7] "The city is a victim of its own tourism success,"
write professors Dario Bertocci and Francesco Visentin.[8] The same authors
have called Venice "the poster child of over-tourism."[9]

One major consequence of this over-tourism is that many Venetians
leave and seek homes elsewhere. Bertocci and Visentin recently conducted

an online survey that crossed all the districts of the city. They received responses from 12 percent of the population and found that everyone, from those living in the tourist bull's-eye of San Marco, to those living in the more remote area of Castello, felt highly stressed from over-tourism. Between 20 to 50 percent of these Venetians have also considered moving away due to mass tourism, and the same percentage cite the fact that Venice has become too expensive for normal people. Residents also say there are no jobs and that the city is inconvenient, but it is telling that tourism, in particular, is so much a part of their thoughts about abandoning their city.[10] Also, those who remain in the city of Venice are often older people, since the young move out for opportunities and convenience. When those older people die, they are not replaced by a corresponding number of births, because the younger people have left.[11] And so the resident population decreases even further.

The visual evidence of over-tourism in Venice is at hand every minute of every day as both residents and tourists push each other out of the way to walk down narrow *calli*, as residents are stopped short while crossing small bridges on their way to work or home, or when they are blocked by a crowd staring into canals. The conflict continues when tourists stop abruptly to look at a shop window with no awareness of the people behind them or when it becomes impossible to order a morning coffee because of the crowds. Just seeing the lines of hot, cranky people snaking out into the piazza as they wait hours to enter the basilica or the Palazzo Ducale makes the tourist life look so unpleasant, and not exactly a vacation moment to remember.

One might also wonder why a working city (as opposed to an abandoned ruin or a people-free tropical forest or desert) would want to attract such attention. It's about money, of course, and since Venice is a city built on mercantile business, its citizens are certainly attuned to economic opportunity. Indeed, Venice has a long history of branding itself as a way to bring in tourists and their wallets—just think of the Crusades. But as anthropologist of tourism Sharon Bohn Gmelch points out, "... international tourism is a fickle form of development" because those tourist dollars can vanish overnight without warning if the destination becomes unpopular.[12] Natural disasters, the outbreak of war and disease, and political unrest dramatically affect tourist numbers. The international press coverage of crowding and anti-tourist sentiment in Venice, combined with now frequent bouts of *acqua alta*, are very real threats to the ongoing popularity of Venice as a must-see destination. It's a risky business, at best.

The combination of cruise ships, bus tours, package tours, and such has grown exponentially because of great marketing and also because more people with money can afford to travel abroad.[13] The tourism business is going gangbuster at the moment; it represents 12 percent of worldwide GDP and 13.2 percent of Italian GDP,[14] while European cruise lines report they make 16.6 billion dollars each year.[15] The industry also claims one out of every twelve jobs on earth has to do with tourism.[16] Admittedly, these figures come from inside the very industry spouting them, but such pronouncements also project the idea that mass tourism is a good thing for the global economy, so it should be allowed to grow unchecked without any consideration of the consequences beyond making money.

Tourism is, in the words of anthropologist Davydd Greenwood, "The largest scale movement of goods, services, and people that humanity has perhaps ever seen."[17] But it's an odd sort of business, because it's about buying an experience, not a thing, the "seeing" of something rather than engaging in something. The tourism business doesn't seem to understand that jam-packed tours to cities and sites that don't have the infrastructure, or the patience, to survive the onslaught might make for lousy experiences. Tour operators focus narrowly on their customers, convinced they know what people want. And they seem to have no regard or consciousness about the people or places visited.[18] Even during so-called eco-tours, someone is being exploited economically, culturally, and psychologically. When mobs infringe on a group's identity, they change it, even obliterate it. There might be an economic gain in there for some of the locals, but the tourism industry is making most of the money.

One analysis of the rise of mass tourism compares its growth to a life cycle.[19] At first, travelers discover a new and interesting place, then the locals take note and realize money could be made by encouraging tourism, and they begin to put tourist infrastructure in place—hotels, restaurants, day tours, docking for cruise ships and such. Then come the big time "holiday-conglomerates" with their massive hotels and ships which invite huge numbers of people at a time. At some point, this scenario suggests, everyone who wanted to visit a place has done so and then the numbers decline. There is no sustainable balance of tourists and locals because there is no control of this cycle; it's a heedless process that assumes everyone has a right to see everything no matter the effect on the chosen spot. And yet, throughout this vacation destination life cycle, there is a concomitant reaction by the locals. At first, they think about the money, but as soon as the tourist infrastructure phase begins, the locals start to feel left out. By the time the holiday conglomerates arrive, the locals are fully pushed out and furious. Making money no longer seems worth it. Visitors become the target of their rage as they clog the streets, leave trash, and spend money on cheap imported goods that have nothing to do with the destination.[20] Also, money might be flowing in, but it is not, by and large, going to locals but to the tourism industry and the holiday conglomerates. In other words, the promise of tourism saving a place turns out to be a lie, because there is so much investment required to support the industry and the bulk of the income ends up in the hands of outsiders. With many tourists, such as backpackers or those who come for just a day, there is no money coming in at all. According to this predictive path, the next stage is the failure of holiday infrastructure and a decline in services and comfort. Then fewer people visit and the tourism business at that spot dies out. Venice is on the cusp of that final stage. It remains to be seen if the tourism business in Venice will stabilize or die out.

In another analogy, a place that relies solely or largely on tourism is much like a culture that depends on one crop for existence. If that crops fails or even dips, the culture is at risk of famine and devastation. And as that culture continues to fixate on the one crop, be it rice or yams or anything else, its people become malnourished, with the result that children experience impaired growth, overall sickness increases, and people die much younger. There are also social effects. In Venice these days it's hard to find a job outside of serving tourists in some way, so the society has become more of a monoculture, potentially losing its complexity and vibrancy.[21]

Other studies focus on residents' reactions toward tourists. Having hordes of outsiders come into your city every day triggers the tribal instinct that comes naturally to our species. The influx of strangers makes for an "us" and "them" mentality, with residents trying to ignore all the bad tourist behavior while muttering about all these people in the way.[22] In a sense, mass tourism is a microcosm of the global immigration crisis. Tourists aren't perceived as a wave of needy or scared outsiders who might want local jobs, but they are still non-Venetians trampling the city and changing its identity. Tourists are marked as foreigners who know little or nothing about the city and its history and have no concern about the city's future. Venetians also know that many tourists see Venice as a Disneyland put there for their personal entertainment rather than as a sacred place deserving of respect and honor and a place where people actually live. For example, on the island of Burano, where houses are traditionally painted bright colors, tourists regularly stand in residents' doorways to have their pictures taken. These tourists are oblivious to the fact that they are blocking someone's private front door, and they're surprised and offended when a home owner tells them to go away. That rage against outsiders is fueled, and made manifest, when a tourist jumps into the Grand Canal for a swim (which is not allowed and no Venetian would do so), or cooks his breakfast on a camp stove on the iconic Rialto Bridge, or when packs of young people roam about the *campi* drunk, throwing glass bottles or leaving drink glasses on residents' doorsteps. As tourism scholar Papathanassis Alexis puts it, "Stating it simply, tourists are visible, different, not politically represented and therefore, a convenient projection surface for disenchantment and populistic/demagogic motives."[23] And so, the antagonism grows in Venice. Given that Venice is a small walking city, for tourists and residents alike, these mental and physical confrontations happen every minute of every day, and it would take the patience of a saint to put up with it.

NO GRANDI NAVI

On June 2, 2019, a 275-foot-long 65-ton cruise ship lost control and crashed into a much smaller river cruise boat docked at the quayside of the southern edge of Venice. People on the river cruise boat were eating breakfast. As the giant ship bore down on them, they fled into the water and onto land, screaming. Five people were injured. Ironically, this accident occurred just as the rest of Venice was celebrating the *Festa della Sensa* at the mouth of the Adriatic, where the rowing clubs were bringing the mayor out to toss a gold ring into the sea, thus marrying Venice to the water. The ceremony was delayed that year, as the rowed boats had to bypass the crash site. By the next day, posters went up all over Venice calling for a spontaneous protest against the giant cruise ships. The following weekend, between eight and ten thousand marchers with flags reading NO GRANDI NAVI (No Giant Cruise Ships) processed from the Zattere promenade on the southern shore of Venice through the city to Piazza San Marco. These *No Grandi Navi* protests against mega cruise ships have been going on for a few years in Venice. During a previous demonstration, a cruise ship happened to enter the lagoon and pass along the Giudecca Canal, right in front of a fleet of local boats bravely flying the *No Grandi Navi* flag. The protestors shouted, "Go home" and "We don't want you here" in English at the passengers, who were so high up on their cruise ship decks they probably didn't even hear.[24] But surely they saw the antagonism.

The World Tourism Organization classifies cruise ships as floating resorts, not just transportation, and they are always a package deal.[25] Between 1997 and 2010, cruise ship passengers disembarking in Venice increased by 440 percent, and over those same thirteen years, 200 percent more ships docked at Venice.[26] Today, about 2.5 million tourists arrive each year on cruise ships into Venice, and that number is growing.[27] Although cruise passengers make up a small fraction of the yearly total visitors to Venice, they come all at once, disembarking thousands at a time for a quick collective walk to Piazza San Marco and back.

Worse, these giant ships are disproportionally large for such a small city. Venetians and tourists who have arrived by land have a very strong reaction to these giants because they are floating skyscrapers that overshadow, literally, the beauty of the city. Looking out from the imposing Piazza San Marco and watching one pass, seeing it tower over the biggest building in Venice while obliterating the view of the lagoon beyond, is just plain shocking. These ships are often sixty meters high or more, while the buildings in Venice, on average, are fifteen meters high. The ships dwarf the city and literally block out the sun. In 2019 a new protest sign of this incongruity showed up around town. It's a stencil of a man holding a boy's hand and under that drawing are the pointed words "We Demand To See The Horizon."

Beyond how disproportionate and out of place they look, cruise ships are also causing environmental damage at a level their passengers are probably unaware of. In Venice, waves created in the wake of grand ships put smaller boats, especially rowed boats, in danger. Those waves also undermine the foundation of the city itself. Even worse, the engines from cruise ships belch out continuous clouds of fine dust that fills the air.[28] The ships also dump toxic benzopyrene, among other chemicals into the lagoon.[29] One estimate puts the air pollution produced by a resting cruise ship at the same level of twelve thousand cars idling (and how ironic that a city without cars now has a car pollution problem).[30] For people living in the area of Santa Marta, where the ships dock, this situation has become an environmental nightmare because the ships run their engines all night, puffing out pollution and black smoke; citizens of Santa Marta now have to keep their windows closed at all times. And then there is the noise and vibration of these huge boats, creeping through the city, disturbing its ancient peace. "Cruise tourism is a selfish and vague commitment," writes one critic.[31]

Some believe the arrival of cruise ships is necessary because of the potential money they bring to a place. But it's hard to know if cruise ship passengers' excursions into Venice are any benefit to the economy beyond docking fees. Passengers sleep on the ship, eat on the ship, and are usually around for only one day. One estimate (from the cruise ship organizations) claims that passengers spend between 120 and 160 euros each while in port, but that figure is difficult to track. Cruise ships themselves are designed to be shopping malls that entice passengers with a constant stream of consumer goods, and when in port passengers don't even have time to seek out locally made souvenirs. Most cruise ship passengers just walk around a bit and buy nothing, or pick up cheap items such as masks and T-shirts sold at outdoor kiosks around Piazza San Marco; these souvenirs are made in China, not Venice.[32] On the other hand, the ports also employ many people who restock the ships, and if they move the cruise ship port outside the lagoon, those jobs will go to others.

After the crash of June 2019, there was a flurry of demands for action. Various proposals have been floated about, but it remains to be seen what Venice, and the Italian State, will do about this problem. One idea is to build new docking infrastructure on the Lido, on the Adriatic side, then ferry passengers into Venice on smaller boats. Another is the possibility of redredging the Vittorio Emanuele canal put in place to allow oil tankers to enter the southern end of the lagoon at Chioggia and pass to the refineries at Marghera on the mainland while staying reasonably far away from Venice proper. Or they might bring cruise ships to the city of Trieste at the northern apex of the Adriatic and somehow bring people to Venice by train or bus. All of these plans are inconvenient for tourists, and they will cost a fortune. Tourists also lose that moment when they look over the ship's deck into the glory of Piazza San Marco while spoiling the view for everyone on land. If that iconic moment is lost to those who pay for a visit to Venice, the number of cruise ship passengers into Venice might decline on its own.

ACQUA ALTA, CLIMATE CHANGE, AND THE FLOODING OF VENICE

Flooding in Venice comes and goes from the Adriatic Sea with the twice-daily tide. It flows in through three breaks in the barrier islands on the east side of the lagoon. The northernmost break is at the tip of the Lido island, the second south of that island, and the third near the city of Chioggia on the southeast tip of the lagoon. Usually, the tide comes and goes as part of the rhythm of the Venetian day, and no one pays much attention except for fishermen and boatmen. But with a combination of unpredictable factors, mostly during November and March when unusually high and forceful tides rush in, they overrun canals and flow into the area under the city, bubbling up through the drains. Venice was initially built at sea level and these tides have been part of its history, but various complications in the last hundred years have made these tides a destructive phenomenon.

In the 1920s, with the development of the industrial port of Marghera— now virtually abandoned—on the mainland (those are the ugly stacks one sees on the left when leaving Venice), huge quantities of water were sucked out of the underground reservoirs in the lagoon. That groundwater had acted like a pillow, buoyantly holding up the stratum of stable clay that makes up the floor of the lagoon. Venetian builders depend on that layer; they drive fields of pylons into the dense clay and that's what holds up the city. Venice is basically a city built on very secure stilts that are out of sight. As groundwater under the lagoon was sucked out over fifty years from 1920 to 1970, that clay layer slowly compacted and the city sank ten centimeters (four inches), more than it should have.[33] Once the removal of groundwater was halted, the rate of annual sinking became minuscule; it is still ongoing at the rate of about 0.5mm (0.02 inches) per year. But the loss of those initial four inches made the city increasingly, and permanently, more vulnerable to the weather conditions that bring in *acqua alta*.

Extreme tidal wash is a result of unpredictable factors that combine to make for high tides that barrel into the lagoon. Those factors include low-pressure weather systems and forceful winds that drive the currents of the Adriatic dramatically up toward Trieste at the apex of that sea and then down the east side of Italy. At the same time, a cold bora wind from the northeast and a warm sirocco wind from the Sahara Desert in Africa add to the mix, pushing the waters of the Adriatic even more forcefully. Rains also dump more water. Add to those uniquely Adriatic ecological forces the precipitously rising seas levels from the recent effects of climate change. With rising global temperatures, the seawater volume expands and polar ice caps and glaciers melt, filling the oceans and seas with more water.[34] And so, global warming makes any waterside city frighteningly vulnerable to flooding. In the end, this combination of wind, rain, and water arrives on the incoming tide in about forty-five minutes from the Adriatic to the Grand Canal, carving a path of destruction to land and buildings. After a few hours, it recedes, and then it comes again on the second daily tide. Records show that the number of tides classified as *acqua alta*, defined as water over a tolerable level, has increased, and many suspect that increase is due primarily to global warming.

The city can stand some annual flooding, but the height, rate, and force of the extreme tides are increasing and causing irreparable damage to the infrastructure of the city. The most recent and very destructive flooding came in the fall of 2019; it was the worst high tide since 1966.[35] On the morning of November 12, everyone in Venice woke up to the long siren. The water had begun to arrive during the night, and by morning pretty much every part of town was covered with water. The "duck walks" were put out across Piazza San Marco, and the tourist crowds wandered about confused but determined to sightsee. The flooding, from a touristic point of view, seems like fun because they have no houses to protect from rising water, no businesses to close with a flood gate, no money lost as the city shuts down, and they can leave for their next destination. That November, tourists did indeed begin to retreat and to cancel their reservations, leaving a devastated Venice to cope with five more days of flooding and a major loss of tourist dollars. Shop owners in every part of the city inserted extra-tall gates into their front doors and reworked them every day after the force of one tide leaked in anyway. Many people used cans of spray foam insulation to temporarily fill the cracks in those gates. Grocery stores shut their doors as they coped with soggy stock and broken refrigerated cases that fizzed out as water hit electrical wires. Pretty much everything was closed for five days; it was almost impossible to find an open bar or restaurant and get a cup of coffee, spritz, or a meal out and the city became a social wasteland. The force of those tides had been so strong that they also ripped out pilings for mooring boats and undermined public waterbus stops, forcing the closure of many embarkation platforms on the Grand Canal. The waves demolished a long and ancient brick wall on Giudecca island and upended bushes and plants in a famous garden and the Franciscan monastery next door. It's impossible to calculate what it will take to rebuild those walls and gardens and bring them back to their former beauty. After about a week, the worst flooding receded. Shops and eating places that had been pumping out the water every day just to get in front of the damage began the hard work of evaluating the state of their stock and washing floors and counters with disinfecting soap in preparation for opening up again. Many places put their water-damaged stock on tables outside hoping for some sales.

During that November, *acqua alta* was also the only thing anyone talked about. There were many shaking heads, but also a sense of community as groups of young people spontaneously showed up to help the elderly escape ground-floor apartments or clean up shops and community venues. In the last few years, citizens have moved from knee-high to thigh-high boots during *acqua alta*—a visual confirmation of the rising height of floods over time. Every Venetian and long-term resident now has an app or two on their phones that predict *acqua alta* three days ahead. These apps also give warnings about spots around the city and what footwear is necessary—regular shoes, boots, or if it's best not to even go near that area.

The point is that Venice is not exactly sinking. It is being submerged by high water from climate change, including the increase in snowmelt flowing down from the Alps into nearby rivers and into the lagoon; the Adriatic's increased sea level rise; the rise in global sea levels; and the atmospheric tumult that affects winds, storms, and rain. Venice is a ground zero for the global effects of climate change.

Tourism, in general, is also instigating climate change as tourists demand to enter remote areas or to be entertained in ways that destroy natural areas. The tourism business also relies on the burning of fossil fuels to get passengers to their destinations by ship, train, bus, or car. And tourism also uses up natural resources such as clean water, locally sourced food, firewood, and gasoline.[36] Many people want to travel with the same comforts they have at home, so there is importation of major infrastructure and goods necessary to provide that comfort—air-conditioning, imported foods, plumbing for long showers and flushing toilets. Delivering that level of care often overwhelms local economies, especially in non-Western cultures that do not normally have those comforts in place. Presumably, most tourists have no idea that their requests are damaging. But often, providing these comparative "luxuries" takes resources away from locals and creates an unequal distribution of wealth that is obvious to both the tourists and the locals.[37] The exploitation of locals by tourists reaches a peak with "sex tourism," where men arrive to prey on local women and girls for money, and procurers encourage their behavior.

In Venice, the expectations and demands of tourists specifically influence the speed of taxis and boats and causes damage via water. Tourists love to speed down the Grand Canal, but the wakes of those boats damage the foundations of Venetian buildings. And speeding tourist boats interrupt the normally busy but calm Venetian canal traffic. Recently, the rowing clubs of Venice and surrounding areas banded together for a demonstration. They rowed down the Grand Canal in a fleet, demanding less speed and no waves from motor-powered boats. These clubs are refusing to participate in the annual first day of carnival boat parade down the Grand Canal. Instead, they are all rowing to the abandoned island of Poveglia to clean up after the devastation of last November's *acqua alta*. Their protest might be disappointing to tourists but they are trying to save their city.

Venice has been dealing with *acqua alta* since its founding, but now it has to do something before it's too late. In 2003, the city received the go-ahead to build the MOSE gates (*Modulo Sperimentale Elettromeccanico*), three undersea mobile "gates" that will be permanently placed on the seafloor at the three places where the Adriatic meets the lagoon—Lido, Malamocco, and Chioggia. If a tide is exceptionally high, the wing-like gates are supposed to rise and close off the Adriatic from the lagoon. Some have protested that such action will upset the natural ecology of the lagoon, but since the gates would only be used in dire times, the objection is not taken particularly seriously.[38] The bigger problem is the cost and the endless delays that curse the MOSE project. To this date, seventeen years after the plan was approved, none of the gates work. After the most recent *acqua alta* in November 2019, the director in charge of MOSE claimed that the gates would be in operation in a year, but no one believes her. Corruption of funds is a major issue, and people in town claim the project went to mafia interests and that's why the gates are a bust. Additionally, the MOSE gate project has siphoned off much of the funding that could have gone into other projects aimed at restoring or saving the city.[39]

HOW DOES VENICE COPE?

Venice doesn't cope very well, at least at this point. None of the historical towns in Europe overwhelmed with tourists seem to cope. Prague, Amsterdam, Dubrovnik, Bruges, and other cities are overwhelmed each summer with crowds that are so very oddly out of place in small historical towns; obviously, no city born centuries ago is designed for massive crowds.[40] These towns are at risk because they have been "heritageized," given some historical meaning that everyone agrees should be preserved, but that very designation also makes them tourist destinations. The label World Heritage Site puts that place under a spotlight and emphasizes that it should be seen soon because it might evaporate in our lifetime. No city so far has figured out how to control what happens after that spotlight makes it a precious destination, attracting the curious and those who like bucket lists. Fencing and entry gates seem the only solution, but no city yet has been willing to erect excluding barriers.

Venice, in particular, also has philosophical reasons for not bursting forth with solutions to curb the tide of tourists. Venice is a city that historically and economically believes in the free market system. It also spent over a century promoting and embracing economic and personal freedom. That underlying belief system was an exuberantly successful combination until now. As I wrote in chapter 8, for example, the printing business flourished in Venice because writers and publishers knew it to be a place with freedom of the press and the freedom to read anything. In the 1700s, as the world economy changed and Venice was hit with competition in the global marketplace, those values were less important, but they are still part of the identity of the city and its people. Today, tourism is seen not only as a business opportunity but also as a tourist right. That right is echoed in laws that give everyone open access to the cities of Italy,[41] and to most Venetians it would be anathema to close off the city. Instead, vague influences have come into play, most of them ineffectual. For example, the Italian Committee for the Safety of Venice includes a UNESCO official who must approve any public or private work that might impact any territory within the lagoon. Who knows if this bit of bureaucracy makes a difference? Also, there are Italian laws that allocate money for public works and help residents. But these monies are reallocated each year, and of course they have been reduced every year.[42] After bouts of major *acqua alta,* the Italian state says it will allocate recovery money, but that money, so Venetian shop owners say, never reaches the residents who need it for cleanup or to save their businesses.

At the same time, Venice is trying to introduce taxes that might slow down the number of people who come each day. In 2011 the city introduced a "tourist tax" on all hotel rooms and apartments for rent (including Airbnb). When checking in, a guest must pay the hotel or landlord a few euros per night. This tax is then passed on to the government. The rate varies with season, type of accommodation, length of stay, and number of guests, but at the high end it's 4.5 euros per night for a luxury hotel stay. In July 2020, a new tax is aimed at day-trippers, who don't spend the night and for whom, before this, there was no contribution to Venice. It will begin at 3 euros a day for off season and 8 euros a day for high season and will rise each year until 2022. The aim is to use that money to maintain the city and keep it clean.[43] Another initiative will require high season tourists to book their entry, thus treating the whole city as a civic museum of sorts. And the tax will extend to other islands such as Murano and Burano. So far, there is no notice of exactly how this tax will be collected. In 2018 the city tried to regulate Carnival with turnstiles into Piazza San Marco, but it didn't go well. Just try to keep a bunch of dedicated partygoers away from the epicenter of the action.

In 2017 the city decided to address the impoliteness of visitors head-on with a simple plea to tourist behavior. Called "Enjoy Respect Venice," the campaign consisted of pasting up posters that tell tourists they will be fined for misbehavior. The rules include no sitting, eating or drinking on the steps of Piazza San Marco or any bridge. It forbids visitors to feed the pigeons, jump into the canals or the lagoon for a swim, bring bikes into the city, and drop their trash. Also, since Venice is a historical town and not a beach resort, visitors are asked to wear something other than a bikini around town. The most charming, or perhaps forbidding, rule of the campaign is to "not stand without motivation" anywhere in Venice, but then the city is so captivating that one pretty much has lots of motivation to just stop and stare anywhere. These regulations highlight how insensitive some tourists have been and are, but it's impossible to confirm if this citywide campaign has had any effect. People still swim in the Grand Canal for a lark, and they certainly sit where they like while eating, drinking, or just resting. Fines are sometimes imposed, but enforcement relies on police and there aren't many local officers around. And tourists seem oblivious to the posters.

And so, for now, anti-tourism is festering in Venice, as international news-paper articles declaim.[44] Citizens are angry that all those tourists put their World Heritage Site designation at risk and make sustainability impossible. They point out that even the so-called economic boost is a myth that only touches some while bringing economic hardship to others.[45] The only thing the city has going for it is international attention and very active grassroots movements. In 1973 UNESCO designated Venice a World Heritage Site, and in 2016 the organization added the entire lagoon to that designation. Recently, UNESCO has also been considering moving Venice into its "cities in danger" category. In June 2019, the mayor of Venice directly asked UNESCO to include Venice on its blacklist when the Italian Minister of Transport did not come through with a promised ban on cruise ships in the lagoon.[46] That request highlights one of the many conflicts between the city of Venice and the Italian government and fuels Venetian anti-Italian sentiment.

On the upside, private nongovernmental organizations such as Venice in Peril[47] or Save Venice[48] have doggedly gathered donations to save specific buildings and works of art, restoring them in the process. And Venetians have stepped up to save their city. Grassroots organizations include, among many others, *Comitato No Grandi Navi-Laguna Bene Comune*,[49] which is working to implement the removal of cruise ships over forty thousand tons; *Venezia Autentica*,[50] a site that promotes local Venetian artisans and tour guides; *Generazione 90*,[51] which gathers young Venetian volunteers to clean up after *acqua alta* and help with anything community members might need; Venice Calls also organizes young people to help out during city crises.[52] They want to make plans for a better Venice in the future. All these organizations are there because Venetians love their city and want it to survive and thrive. And they are willing to work for it.

CAN VENICE BE SAVED?

One analysis has suggested it takes more than forty-one million euros annually to maintain the unique water-bound life of Venice, to keep the buildings safe and the city clean and functional.[53] Preserving Venice and its heritage is under-scored by Italian law and its designation as a World Heritage Site.[54] The city is indeed described by UNESCO as "one of the most extraordinary built-up areas of the Middle Ages," "one of the greatest capitals in the medieval world," and "one of the most extraordinary architectural museums on earth."[55] But that designation has a downside. Called the "heritization" of a place, the UNESCO label also reinforces the idea that the whole city is a precious museum rather than a living, breathing city needing a preservation plan that incorporates the populace. According to their mandate, only historically designated buildings, including those tourists want to see, are worth saving, while the buildings used by the residents receive none of the money or attention they deserve. Planning to revitalize Venice rather than "preserving" it means thinking about those who live there, because they are the lifeblood of the city. Just look at a map. The homes, shops, schools, and businesses of Venetians *are* Venice. If their homes fall apart and the populace leaves, the city as a whole is dead.

Also, the emphasis on saving the cultural heritage of Venice for tourists has made it impossible for the city to change its economic base; it's hard to imagine that the heritage regulations would allow for the building of new factories, and even finding an open space for a new factory would be difficult. Focusing on tourism has also led many Venetians to move away and rent their family places to visitors because renting to tourists is so lucrative. And who can blame them? Rental income maintains the building in ways that could not be accomplished with a normal salary.

At this point, Venice has been unable to balance the needs of tourists and residents, and it has become one of the most tourist-dependent sites on earth. So many Venetians have left the city that the loss of population is reminiscent of the great plague of 1630. Ironically, this city is the father of capitalism, yet it feels like end-stage capitalism now.

Humans have a long history of killing off or abandoning cities—we owe our knowledge of these disappeared places to archaeologists—but this time it is happening right in front of our eyes. We are all culpable because of climate change and mass tourism.[56] Should we, as a human collective, halt this downward slide or just let Venice fade away? Even if we want to save it, how would that happen, and who, exactly, is responsible for pulling this unique city away from the brink of destruction? Venetians invented so much; they formed the very civilization in which we now operate. Perhaps it's time to just say thank you and walk away. But what a shame that would be. This loss would not be like one of those ancient disappearing cities that were destroyed by local conditions and politics. Instead, today's climate change is a global problem, so letting Venice go because of high water only means many other cities around the world will also be abandoned for the same reason. Mass tourism is more limited in its global scope—only some cities and countries are affected—but how tragic to destroy humanity's treasures simply out of unregulated commercialism and consumerism. Even though Venice invented commerce, that contribution was leavened by its many contributions to the arts, entertainment, printing, and health care. To abandon Venice would be an act that not only disavows our own history but also nullifies the vitality of the living city of today.

Another view is that we all owe Venice a great debt and maybe those of us who appreciate what the people of this city did for our history should band together in protest and action—and help innovate ways to manage the two destructive forces that threaten this city. Perhaps now is the time to recognize not only what Venice has given the world but also what it continues to provide—a way of life on the water that shows, by example, what humankind in its inventive glory can accomplish.

Acknowledgments

As Casanova said at the beginning of his twelve-volume autobiography, *Historie De Ma Vie*, "I expect the friendship, the esteem, and the gratitude of my readers. Their gratitude, if reading my memoirs will have given them instruction and pleasure. Their esteem if, doing me justice, they will have found that I have more virtues than faults; and their friendship as soon as they come to find me deserving of it by the frankness and good faith with which I submit myself to their judgment without in any way disguising what I am." I have done my best to cross-reference all the inventions in this book, but I am sure others will write to me with Venetian firsts that I haven't yet come across. Feel free to email and let me know.

I am indebted to my literary agent, Wendy Levinson, at the Harvey Klinger Agency, for her enthusiasm for this project from start to finish. Wendy's support has been very important as I worked on the proposal and the book; I never felt like I was out here working alone. Wendy is also a fabulous editor, and I thank her profusely for volunteering to be the first reader and editor. I also thank Wendy for always having my best interests in mind, and for her kind comments about my writing and the structure of the book.

My publisher, Pegasus Books, has been fabulous to work with. I so appreciate the comments on the manuscript by Deputy Publisher Jessica Case, the wonderful cover art, and the title, which perfectly sums up the book. But mostly I am lucky enough to have a publishing house that has been behind this book from proposal to the final product, and they have worked with me in a congenial and supportive partnership to make this happen.

I am grateful to Francesca Merrick for construcing my website (meredithfsmall.com) and working diligently on all aspcts of promoting the book. I simply would not have done all that without her guidance and encouragement. I am also grateful to Australian graphic designer Alex Darbyshire for the compass rose logo that he created for my website and my so-called brand. It completely sums up my interests and my approach to work and life.

I owe special thanks to my Italian teacher, Stefano Chiaromonte, who has so improved my Italian that I can have coherent conversations in Venice and with Italian tourists in Philadelphia. One of the greatest pleasures of writing this book was working to improve my ability to read and speak in Italian. That never would have happened without Stefano, who is such a great teacher. Given his fluent English, he understands why, as an English speaker, I get things wrong, and how to fix them. Under Stefano's guidance, I have also deepened my understanding of Italian grammar, which informs my casual conversations in Venice. Recently, during an over-the-counter chat, a Venetian barista asked me where I was from, and when I said "Filadelfia," she responded, "Oh, I thought you were Italian." I owe that moment to Stefano. *Mille grazie.*

Over the years, I also took many courses at the Instituto di Venezia language school in Venice, and I am grateful to the teachers and fellow students for making conversation more a pleasure than a test.

I thank the Department of Anthropology, the University of Pennsylvania, for appointing me Visiting Scholar, giving me a space to work on this book and access to the vast libraries of the University of Pennsylvania. I am also grateful to the staff of the *Dipartimento di Studi Umanistici, Università Ca' Foscari* (The University of Venice) for appointing me Visiting Researcher and for working on my behalf trying to obtain a longer-term visa to Italy.

Various American friends have been interested in the book and very supportive. Giuliana Pierson helped me write (or she rather rewrote) official letters in proper Italian; Helen Cunningham, Nancy Zahler, and Judy Berringer were always interested in how the chapters were going; Steve Cariddi and Tamara Loomis managed to make me laugh about this book (and everything else). Robert Urbansky also provided laughs and took care of my apartment and various tenants in Philadelphia while I was away oh so many times. My Australian friend Deb Elkington came to Venice in the fall of 2019 and braved severe *acqua alta* in cheap plastic galoshes so I could drag her about town looking for an open coffee bar. We saw about half of what I had planned, so Deb has to come back.

About forty years ago, Sandra Quinn and I traveled across Europe for two months. We experienced Europe, dirty hair, sleeping in parks and on beaches, interacting with all sorts of characters, and during those fun months of our youth Sandy also taught me about art. After over four decades of mutually supportive friendship, Sandy came to Venice in the fall of 2018, and I ran her ragged showing off the city. She was always interested in what I had to say and never once asked me to please stop talking about Venetian inventions. Sandy has also been consistently interested and supportive of this book and my writing.

My sister Andrea Small Perkins listened endlessly to details of every chapter and asked insightful questions. Andrea also acted as my "stringer," passing along bits of news about Venice that she came across in the foreign press. This book is not just for Venetofiles, and the interest of Deb, Sandy, and Andrea proves that even those who don't know the city can be entertained by how Venetians have influenced our lives.

Many friends and acquaintances in Venice have made my stays there memorable. I thank the hundreds of shopkeepers and baristas on whom I practice my Italian. They have all been encouraging and complimentary as I stand there chatting and butchering their lovely language (or so it seems to me). In Venice, I am especially grateful to the bakers and counter staff of Pasticceria Tonolo, the chocolate makers of Virtu, the "coffee ladies" at Bar Adagio, the vegetable boat in Dorsoduro, and all the various *vino sfuso* places around town, especially my friends at the one near the Church of the Carmini. It was at that *vino sfuso* that I saw Venetians playing cards of some sort with a very oddly shaped deck, which turned me on to the longevity of card playing in Venice.

I also thank Venetian historians Alberto Toso Fei and Andrea di Robilant. Unbeknownst to them, these Venetian writers were an inspiration and a resource. Besides Toso Fei's intriguing guidebooks for walks around Venice, I just happened to be in the city when two volumes of his *Ritratti Veneziani* (Venetian portraits) came out on the newsstand. Those books were good fact-checking resources, and I picked up a few more firsts while reading them. I relied on Di Robilant's book, *Irresistible North,* to write about the Zen brothers' probable adventure in the North Sea.

Special mention goes to the *Reale Società Canottieri Bucintoro,* for welcoming me into the rowing club and teaching me Venetian rowing. Silently rowing for hours across the Venetian lagoon (with a break for prosecco) is a life-changing experience. I am especially thankful to my rowing teacher, Paula Landart, a woman of great patience and skill, who cheerfully takes me out on the water after months away and teaches me all over again how to row like a Venetian. I was introduced to the Bucintoro, and encouraged to become a rower, by fellow language student Oliver Loeb-Mills. Oliver and I have had many adventures walking around Venice, and it was always great to explore the city with someone who walks as fast as I do, talks as much about Venice as I do, and can be easily persuaded to sit down for a spritz and a chat at any hour of the day.

Initially, I was concerned this book would be of no interest to Venetians, people who know their history well. But the Venetians I spoke to about the book over the years were all interested and enthusiastic, and they sometimes added new firsts to my list. One landlord asked why I was staying so long in Venice, and I bashfully talked about the book, expecting a negative reaction. Instead, he looked me in the eye and said, maybe for all Venetians, "It needs to be said." I was so reassured by that comment. And now it is said.

This book is dedicated to my daughter Francesca Deane Merrick because she has been involved and supportive from the minute I came up with the idea to the finished manuscript. Francesca also pushed me, in no uncertain terms, to go forward. She demanded I propose this book to the Harvey Klinger Agency. Her pivotal and insightful comments and questions on the early proposal for this book made me think long and hard about what I was trying to accomplish. And Francesca's last comments about that early proposal were exactly what every unsure writer needs to hear: "You should write this book. I would read it." She also read drafts of the first few chapters and let me know I was on the right track, and lucky for me, she is a master editor. Francesca's intelligence and whit have also informed and entertained me during the process of working on this book. And she has been very patient (OK, some eye-rolling) when I have repeatedly inserted the words "Well, yes, but Venetians invented…. (fill in the blank)" into any conversation. Francesca also put up with my many trips to Venice, being far away from her in body but not in soul. Love is a given between us, but I am also deeply touched and eternally grateful for Francesca's belief in me as an anthropologist and a writer.

Meredith F. Small
Philadelphia, PA, USA
March, 2020

Chronology of
Venetian Inventions

675   Lagoon islands in the Venetian band together and form a government that is based on a system of checks and balances.

1072   Greek princess Teodara marries Venetian Doge Selvo and eats with a gold fork, thus introducing the fork (*priòn* in Greek and *veneziano*) to Europe.

1104   Construction of the Arsenal shipyard, the largest industrial complex in the world before the Industrial Revolution.

1128   First city ordinance to require street lighting at night.

1157   First national bank (guaranteed by the state) established in Venice.

c. 1160   Nicolò Starantonio Barattiero invents a system of wood boxes and pulleys to bring materials up to the bell tower of Piazza San Marco, thereby inventing the elevator.

1171   The Venetian state forces loans on citizens to pay for a war, and the citizens start to trade these loans among themselves, inventing government bonds.

c. 1200s   Venice invents the first bill of exchange to allow international trade without the need to haul loads of gold or silver.

c. 1200s   Venetian merchants trade government securities.

c. 1200s   Venetians infuse fragrances into soap, inventing perfumed soap and soap solely intended for personal hygiene.

1227   First brick hearth with a chimney breast and flu constructed in Venice.

1260   Venetian merchants Maffeo and Nicolò Polo are the first Europeans to travel to East Asia.

1260   First *Scuola Grande*, powerful civic confraternity organization for non-nobles.

1271   Venetians make giant wooden dolls called *Marione* (big Maria) to use during a traditional festival from 944 celebrating twelve captured brides who were returned to the city. The dolls were also copied as miniatures and sold as souvenirs called "marionettes."

1272	Venice offers foreign wool weavers a free house and ten years free of taxes if they move to her islands, thus inventing the economic tax incentive.
1284	Venetians enact the first child labor laws to protect children from working in the more dangerous areas of glass production on Murano.
1284	Venetians mint the first gold ducat, called a *zecchino*, the most stable currency in the Mediterranean in the Middle Ages, which was accepted until modern times.
c. 1285	Double magnifying lenses for reading set in bone, metal, or leather frames and balanced on bridge of nose.
1291	City of Venice orders glassblowers to move to Murano because of the threat of fire. This is the first forced exodus of a particular trade to protect a city.
1295	Marco Polo, on his way back from China, brings diamonds from India to Venice, making Venice the European center of the diamond trade.
1295	Marco Polo brings fireworks to Venice from China.
1297	The Venetian government passes a law regulating the making and sale of medicines to state-run shops.
1300	First boat race.
c. 1300	Venetian Marco Polo publishes a handwritten, four-volume book of his travels and invents travel writing, explains Asian topography, and sends Columbus to America looking for Asia.
1309	The Venetian Della Borsa family invents the first stock exchange in Bruges, Belgium.
c. 1300s	Venetians invent mirrors of glass backed by tin and mercury.
1310	Cartographer Pietro Vesconte draws the oldest surviving signed and dated nautical map, commonly called a portolan.
1320	The factory line is invented at the Arsenale shipyard.
c. 1320	An unknown Venetian publishes *Zibadone da Canal*, the first extensive merchant's manual.
1339	Lace-making style using a needle, called *punto in aria,* is invented on Burano, a Venetian lagoon island, and alters the style of clothing embellishment in Europe.
1347	The Republic of Venice imposes quarantine for ships entering its trading post Ragusa (Dbrovnik, Croatia). That quarantine lasted thirty days but was extended to forty days (*quaranta*).
1348	Venice is the first city to codify quarantine for incoming ships.

1348	Venice establishes rules for burial of plague victims in remote locations to protect its citizens.
1348–1364	Giovanni Dondi dell'Orologia builds the Astrarium, the first astronomical clock and planetarium.
1351	Venetian government outlaws spread rumors intended to lower the price of government funds, thus instituting state control of fluctuating financial markets.
1364	Giovanni Dondi dell'Orologia publishes *Tractus Astarri*, the first written description of clockworks.
1374	A Venetian state committee proposes the formal establishment of the first public bank but it does not happen.
1378	Venetian navy mounts gunpowder weapons on ships for the first time during the Battle of Chioggia.
1396	Child labor laws are extended to exclude children under the age of thirteen from certain dangerous trades, the last child labor laws in Europe until the mid-19th century.
1402	Pietro Paolo Vergerio publishes *On the Conduct of Honorable Men,* instructions to young men on how to become an educated person, and invents humanism at Padua University.
1423	First permanent quarantine hospital is built on the lagoon island Santa Maria di Nazaret. It's called Lazzaretto, a name then used for quarantine hospitals throughout Europe.
1443	The republic states it will provide an attorney for those who cannot afford one, making this the first city or state to provide public defenders.
1452	Venetian Giovanni Leone translates Alberti's *De Pictura* from Latin to English, thereby giving the English-speaking world the mathematics and optics of art and science.
c. mid-1400s	Venice is first place where fumigation is attempted on goods exposed to contagion.
1455	Alvise Cadamosto first describes and draws the Southern Cross while at the mouth of the Gambia River, calling it the southern chariot (*carro dell' ostro*).
1455	Glassmaker Angelo Barovier of Murano invents *cristallo* glass.
1456	Alvise Cadamosto and colleagues discover the Cape Verde Islands, then uninhabited.
1459	Venetian monk and cartographer Fra Mauro finishes oldest surviving, most accurate and detailed map of the world drawn in medieval times. His map is based on geography and oral accounts, moving cartography from a religious,

faith-based approach to an accurate science. Mauro's map also argues that passage around the Cape of Good Hope is feasible, and that the Indian Ocean is an ocean and not an inland sea. In doing so, Mauro initiates globalism.

1460s	Venice is the first government in Europe to make survey maps of its territories.
1461	The first printed book in Italian, *Decor Puellarum*, is printed by Nicholas Jenson in Venice.
1468	Grazioso Benincasa draws a portolan chart that includes West Africa, based on descriptions of explorer Alvise Cadamosto.
1468	Grazioso Benincasa's atlas includes the first map of Ireland as a stand-alone map.
1469	Venice awards the first printing privilege granted by a European government, and it conceded a commercial monopoly over the art of printing.
1470	Nicholas Jenson, a French printer working in Venice, invents the first Roman type based on typographic principles rather than manuscript models.
c. late 1400s	Venetian printers apply page numbers to books for the first time.
1470	Venetians are the first to add ground glass to oil paint.
1472	Paolo Bagellardo publishes *De infantium aegritudinibus et remediis*, the first printed book on pediatrics, which is also the first printed medical book.
1472	The government of Venice mints the *lira* coin which is the first *lira* in Italy.
1474	First codified patent system and introduction of the concept of intellectual property.
1474	The first printed cookbook, *De honesta voluptate et valetudine* (On honest indulgence and good health), by Bartolomeo Sacchi, is published in Venice. His book also includes ideas about healthy living.
1475	Father Clement of Padua is the first Italian to print a book and he does so in Venice.
1480	Marietta Barovier invents the rosette or chevron bead, which is then used worldwide for trade.
1481	First women's public regatta.
1486	First book copyright given to Marco Antonio Sebellico for his book *Decades Rerum Venetarum*, a history of Venice.
1488	Laura Cereta assembles her feminist writings; their publication forms the basis for modern feminism.
1493	Venetian health authorities are the first to attempt disinfection of post by dipping letters into vinegar.

1494	Luca Pacioli publishes first book of algebra and accounting, and codifies double-entry bookkeeping.
1496–1533	Over four decades, Venetian Marino Sanudo writes a fifty-eight-volume diary of daily life in the Venetian government, thereby inventing the chronicle of a nation.
1497	John Cabot, who learned navigation in Venice and was a Venetian citizen, "discovers" Canada, thinking it is Asia and plants Venetian and British flags at Newfoundland.
1498	First patent for using double-impression technique for printing polyphonic music given to Ottaviano Petrucci; one of the earliest known patents, it is for twenty-one years.
1498	Aldus Manutius prints a list of all the books available from the Aldine Press, inventing the catalog.
1499	Aldus Manutius publishes the romance tale *Hypterotamachia Poliphili* with lavish and large illustrations, making it the first printed picture book.
c. 1500	Aldus Manutius invents the semicolon, hooked comma, apostrophe, and accents, and figures out how to print them.
c. early 1500s	Venetians invent *trappola*, the first trick-taking card game and the first set of cards with mirror images of the figure.
1500	Aldus Manutius and Francesco Griffo invent italic typeface.
1500	Venetian artist Jacopo de' Barbari finishes his massive aerial map of Venice. This map and other prints by him initiate the medium of large detailed woodcut print in Italy. It is also the first woodcut map based on the "scientific principles" of surveying and scale.
1500	Leonardo DaVinci invents the wet suit and diving suit for Venice.
1501	*Le Cose Volgari*, by Petrarch, is the first book in Italian printed in italic type.
1503	The first printed Venetian merchant manual, *Tariffa de pexi e mesurei*, written and published by Venetian Bartolomeo Paxi.
1504	Jacopo de' Barbari paints *Still-life with Partridge and Gauntlets*, the earliest European still life.
c. 1500	Venetians invent pants. Pantalone, the famous *Commedia dell'arte* character, typically wears tight ankle-length trousers, unlike nobles, who wear knee breeches. Peasants wear them in Venice, and the French call the peasant trousers *pantaloni*. Pants as we know them are born.
c. 1500	Venetian ceruse white makeup is invented and exported.
c. 1500	Venetians invent the galleon, a multi-deck sailing ship designed to fight piracy and fire cannon broadside.

c. 1500s	Anatomist Realdo Colombo describes the clitoris as an organ of female sexual pleasure.
1500s	Gabriello Fallopio, anatomist of Padua, is first person to describe the tubes that connect the ovaries to the uterus.
1500s	Gabriello Fallopio, anatomist of Padua, is first person to use an aural speculum to examine the inner ear.
c. 1500s	Anatomist Realdo Colombo is first Westerner to describe pulmonary circulation.
c. 1500s	Girolamo Fabrici d'Acquapendente first describes membranous folds inside veins, which he names valves.
c. 1500	Election of the doge is made by a complicated secret method using gold and silver balls, called *balote*. They're dropped into a secret box and then chosen randomly by senators, thus inventing the ballot method of voting, which was then copied in France and the United States in the 1700s.
1501	Aldus Manutius invents the pocket book (paperback books known as *libelliu portatiles*).
1506	Giovanni Soro is employed by the Council of Ten to decipher code, making the Council of Ten the first Secret Service to use code breaking. Soro is the first cryptologist in Western history and the father of modern cryptology.
1516	Creation of the first Jewish Ghetto. *Ghetto* is a Venetian word that is then, and now, a description of places where people are forced to live with others of the same identity.
1516	First literary censor, Andrea Navagero, appointed by a state (i.e., the Council of Ten).
1520	Venetian Pietro Bembo publishes *Prose della Volgar Lingua* (Writing in the vernacular tongue) and establishes Tuscan dialect as the language of Italy.
1522	Antonio Pigafetta of Vicenza is one of the remaining 18 sailors (out of 240) who completed Magellan's circumnavigation of the world.
1522	Antonio Pigafetta is also the first mariner to keep a journal (not just a logbook) of a long voyage. He collected information on geography, climate, plants, and animals and drew maps. Handwritten copies of his book, *Relazione del Primo Viaggio Intorno al Mondo* (in Italian), were given to monarchs throughout Europe.
1522	Venice conducts the first public lottery in Italy.
1518–1520	Venetian Pietro Coppo publishes the first atlas (comprising twenty-two maps of the world), called *De Toto Orbe*.

1523   Daniel Bomberg is the first printer and publisher of the Babylonian Talmud, which is the current standard page layout and pagination of the Talmud. Bromberg was also the first printer of Hebrew books and the first non-Jewish printer of Hebrew books.

1528   Pietro Coppo publishes *Portolano*, one of the first rutters, or mariner's handbook, a geography of information for maritime navigation.

1528   Benedetto Bordone of Padua draws the first oval map of the world, which becomes the equal-area elliptical Mollweide projection map three hundred years later.

1528   Benedetto Bordone writes the first description of Pizarro's conquest of Peru in his book *Isolario*.

1528   Benedetto Bordone draws the first European map of Japan.

1528   Aldine Press publishes Baldassare Castiglione's *Il Cortigiano* (The book of the courtier), which instructs gentlemen in Europe how to behave.

1530   Girolamo Fracastoro of Padua names the disease syphilis and suggests mercury for the cure.

1531   Niccolò Tartaglia applies mathematics to the path of a cannonball and invents the science of ballistics.

1534   First art supply shops selling pigments and other art supplies.

1537   Sebastiano Serlio publishes the first architectural treatise with illustrations and mathematical relationships between columns and bases, thereby laying the foundation for Renaissance architecture and architectural scholarship.

1537–1538   Venetian printers Paganino and Alessandro Paganini publish the first printed version of the Koran, in Arabic.

1539   Gian Battista da Monte is the first European to teach clinical medicine. He required his students to attend the bedside of the sick and actually touch them.

1543   University of Padua anatomist and physician Andreas Versalius publishes the first illustrated anatomy book, *De Humini Corpus Fabrica*, with accurate and finely drawn illustrations based on personal observation.

1545   The Orto Botanico di Padova is built by the Venetian Republic as the second botanical garden in the world (Pisa is first in 1544) but is the oldest in its original location (Pisa moved three times).

1545   Venetian Daniele Barbaro builds the first conservatory in the Botanical Garden of Padua.

1545   Establishment of *Commedia dell'arte* acting troupe in Padua, which is the first legitimization of acting as a paid profession.

1545   First time authors are deemed to have an artistic personality and rights, and are given rights to compensation. Reluctantly. Could be considered first copyright.

1546   Girolamo Fracastoro identifies the germ theory of infectious disease and describes typhus for the first time.

1548   Giacomo Gastaldi publishes the first pocket atlas.

1555   Giovanventura Rosetti publishes his two-volume *Secreti nobilissimi dell'arte profumatorio* in Venice which contain the first recipe for dying fiber.

c. mid-1500s   Venice conquers land, but instead of building a standing army hires mercenary soldiers and has them report to members of the Venetian non-military elite, thus introducing a new form of military organization that combines soldiers and politicians, the system used globally today that gives politicians control of military decisions.

c. mid-1500s   Venice produces and sells more books than any other city in Europe; it had about two hundred active publishers, thereby inventing the publishing business as we know it.

c. mid-1500s   Female poetess Gaspara Stampa becomes the most famous poet of Italy.

mid-1500s   Venetian perfumers switch from oil- and fat-based perfume to alcohol-based perfumes, thereby revolutionizing the perfume industry and creating the now standard formula for perfume.

1550   Giovanni Ramusio publishes *Delle Navigationi et Viaggi*, a collection of explorers' accounts, which includes the first mention of tea in European literature.

1550   Antonio Pigafetta of Vicenza publishes his account of the Philippines and Mariana Islands in Ramusio's book and includes the first known use of the words *Oceano Pacifico* (Pacific Ocean) on a map.

1551 and 1553   Giovanni Straparola publishes the first books of fairy tales.

1558   Luigi Alvise Cornaro, born in Venice but living in Padua, at age eighty-three writes *Della Vita Sobria*, the first treatise on how to live a long life, thereby inventing the self-help book.

1558   Luigi Alvise Cornaro invents the concept of food allergies.

1558   Aldus Manutius the Younger publishes *Eleganze della Lingua Toscana e Latina* and invents the modern thesaurus.

1561   Aldus Manutius the Younger, grandson of Aldus, produces the first book on the principles of punctuation, called *Orthographiae Ratio*.

1561   Alchemist Isabella Cortese publishes in Venice *I secreti della signora Isabella Cortese; ne' quali si contengono cose minerali, medicinali, profumi, belletti, artifitij,*

	& alchimi (*The Secrets of Lady Isabella Cortese*), the first book on cosmetology written by a woman.
1563	Newspapers that began as a hand-written government monthly called *avise* and *gaxeta*, the name of the two coins needed to buy it, thus introducing the word *gazette*.
1564	Gabriello Fallopio, anatomist of Padua, is first person to describe a condom and advocates their use to prevent syphilis (published after his death).
1569	First women in a boat race (regatta, which is a Venetian word).
1570	Andrea Palladio, born in Padua, publishes his *Four Books of Architecture* and gives the Western world a building style that remains today.
1571	The Venetian navy mounts a continuous gun deck on large ships and invents the first broadside.
1575	First printed explanation of how to use tarot cards was printed in Venice.
1576	Venetian Isabella Andreini becomes the first diva.
1579	Aldus the Younger adds a list of still-available books from Aldine Press to the back of books.
1581	Venetians invent exclusive and expensive box seats in theaters to increase profits and hide bad behavior.
1591	Paduan botanist Prospero Alpini publishes in Venice the first description of the coffee plant and coffee as a drink in *De Medicina Aegyptiorum*.
1592	Botanist Prospero Alpini first recognizes the sexual division of plants and the need for fertilization, which becomes the basis for Linnean taxonomy.
1592	Galileo Galilei and Santorio Santorio, working collaboratively at University of Padua in the Venetian Republic, improve the thermascope and add numbers, thus inventing the thermometer.
1593	Venetian Pietro Cellini invents the high-stakes card game *basetta,* which is introduced to France (where it is called *basset*) by the Venetian ambassador in 1674.
1595	First permanent anatomical theater is built at University of Padua.
1595–1598	Galileo Galilei invents the sector, a military compass to solve mathematical problems in gunnery, ballistics, surveying, and navigation.
c. late 1500s	Paolo Sarpi explains the contraction of the iris.
c. late 1500s	First perfume shops.
1600	After her death in 1592, writer and poet Modesta da Pozzo de Forzi's (Zorzi) husband publishes her *Worth of Women Wherein Is Clearly Revealed Their Nobility and Their Superiority to Men* counters arguments by men that women are inferior, thereby presaging feminism.

c. early 1600s Venetians invent the galleon, a new class of military ship also used to hunt down pirates.

c. 1600 Santorio Santorio invents the wind gauge.

c. 1600 Santorio Santorio invents the water current meter.

c. 1600 Santorio Santorio invents the *pulsilogium*, a device to measure pulse rate, which is the first human-machine interaction in medicine and the first instrument of medical precision.

c. 1600 Santorio Santorio weighs himself, his food, drink, urine and feces for thirty years, thereby inventing the study of metabolism and introducing weight measurement to medicine. Santorio is considered the father of experimental physiology.

1600s The Miotti glassmaking family infuse glass with copper flecks and invent goldstone glass, also called aventurine.

1601 Lucrezia Marinella publishes *The Nobility and Excellence of Women and the Defects and Vices of Men*, a treatise in response to a negative book about women by Giuseppe Passi. This is the first time a woman argues with a man in print.

1606 Paolo Sarpi, a Venetian monk and intellectual, defies the pope and makes the case for a separation of church and state, which the United States follows 175 year later.

1609 Galileo improves the refracting telescope and presents it to the Venetian doge.

1614 Giacomo Castelvetro publishes *A Brief Account of the Fruits, Herbs and Vegetables of Italy*, the first health cookbook.

1614 Santorio Santorio publishes *De Statica Medicina* and invents evidence-based medicine.

1615 Venetian traders introduce coffee to Europe.

1619 Invention of the first bank-to-bank transfer, called a *banco del giro* (*banco del ziro*).

1635 Pietro Cesare Alberti, a citizen of Venice, is the very first Italian-American immigrant into the New World. He lands in the Dutch colony of Manhattan and works a farm in Brooklyn.

1637 Teatro San Cassiano opens as first opera house for the paying public, rather than nobles only, and it stays open for two hundred years.

1638 First casino in Europe, called the Ridotto, for controlled gambling. Casinos, small houses for fun, were common in Venice.

1645 First coffeehouse in Western culture opens in Venice.

1645 First regatta, a series of boat races, which was initially done with gondolas.

1652	Archangela Tarabotti publishes *La Semplicità Ingannata*, a scathing antimale denunciation of fathers who banish their daughter to convents.
1678	Venetian noblewoman Elena Cornaro Piscopia is first woman in the world to be awarded a Ph.D.
c. 1680	Venetian diamond polisher Peruzzi invents the brilliant cut diamond, which is the most popular diamond cut today.
1681–1747	Venetian painter Guilia Lama is the first woman to draw both male and female nude figures from real life, as evidenced in two hundred of her drawings.
1698	Venetian expatriate Niccolao Manucci publishes his four-volume work, *Storia do Mogor*, about life in the Mughal court of India and gives the first Western description of a harem.
c. late 1600s to early 1700s	Venetian mathematician and astronomer Bernardo Facini builds the goniometer for surveying with a compass and other mathematical instruments such as the logarithmic calculator. He also built a complex astronomic clock called a *planisferologio*.
c. 1700	Venetian female artist Rosalba Carriera is the first painter to use ivory rather than velum for miniatures painted on the tops of snuff boxes.
c. 1700	Artist Rosalba Carriera is the first artist to exclusively use pastels which had previously been used only for sketches.
c. 1700	Artist Rosalba Carriera initiates the rococo style and is the first female painter to initiate a new artistic style.
1703	Physician and medical professor at Padua Bernardino Ramazzini publishes the second edition of *De Morbis Artificium Diatriba* (Diseases of workers; first edition published in 1700 in Modena) and establishes the discipline of occupational medicine.
c. 1705	Vincenzo Coronelli, a Franciscan friar and official "Cosmographer of the Republic of Venice," establishes the first geographical society, l'*Accademia Cosmografica degli Argonauti*.
1709	Eau de cologne, invented by Giovanni Paolo Farina (Feminis) and his brother Giovanni Battista (Farina), who were grandsons of the famous Venetian perfumer Catarina Genarri.
1709	Paduan Bartolomeo di Francesco Christofori invents the piano.
1709	Physicist and mathemetician Giovanni Poleni constructs a pinwheel counting machine and presages the modern calculator.

1720	Alla Venezia Trionfante coffeehouse (known as Café Florian) opens in Piazza San Marco and is first coffeehouse in Italy to allow women. Florian's is also one of two longest-running coffeehouses in the world (the other is in Paris).
1724	Giuseppe Briati produced the first *ciocca* (bouquet of flowers) chandelier with polychrome glass flowers, which then becomes the model Venetian chandelier.
1730	Hydraulic engineer Bernardino Zendrini invents the first stone sea walls, or *murazzi*, to keep the Adriatic from overflowing barrier islands into the lagoon.
1738	Admiral Angelo Emo invents the floating battery, a raft of cannonballs that floats toward the shore of an enemy.
c. 1750	Invention of sunglasses, called "Goldoni glasses."
1757	Benedetto Civran invents the *camello*, a system of pontoons that will raise a ship off of and then through mud.
1760	With the *Gazzetta Veneta*, Gasparo Gozzi designs the newspaper layout we use today.
1761	Anatomist and pathologist at Padua Giovanni Battista Morgagni publishes his five-volume *De sedibus et causis morborum per anatomen indagatis* (Of the seats and causes of diseases investigated through anatomy) and established the field of pathology.
1777	Elizabetta Caminèr Turra became director of the literary magazine *Nuovo Giornale Enciclopedico*, and when no one would publish her groundbreaking magazine, she opened her own print shop.
1786	Gioseffa Cornoldi publishes *La Donna Galante ed Erudita* (The elegant and educated woman), the first women's magazine.
1792	Venetian Casanova finishes the first draft of his 12-volume memoir, inventing the image of the Latin lover and acknowledging female sexuality.
1798	Venetian Giustina Renier Michiel is the first translator of Shakespeare from English to Italian.
1829	Venetian geographer Adriano di Rodolfo Balbi publishes (with Andre-Michel Guerry) the first comparative choropleth map depicting rate of crime in relation to education and begins the discipline of graphing types of human behaviors and decisions geographically.
1849	Venice is the first city in the world to be attacked from the air. The Austrians launched armed aerostatic balloons to rain bombs on the city, but they went off target and the crowd below cheered.
1856–60	Felice Beato documents a military campaign as it unfolded during the Second Opium War, making him the first photojournalist.

1858	Luigi di Lucia restores the clock of Piazza San Marco and adds numbers on drums that flip forward every five minutes, thus inventing the first digital clock and the first public digital city clock.
1859	Adele Della Vida Levi opens the first kindergarten in Italy and promises an education free of economic or religious discrimination.
1895	Venice holds the first Biennale of Art, a form of art festival that is then copied around the world.
c. 1900s	The first open-air swimming pool opens on the Lido.
1919	Barbieri brothers in Padua invent Aperol and thus the Aperol spritz.
1932	First international film festival is founded in Venice.
1945	Students at the University of Padua are awarded the Gold Medal of Military Valor, the only university students to be so awarded, for their active and dangerous resistance to German occupation.
1960s or 1970s	Tiramisu is concocted for the first time in the Veneto city Treviso.
1971	Fererico Faggin, who was born in Vicenza and attended the University of Padua, leads the team that invents the first singlechip processor.
1987	Venice is first Italian city to adopt an Animal Rights Act.
1997	Massimo Marchiori, professor of computer science, the University of Padua, invents the HyperSearch, which ranks the relationship among several web pages and paves the way for Google.
2013	Venice opens the first museum of perfume in Italy in the civic museum Palazzo Mocenigo.

Bibliography

Adams, J. Q. (1786). *In Defense of the Constitutions of the United States* (Vol. 1).

Agee, R. J. (1983). The Venetian privilege and music-printing in the sixteenth century. *Early Music History, 3,* 1–42.

Alexis, P. (2017). Over-tourism and anti-tourist sentiment: An exploratory analysis and discussion. *Economic Sciences Series, 17*(2), 288–294.

Altshuller, G. S., & Shapiro, R. B. (1956). On the psychology of inventive creativity. *Problems of Psychology, 6,* 37–49.

Ambrose, C. T. (2005). Osler and the infected letter. *Emerging Infectious Diseases, 11*(5), 689–693.

Ambrosini, F. (2000). Toward a social history of women in Venice: From the Renaissance to the Enlightenment. In *Venice Reconsidered: The History and Civilization of an Italain City-State, 1297–1797* (pp. 420–453). Baltimore: The Johns Hopkins University Press.

Ammerman, A. J. (2001). Venice. In *Routledge Revivals; Medevil Archaeology.* New York; Routledge.

Ammerman, A. J. (2003). Venice before the Grand Canal. *Memoirs of the American Academy in Rome, 48,* 141–158.

Ammerman, A. J., DeMin, M., & Housley, R. (1992). New evidence on the origins of Venice. *Antiquity, 66,* 913–916.

Ammerman, A. J., DeMin, M., Housley, R., & McClennen, C. E. (1995). More on the origins of Venice. *Antiquity, 69,* 501–510.

Ammerman, A. J., McClennen, E., DeMin, M., & Housley, R. (1999). Sea-level change and the archaeology of early Venice. *Antiquity, 73,* 303–312.

Ammerman, A. J., Pearson, C. L., Kuniholm, P. I., Selleck, B., & Vio, E. (2017). Beneath the Basilica of San Marco: New light on the origins of Venice. *Antiquity, 91*(360), 1620–1629.

Anselme, P. (2013). Dopamine, motivation, and the evolutionary signifigance of gambling-like behavior. *Behavioral Brain Research*, *1*(256), 1–4.

Anselme, P., & Robinson, M. J. (2013). What motivates gambling behavior? Insights into dompamine's role. *Frontiers in Behavioral Neuroscience*, *7*, 182–199.

Apellániz, F. (2013). Venetian trading networks in the Medieval Mediterranean. *Journal of Interdisciplinary History*, *2*, 157–179.

Argardy, T., & Alder, J. (2005). Coastal systems. In Millennium Ecosystem Assessment (Ed.), *Ecosystems and Human Well Being* (pp. 513–549). Washington, D.C.: Island Press.

Armstrong, L. (1996). Benedetto Bordone, "miniator," and cartography in early sixteenth-century Venice. *Imago Mundi*, *48*, 65–92.

Barolini, H. (1992). *Aldus and His Dream Book: An Illustrated Essay*. New York: Italica Press.

Barone, A. (1989). *Italians First! An A to Z of Everything First Achieved by Italians*. Folkstone, Kent, U.K.: Renaissance Books.

Barzini, L. (1964). *The Italians*. New York: Touchstone.

Bassi, S. (2019). Waters close over Venice. Retrieved from https://www .nytimes.com/2019/11/15/opinion/venice-flood-climate-change .html?searchResultPosition=1.

Beard, M. (2015). *SPQR; A History of Ancient Rome*. London: Liveright.

Becker, E. (2013). *Overbooked: The Exploding Business of Travel and Tourism*. New York: Simon and Schuster.

Bedini, S. A. (1985). *Clockwork Cosmos: Bernardo Facini and the Farnese Planisferiologio*. Vatican City: Biblioteca Apoltolica.

Belzoni, G. (2001). *Belzoni's Travels: Narrative of the Operations and Recent Discoveries in Egypt and Nubia*. (A. Siliotti, Ed.). London: The British Museum Press.

Bembo. (n.d.). Retrieved from https://www.fontshop.com/families /bembo-adobe.

Benazzi, S., Doak, K., Fornai, C., Baur, C., Kullmaer, O., Svoboda, J., Weber, G. (2011). Early dispersal of moden humans in Europe and implications for Neanderthal behavior. *Nature*, *479*, 525–528.

Benedict, R. (1934). *Patters of Culture*. Boston: Houghton Mifflin Company.

Berendt, J. (2006). *City of Falling Angels*. New York: Penguin Books.

Bergreen, L. (2007). *Marco Polo: From Venice to Xanadu*. New York: Alfred A. Knopf.

Bernstein, J. A. (1998). *Music Printing in Renaissance Venice: The Scotto Press 1539–1572*. Oxford: Oxford Univeristy Press.

Bernstein, J. A. (2001). *Print Culture and Music in Sixteenth-Century Venice*. Oxford: Oxford University Press.

Berrie, B. H., & Matthew, L. C. (2006). Venetian "colore": Artists at the intersection of technology and history. In D. A. Brown & S. Feino-Pagden (Eds.), *Bellini, Giorgione, Titian and the Renaissance of Venetian Painting* (pp. 301–309). New Haven: Yale University Press.

Bertocchi, D., & Visentin, F. (2109). "The Overwhelmed City": Physical and social over-capacities of global tourism in Venice. *Sustainability*, *11*(6937).

Black, C. F. (1989). *Italian Confraternities in the Sixteenth Century*. Cambridge: Cambridge Univerity Press.

Blumenthal, J. (1973). *Art of the printed book, 1455–1955: masterpieces of typography through five centuries, from the collections of the Pierpont Morgan Library, New York. With an essay by Joseph Blumenthal*.

Boelhower, W. (2018). Framing a new ocean geneology; The case of Venetian cartographers. *Atlantic Studies*, *15*(2), 279–297.

Bonfil, R. (1994). *Jewish Life in Renaissance Italy*. Berkeley: University of California Press.

Brainerd, P. (1992). Foreward. In *Aldus and His Dream Book; An Illustrated Essay* (pp. xvii–xix). New York: Italica Press.

Brambati, A., Carbognin, L., Quaia, T., Tiatini, P., & Tosi, L. (2003). The lagoon of Venice: Geological setting, evolution, and subsidence. *Episodes*, *26*(3), 264-.

Braudel, F. (1982). *Civilization and Capitalism, 15th to 18th Centuries*. New York: Harper & Row.

Brosnan, S. F., & DeWaal, F. B. M. (2005). Response to a simple barter task in chimpanzees, Pan trolodytes. *Primates*, *46*, 173–182.

Brown, H. F. (1891). *The Venetian Printing Press: An Historical Study Based on Documents for the Most Part Hitherto Unpublished*. New York: G.C. Puntam's Sons.

Brown, H. F. (1893). *Venice; An Historical Sketch of the Republic*. New York: G.P. Putnam's Sons.

Brown, P. (1991). Self-definition of the Venetian Republic. In A. Mohlo, K. Raaflaub, & J. J. Emlen (Eds.), *City-States in Classical Antiquity and Medieval Italy* (pp. 511–548). Ann Arbor: University of Michigan Press.

Brown, P. F. (1977). *Art and Life in Renaissance Venice*. Upper Saddle River, NJ: Prentice Hall.

Brown, P. F. (2000). Behind the walls; The material culture of Venetian elites. In J. Martin & D. Romano (Eds.), *Venice Reconsidered: The History and Civilization of an Italain City-State, 1297–1797* (pp. 295–328). Baltimore: The Johns Hopkins University Press.

Brown, P. F. (2004). *Private Lives in Renaissance Venice*. New Haven: Yale University Press.

Brown, R. G., & Johnson, K. S. (1963). *Pacioli on Accounting*. New York: McGraw Hill.

Brown, V. (2015). *Cool Shades*. London: Bloomsburg Academics.

Bryer, R. A. (1993). Double-entry bookkeeping and the birth of capitalism; Accounting for the commercial revolution in Medieval Northern Italy. *Critical Perspectives in Accounting, 4*, 113–140.

Buckley, J. (2017). Venice in the summer is "like war," according to tourism chief Paola Mar. Retrieved from https://www.independent.co.uk/travel/news-and-advice/venice-enjoyrespectvenezia-tourism-campaign-overcrowding-paola-mar-litter-responsible-travel-a7863041.html.

Calabi, D. (2017). *Venice and Its Jews; 500 Years Since the Founding of the Ghetto*. Milano: Officia Libraria.

Calliari, I., Canal, E., Cavazzoni, S., & Lazzarini, L. (2001). Roman bricks from the Lagoon of Venice; A chemical characterization with methods of multivariate analysis. *Journal of Cultural Heritage, 2*, 23–29.

Calvani, A. (2011). Translating in a female voice; The case history of Giustina Renier Michiel. *Translation Journal, 15*(3), 1–16.

Campbell, T. (1987). Portolan charts from the late thirteenth century to 1500. In J. B. Harley & D. Woodward (Eds.), *The History of Cartography, Volume 1* (pp. 371–463). Chicago: University of Chicago Press.

Canal, E. (2013). *Archaeloggia della Laguna di Venezia 1960-2010*. Verona: Caselle di Sommacampagna.

Canal, E., Fersuoch, L., Spector, S., & Zambon, G. (1989). Indagini archeologiche a San Lorenzo di Ammiana (Venezia). Venezia: Archeologia Veneta XII.

Caracausi, A. (2014). Textile manufacturing, product innovation and transfer of technology in Padua and Venice between the 16th and 18th centuries. In K. Davids & B. DeMunk (Eds.), *Innovations and Creativity in Late Medieval and Early Modern European Cities* (pp. 131–160). Surrey: Ashgate.

Carlton, G. (2012). Making an impression; The display of maps in sixteenth century Venetian homes. *Imago Mundi, 64*(1), 28–40.

Carpo, M., & Benson, S. (2001). *Architecture in the Age of Printing: Orality, Writing, Typography, and Printed Images in the History of Archtectural Theory*. Boston: MIT Press.

Casagrande, M. (2016). Heritage, tourism, and demography in the island city of Venice: Depopulation and Heritazation. *Urban Island Studies, 2*, 121–141.

Cattaneo, A. (2003). God in his world; The earthly paradise in Fra Mauro's Mappamundi illuminated by Leonardo Bellini. *Imago Mundi, 55*, 97–1902.

Cattaneo, A. (2011). *Fra Mauro's Mappa Mundi and Fifteenth Century Venice*. Turnhout, Belgium: Brepols.

Cattani, A., & Fazzini, G. (2011). Venezia e la peste; Lettere, decreti e fede di sanità. *ArcheoVenezia, 21*(1), 1–4.

Cheney, D. L., & Seyfarth, R. M. (1990). *How Monkeys See the World; Inside the Mind of Another Species*. Chicago: University of Chicago Press.

Cheney, D., Seyfarth, R., & Smuts, B. (1986). Social relationships and social cognition in nonhuman primates. *Science, 234*(4782), 1361–1366.

Chiapello, E. (2007). Accounting and the birth of the notion of capitaliam. *Critical Perspectives on Accounting, 18*(3), 263–296.

Chojnacka, M. (2001). *Working Women of Early Modern Venice*. Baltimore: The Johns Hopkins University Press.

Chojnacki, S. (2000). Identity and ideology in Renaissance Venice. In J. Martin & D. Romano (Eds.), *Venice Reconsidered: The History and Civilization of an Italain City-State, 1297–1797* (pp. 264–294). Baltimore: The Johns Hopkins University Press.

Clark, J. (2018, January 3). The hunt for centuries old books relveals the power of the printed word; Oxford researchers have begun building a data base of a half-million volumes to map the spread of knowledge. *The Wall Street Journal*.

Clarke, P. C. (2015). The business of prostitution in Early Renaissance Venice. *Renaissance Quarterly, 68*, 419–464.

Coldwell, W. (2017). First Venice and Barcelona; Now anti-tourism marches spread across Europe. Retrieved from https://www.google.com/url ?client=internal-element-cse&cx=007466294097402385199:m2ealv uxh1i&q=https://www.theguardian.com/travel/2017/aug/10/ant i-tourism-marches-spread-across-europe-venice-barcelona&sa=U&ve d=2ahUKEwi3y9DTms3pAhVohXIEHWwWCgEQFjAAegQI ABAC&.

Colorito. (n.d.). Retrieved from http://www.visual-arts-cork.com/painting /colorito.htm.

Cortese, I. (2017). *I secreti della signora Isabella Cortese; ne' quali si contengono cose minerali, medicinali, profumi, belletti, artifiti, & alchimia.* Rome: Forgotten Books.

Cosgrove, D. (1992). Mapping new worlds: Culture and cartography in sixteenth century Venice. *Imago Mundi, 44,* 65–89.

Cowan, J. (2007). *A Mapmaker's Dream; The Meditations of Fra Mauro, Cartographer to the Court of Venice.* Boston: Shambala.

Crawshaw, J. L. . (2011). The beasts of burial; Pizzigamorti and public health for the plague in Early Modern Venice. *Social History of Medicine, 24*(3), 570–587.

Crawshaw, J. L. . (2012). *The Plague Hospitals; Public Health for the City in Early Modern Venice.* New York: Routledge.

Crone, G. R. (2016). *The Voyages of Cadamosto and Other Documents on Western Africa in the Second Half of the Fifteenth Century.* London: Routledge.

Crouzet-Pavan, E. (1999). *Venice Triumphant; The Horizons of a Myth.* Baltimore: The Johns Hopkins University Press.

Crouzet-Pavan, E. (2000). Toward an ecological understanding of the myth of Venice. In J. Martin & D. Romano (Eds.), *Venice Reconsidered* (pp. 39–64). Baltimore: The Johns Hopkins University Press.

Dacome, L. (2012). Balancing acts: Picturing perspiration in the long eighteenth century. *Studies in the History and Philosophy of Biological and Biomedical Sciences, 43,* 379–391.

Davies, M. (1995). *Aldus Manutius: Printer and Publisher of Renaissance Venice.* London: The British Library.

Davis, R. C. (1997). Venetian shipbuilders and the fountain of wine. *The Past and Present Society, 156,* 55–86.

Davis, R. C. (2000). Slave redemption in Venice. In J. Martin & D. Romano (Eds.), *Venice Reconsidered: The History and Civilization of an Italian City-State, 1297–1797* (pp. 454–487). Baltimore: The Johns Hopkins University Press.

Dazzi, M. (1969). *Aldo Manuzio e i Dialogo Veneziano di Erasmo.* Vicenza: Neri Pozza.

Dealla Valentina, M. (2006). The silk industry in Venice; guilds and labour relations in the eventeeth and eighteenth centuries. In P. Lanaro (Ed.), *At the Centre of the Old World; Trade and Manufacturing in Venice and on the Venetian Mainland* (pp. 109–142). Toronto: Centre for Reformation and Renaissance Studies.

Demo, E. (2014). New products and technological innovation in the silk industry of Vicenza in the fiftheen and sixteenth centuries. In K. Davids & B. DeMunck (Eds.), *Innovations and Creativity in Late Medieval and Early Modern European Cities* (pp. 82–93). Surrey: Ashgate.

DeRoover, R. (1948). *Money, Banking and Credit in Mediaeval Bruges;* Cambridge: Medieval Academy of America.

Di Robilant, A. (2011). *Irresistible North; from Venice to Greenland on the Trail of the Zen Brothers.* New York: Alfred A. Knopf.

Dickinson, G. (2018). Venice mulls charge for day-trippers—again. Retrieved from https://www.telegraph.co.uk/travel/destinations/europe/italy /veneto/venice/articles/venice-entry-charge-for-tourists/.

Divulgazione; Banca Dati Ambientale sulla Laguna di Venezia. (n.d.). Retrieved from http://www.istitutoveneto.org/venezia/divulgazione/valli/barene.php.

Donadio, R. (2016). Justice Ruth Bader Ginsburg Presides Over Shylock's Appeal. Retrieved from https://www.nytimes.com/2016/07/28/theater /ruth-bader-ginsburg-rbg-venice-merchant-of-venice.html.

Dondi, C. (2018). Printing Evolution 1450-1500: Fifty Years That Changed Europe (*I Cinquant'Anni Che Hanno Cambiato l'Europa*). Venice: Marsilio.

Dondi, C., & Harris, N. (2013a). Best selling titles and Books of hours in a Venetian bookshop of the 1480s: The Zornale of Francesco de Madiis. *La Bibliofilia, 115*(1), 63–82.

Dondi, C., & Harris, N. (2013b). Oil and green ginger. The Zornale of the Venetian bookseller Francesco de Madiis, 1484-1388. In M. Walsby & N. Constantinidou (Eds.), *The Handpress World: Documenting the Early Modern Book* (pp. 341–406). Leiden: Brill.

Dotson, J. E. (1994). *Merchant Culture in Fourteenth Century Venice; The Zabaldone da Canal*. Binghamton, NY: Medieval and Renaissance Texts and Studies.

Dummett, M. (1980). *The Game of Tarot: From Ferrara to Salt Lake City*. London: Duckworth.

Dummett, M. (1993). The history of card games. *European Review, 1*(2), 125–135.

Dunbar, C. F. (1892). The Bank of Venice. *Quarterly Journal of Economics, 6*(3), 308–335.

Edson, E. (2007). *The World Map, 1300–1492; The Persistence of Tradition and Transformation*. Baltimore: The Johns Hopkins University Press.

Eisenstein, E. (1979). *The Printing Press as an Agent of Change, Vols. 1 and 2*. Cambridge: Cambridge University Press.

Eldridge, L. (2015). *Face Paint; The Story of Makeup*. New York: Abrams.

Falchetta, P. (2006). *Fra Mauro's World Map With a Commentary and Translation of the Inscriptions*. Turnhout, Belgium: Brepols.

Farago, J. (2018). 3 days, 150 paintings; A whirlwind Tintoretto Tour. Retrieved from https://www.nytimes.com/2018/11/15/arts/design/tintoretto-venice.html?searchResultPosition=1.

Farrington, L. (2015). "Though I could lead a quiet and peacful life, I have chosen one full of toil and trouble"; Aldus Manutius and the printing history of *Hypnerotomachia Poliphili. Word & Image, 31*(2), 88–101.

Feldman, M. (2000). Opera, festivity and spectacle in "Revolutionary" Venice: Phantasms of time and history. In J. Martin & D. Romano (Eds.), *Venice Reconsidered: The History and Civilization of an Italian City-State, 1297–1797* (pp. 217–260). Baltimore: The Johns Hopkins University Press.

Fenlon, I. (2009). *Piazza San Marco*. London: Profile Books.

Ferguson, N. (2000). *The Ascent of Money; A Financial History of the World*. London: Allen Lane.

Ferreiro, L. D. (2010). The Aristotelian heritage in early naval architecture, from the Venice Arsenal to the French Navy, 1500–1700. *Theoria, 68*, 227–241.

Fiorani, F. (2005). *The Marvel of Maps: Art, Cartography and Politics in Renaissance Italy*. New Haven: Yale University Press.

Fiorin, A. (1989). *Carte, dadi e tavolieri*. In A. Fiorin (Ed.), *Fanti e Danari* (pp. 59–103). Venezia: Arsenale Editrice.

Fischer, M. J. (2000). Luca Pacioli on business profits. *Journal of Business Ethics, 25*(4), 299–312.

Fletcher, C., & Da Mosto, J. (2004). *The Science of Saving Venice.* Turin: Umberto Allemandi & Co.

Florida, R. (2005). *Cities and the Creative Class.* New York: Routledge.

Frankopan, P. (2016). *The Silk Roads; A New History of the World.* New York: Vintage.

Freeland, C. (2012). *Plutocrats; The Rise of the New Global Super-Rich and the Fall of Everyone Else.* New York: Penguin Books.

Freeman, D. E. (1996). "La guerriera amante:" Representations of Amazons and Warrior Queens in Venetian Baroque Opera. *The Musical Quarterly, 80*(3), 431–460.

Friedman, T. L. (2019, May 14). President Trump, Come to Wilmar. *The New York Times.*

Frith, J. (2012). Syphilis—its early history and treatment until penicillin and the debate on its origins. *Journal of Military and Veterans' Health, 20*(4), 49–58.

Frumkin, M. (1945). The origin of patents. *Journal of the Patent Office Society, 27*(143–148).

Fuentes, A. (2017). *The Creative Spark: How Human Imagination Made Humans Excecptional.* New York: Dutton.

Galloway, J. H. (1977). The Mediterranean sugar industry. *Graphic Review, 67*(2), 177–194.

Generazione 90. (n.d.). Retrieved from http://generazione90venezia.it.

Geysbeek, J. B. (1914). *Ancient Double Entry Bookkeeping.* By Author, a translation.

Gibbons, A. (2001). The peopling of the Pacific. *Science, 291*(5509), 1735–1737.

Ginnaio, M. (2011). Pellagra in late nineteenth century Italy: Effects of a deficiency disease. *Population, 66*(583–609).

Giuffrida, A. (2017). "Imagine living with this crap": tempers in Venice boil over in tourist high season. Retrieved from https://www.google .com/url?client=internal-element-cse&cx=007466294097402385 199:m2ealvuxh1i&q=https://www.theguardian.com/world/2017 /jul/23/venice-tempers-boil-over-tourist-high-season&sa=U&ved =2ahUKEwiYzM-PqaznAhW_lnIEHVjlBCEQFjABegQIAxAC&usg =AOvVawoisvL.

Gleeson-White, J. (2011). *Double Entry; How the Merchants of Venice Created Modern Finance.* New York: W.W. Norton & Company.

Glixon, B. L., & Glixon, J. E. (2006). *Inventing the Business of Opera; The Impresario and His World in Seventeenth-Century Venice*. Oxford: Oxford University Press.

Gmelch, S. B. (2010). Why tourism matters. In *Tourists and Tourism A Reader* (pp. 2–24). Long Gove IL: Waveland Press.

Goldthwaite, R. (1987). The empire of things; consumer demand in Renaissance Italy. In F. W. Kent, P. Simons, & J. Eade (Eds.), *Patronage, Art, and Society in Renaissance Italy* (pp. 153–175). Oxford: Oxford University Press.

González, A. T. (2018). Venice: The problem of over tourism and the impact of cruises. *Journal of Regional Research*, *42*, 35–51.

Goy, R. J. (1989). *Venetian Vernacular Architecture; Traditional Housing in the Venetian Lagoon*. Cambridge: Cambridge University Press.

Goy, R. J. (1997). *Venice: the City and Its Architecture*. London: Phaedon.

Grandi, A. (2014). The secret perfume: Technology and organization of soap production in Northern Italy between the sixteenth and eighteenth centuries. In K. Davids & B. DeMunck (Eds.), *Innovations and Creativity in Late Medieval and Early Modern European Cities* (pp. 115–130). Surrey: Ashgate.

Greenwood, D. (1989). Culture by the pound: An anthropological perspective on tourism as cultural commoditization. In V.L. Smith (Ed), *Hosts and Guests: The Anthropology of Tourism* (pp. 171–185). Philadelphia: University of Pennsylvania Press.

Grendler, P. F. (1975). The Roman Inquisition and the Venetian press. *The Journal of Modern History*, *47*(1), 48–65.

Griffin, J. P. (2004). Venetian treacle and the foundation of medicines regulation. *British Journal of Clinical Pharmacology*, *58*(3), 317–325.

Griffo Classico. (n.d.). Retrieved from https://www.myfonts.com/fonts/linotype/griffo-classico/.

Grubb, J. S. (2000). Elite citizens. In J. Martin & D. Romano (Eds.), *Venice Reconsidered: The History and Civilization of an Italain City-State, 1297–1797* (pp. 339–364). Baltimore: The Johns Hopkins University Press.

Haraniya, K. (2018). The Venetian Ducat's Quaint India Link. Retrieved from https://www.livehistoryindia.com/forgotten-treasures/2018/11/16/venetian-ducat-a-gold-standard.

Hardy, P. (2019). Sinking city; How Venice is managing. Retrieved from https://www.theguardian.com/cities/2019/apr/30/sinking-city-how-venice-is-managing-europes-worst-tourism-crisis.

Hargrave, C. P. (1930). *History of Playing Cards and a Bibliogrpahy of Cards and Gaming*. Boston: Houghton Mifflin Company.

Harvey, P. D. A. (1991). *Medieval Maps*. Toronto: University of Toronto Press.

Heller, W. (2003). *Emblems of Eloquence; Opera and Women's Voices in Seventeenth-Century Venice*. Berkeley: University of California Press.

Heritage, U. W. (n.d.). Venice and its lagoon. Retrieved from https://whc .unesco.org/en/list/394.

Hocker, F. M., & McManamon, J. M. (2006). Medieval shipbuilding in the Mediterranean and written culture at Venice. *Mediterranean Historical Review, 21*(1), 1–37.

Honour, H. (1990). *The Companion Guide to Venice*. London: Harper Collins.

Housley, R. A., Ammerman, A. J., & McClennen, C. E. (2004). That sinking feeling: Wetland investigations of the origins of Venice. *Journal of Wetland Archaeology, 4*, 139–153.

Hudson, S. (2015). Aldus Manutius: Innovator of the Pocket Book, and the Semicolon. Retrieved from https://www.proquest.com/blog /pqblog/2015/EEB2015-Aldus-Manutius.html.

Hume, I. N. (2011). *Belzoni: The Giant Archaeologists Love to Hate*. Charlottesville: University of Virginia Press.

Hunt, J. D. (2015). The plot of *Hypnerotomachia Poliphili* and its afterlives. *Word & Image, 31*(2), 129–139.

Irvine, C. (September 17, 2008). Ancestor city of Venice unearthed. *The Telegraph*. Retrieved from https://www.telegraph.co.uk/news/worldnews /europe/italy/2975502/Ancestor-city-of-Venice-unearthed.html.

Italy-Contribution of travel and tourism to GDP as share of GDP. (2018). Retrieved from https://knoema.com/atlas/Italy/topics/Tourism/Travel -and-Tourism-Total-Contribution-to-GDP/Contribution-of-travel-and -tourism-to-GDP-percent-of-GDP.

Jaeggi, A., & Gurven, M. (2013). Natural cooperators; Food sharing in humans and other primates. *Evolutionary Anthropology, 22*, 186–195.

Johns, A. (2009). *Piracy: The Intellectual Property Wars from Guttenberg to Gates*. Chicago: University of Chicago Press.

Johnson, E. J. (2002). The short, lacsivious lives of two Veneitan theaters, 1580-85. *Renaissance Quarterly, 55*(3), 936–968.

Johnson, J. H. (2011). *Venice Incognito: Masks in the Serene Republic*. Berkeley: University of California Press.

Johnson, S. (2010). *Where Good Ideas Come From: The Natural History of Innovation.* New York: Riverhead.

Jonglez, T., & Zoggoli, P. (2018). *Secret Venice.* Paris: Jonglez Publishing.

Jordan, P. (2014). *The Venetian Origins of the Commedia dell'Arte.* New York: Routledge.

Keahey, J. (2002). *Venice Against the Sea; A City Besieged.* London: Thomas Dunne Books.

Kendall, S. B. (1987). An animal analogue of gambling. *The Psychological Record, 37,* 247–256.

Kennedy, P. (2016). *Inventology: How We Dream Up Things That Change the World.* Boston: Mariner Press.

Kerr, R. (2015). The fame monster: Diva worship from Islabella Andreini to Lady Gaga. *Italain Studies, 70*(3), 402–415.

Khan, D. (1996). *The Codebreakers; The Comprehensive History of Secret Communication from Ancient Times to the Internet.* New York: Scribners.

Lace Guild of the United Kingdom (n.d.). A Brief History of Lace. Retrieved from https://www.laceguild.org/a-brief-history-of-lace.

Kittler, J., & Holdsworth, D. W. (2014). Digitizing a complex urban panorama in the Renaissance: The 1500 bird's-eye view of Venice by Jacopo de'Barbari. *New Media & Society, 16*(5), 770–788.

Klestinec, C. (2004). A history of anatomy theaters in sixteenth-century Padua. *Journal of the History of Medicine and Allied Sceinces, 59*(3), 375–412.

Klestinec, C. (2011). *Theaters of Anatomy; Students, Teachers, and Traditions of Disection in Renaissance Venice.* Baltimore: The Johns Hopkins University Press.

Knoops, J. M. (2018). *In Search of Aldus Manutius a campo Sant'Agostin.* Venezia: Bookshop Damocle Edizione.

Konner, M. (2015). *Women After All: Sex, Evolution, and the End of Male Supremacy.* New York: W. W. Norton.

Konstantinidou, K., Mantadakis, E., Falagas, E., Sardi, T., & Samonis, G. (2009). Venetian rule and control of plague epidemics on the Ionian Islands during the 17th and 18 centuries. *Emerging Infectious Diseases, 15*(1), 39–43.

Kostylo, J. (2008). Commentary on the Venetian statute on industrial brevets (1474). Retrieved from http://www.copyrighthistory.org/cam /index.php.

Kostylo, J. (2010). From gunpowder to print; The common origins of copyright and patent. In R. Deazley, M. Kretschmer, & L. Bently (Eds.), *Essays on the History of Copyright* (pp. 21–50). Cambridge: Open Book Publishers.

Kummu, M., de Moel, H., Ward, P. J., & Varis, O. (2011). How close do we live to water? A global analysis of freshwater distance to freshwater bodies. *PloS One, 6*(6), e20578.

Kurlansky, M. (2002). *Salt; A World History*. New York: Walker Publishing Company.

Labalme, P. H. (1981). Venetian women on women: Three early modern feminists. *Archivio Veneto, 117*, 81–109.

Labalme, P. H., & Sanguineti White, L. (2008). *Città Excelentissima; Selections from the Renaissance Diaries of Marin Sanudo*. Baltimore: The Johns Hopkins University Press.

Lall Niham, B. M. (1986). "Bahi-Khata": The pre-Pacioli Indian double-entry system of bookkeping. *Abacus, 22*(2), 148–161.

Lanaro, P. (2006). Reinterpreting Venetian economic history. In P. Lanaro (Ed.), *At the Centre of the Old World; Trade and Manufacturing in Venice and on the Venetian Mainland* (pp. 19–69). Toronto: Centre for Reformation and Renaissance Studies.

Lane, F. C. (1933). Venetian shipping during the commercial revolution. *The American Historical Review, 38*(2), 219–239.

Lane, F. C. (1934). *Venetian Ships and Shipbuilders of the Renaissance*. Baltimore: The Johns Hopkins Press.

Lane, F. C. (1937). Venetian bankers, 1496–1533; A study in the early stages of deposit banks. *Journal of Political Economy, 45*(2), 187–206.

Lane, F. C. (1973). *Venice; A Martitime Republic*. Baltimore: The Johns Hopkins University Press.

Lane, F. C. (1985). Coins and Money of Account, Volume I. In *Money and Banking in Medeival and Renaissance Venice*. Baltimore: The Johns Hopkins University Press.

Larner, J. (1999). *Marco Polo and the Discovery of the World*. New Haven: Yale University Press.

Lavin, M. (2002). *The Virgins of Venice: Broken Vows and Cloistered Lives in the Renaissance Convent*. New York: Viking.

Lazaretto Nuovo. (n.d.). Retrieved from https://www.lazzarettonuovo.com/home-2/.

Leciejewicz, L., Tabaczynska, E., & Tabaczynskis. (1977). *Torcello: Scavi 1961–1962.* Rome: Fondazione Giorgio Cini.

Lennon, J. R. (2014). The accursed items. In *See You in Paradise; Stories.* Minneapolis: Graywolf Press.

Lessage Workshop at Homo faber. (2018). Retrieved from https://2018 .homofaberevent.com/en/house-of-lesage-workshop.

Lessing, L. (2001). *The Future of Ideas.* New York: Random House.

Letizia Battaglia. Photography as a life choice. (2019). Retrieved from http:// www.treoci.org/index.php/it/2013-02-05-10-08-35/in-programmazion e/item/348-letizia-battaglia-fotografia-come-scelta-di-vita.

Levett, J., & Agarwal, G. (1979). The first man/machine interaction in medicine; the pulsilogium of Sanctorius. *Medical Instruments, 13*(1), 61–63.

Library, P. M. (2003). *Art of the Printed Book, 1455-1955.* New York: The Pierpont Morgan Library.

Long, P. O. (1991). Intellectual property and the origin of patents: Notes toward a conceptual history. *Technology and Culture, 32*(4), 846–884.

Lopez, R. S. (1971). *The Commercial Revolution of the Middle Ages.* Cambridge: Cambridge University Press.

Lopez, R. S., & Raymond, I. W. (1955). *Medieval Trade in the Mediterranean World.* New York: Columbia University Press.

Lovric, M. (2016). The Great Venice Boil-Off. Retrieved from https://the-his tory-girls.blogspot.com/2016/11/the-great-venice-boil-off-michelle.html.

Lowry, M. (1979). *The World of Aldus Manutius; Business and Scholarship in Renaissance Venice.* Ithaca: Cornell University Press.

Lowry, M. (1991). *Nicholas Jenson and the Rise of Venetian Publishing in Renaissance Europe.* Cambridge: Basil Blackwood.

Macfarlane, A., & Martin, G. (2002). *Glass; A World History.* Chicago: University of Chicago Press.

Mackenny, R. (2000). "A plot discover'd?" Myth, legend, and the "Spanish" conspiracy against Venice in 1618. In J. Martin & D. Romano (Eds.), *Venice Reconsidered: The History and Civilization of an Italian City-State, 1297–1797* (pp. 185–216). Baltimore: The Johns Hopkins University Press.

Madden, T. F. (2012). *Venice: A New History.* New York: Penguin Books.

Magno, A. M. (2013a). *L'alba dei libri; Quando Venezia ha fatto leggere il mondo (Bound in Venice; The Serene Republic and the Dawn of the Book).* New York: Europa Editions.

Magno, A. M. (2013b). *Murano; The History of Venetian Glass-Blowing*. Milano: VandA.ePublishing.

Maitte, C. (2014). The cities of glass; Privileges and innovations in Early Modern Europe. In K. Davids & B. DeMunck (Eds.), *Innovations and Creativity in Late Medieval and Early Modern European Cities*. (pp. 36–53). London: Routledge.

Mancall, P. C. (2006). *Travel Narratives From the Age of Discovery: An Anthology*. Oxford: Oxford University Press.

Mandich, G. (1948). Venetian patents (1450–1550). *Journal of the Patent Office Society, 30*, 166–224.

Mandich, G. (1960). Venetian origins of inventors' rights. *Journal of the Patent Office Society, 42*, 378–388.

Marchini, N. V. (1995). *Le Legge di Sanità dalla Republica di Venezia*. Vicenza: Neri Pozza.

Marina Vidal. (n.d.). The History of Burano Lace. Retrieved from https://www.martinavidal.com/en/history.aspx.

Marino, J. (1992). Administrative mapping in the Italian state. In D. Buisseret (Ed.), *Monarchs, Ministers and Maps* (pp. 5–25). Chicago: University of Chicago Press.

Martin, J., & Romano, D. (2000). Reconsidering Venice. In J. Martin & B. Romano (Eds.), *Venice Reconsidered* (pp. 1–35). Baltimore: The Johns Hopkins University Press.

Martinelli, A. (1977). Notes on the origian of double-entry bookkeeping. *Abacus, 13*(1), 3–27.

Marx, K. (1867). *Das Kapital; Kritik der Politischen Oekonomie*. Hamburg: Verlag von Otto Meissner.

Marx, K., & Engels, F. (1848). *Manifest der Kommunistischen Partei*. London: Workers' Educational Association.

Mass, J. L., & Hunt, J. A. (2002). The early history of glassmaking in the Venetian lagoon; A microchemical inventigation. *Material Research Society Symposium Proceedings, 712*, II9.3.1–II9.3.11.

Matthew, L. (2010). Painting techniques in Renaissance Europe. *Arts Magazine*, (July 20).

Matthew, L. C. (2002). "Vendecolori a Venezia": The reconstruction of a profesion. *The Burlington Magazine, 144*(1196), 680–686.

Matthew, L. C., & Berrie, B. H. (2010). *Memoria de colori che bisogno torre a venetia*; Venice as a centre for the purchase of painters' colours. In

J. Kirby, S. Nash, & J. Cannon (Eds.), *Trade in Aritsts' Materials; Markets and Commerce in Europe to 1700* (pp. 245–252). London: Archetype Publications.

May, C. (2002). Venetian moment: New technologies, legal innovation and the institution origins of intellectual property. *Prometheus, 20*(2).

McCray, W. P. (1998). Glassmaking in Renaissance Italy: The innovation of Venetian cristallo. *Journal of Metals, 50*(5), 14–19.

McCray, W. P. (1999). *Glassmaking in Renaissance Venice; The Fragile Craft.* Adershot, England: Ashgate.

McGrew, W. C. (1992). *Chimpanzee Material Culture; Implications of Human Evolution.* Cambridge: Cambridge Univerity Press.

McIntyre, J. (1987). The awisi of Venice: Toward an archaeology of media forms. *Journalism History, 14*(2–3), 69–77.

McNeill, W. (2010). *Plagues and People.* New York: Random House.

Mervosh, S. (2019). Carnival cruises to pay $20 million in pollution and cover-up case. Retrieved from https://www.nytimes.com/2019/06/04/business/carnival-cruise-pollution.html?searchResultPosition=1.

Messinis, A. (2017). *The History of Perfume in Venice.* Venice: Lineadaqua.

Milanesi, M. (2016). *Vincenzo Coronelli Cosmographer 1650–1718.* Turnhout, Belgium: Brepols Publishers.

Minuzzi, S. (2016). *L'Invenzioni dell'Autore.* Venezia: Marsilio.

Miozzi, E. (1957). *Venezia Nei Secolo; La Citta.* Venezia: Casa Editrice.

Molà, L. (2000). *The Silk Industry of Renaissance Venice.* Baltimore: The Johns Hopkins University Press.

Mollin, R. (2001). *An Introduction to Cryptogrpahy.* Boca Raton: Chapman and Hall.

Molmenti, P. (1906). *Venice; Its Individual Growth from the Earliest Beginnings to the Fall of the Republic, Volume 1.* Chicago: A.C. McClurg & Co.

Morelli, L. (n.d.-a). Burano Lace; A Brief History. Retrieved from https://laura morelli.com/burano-lace-a-history/.

Morelli, L. (n.d.-b). Murano glass; A brief history. Retrieved from https://laura morelli.com/murano-glass-a-brief-history/.

Morris, J. (1960). *The World of Venice.* New York: Pantheon.

Morris, J. (1990). *The Venetian Empire: A Sea Voyage.* London: Penguin Books.

Mortazavi, M. M., Adeeb, N., Latif, B., Watanabe, K., Deep., A., Griessenauer, C. J., Fukushima, T. (2013). Gabriele Fallopio (1523-1562) and his contributions to the development of medicine and anatomy. *Child Nervous Systems, 29*, 877–880.

Mozzato, A. (2006). The production of woolens in fifteenth and sixteen century Venice. In P. Lanaro (Ed.), *At the Centre of the Old World; Trade and Manufacturing in Venice and on the Venetian Mainland* (pp. 73–107). Toronto: Centre for Reformation and Renaissance Studies.

Mueller, R. C. (1997). The Venetian Money Market, Banks, Panics and the Public Debt 1200-1500, Volume II. In *Money and Banking in Medieval Venice*. Baltimore: The Johns Hopkins University Press.

Muir, E. (1981). *Civic Ritual in Renaissance Venice*. Princeton: Princeton University Press.

Nadeau, B. L. (2015). Aldus; The Venetian Roots of the Modern Book. *Smithonian Journeys, Winter*, 108–112.

Needham, P. (1998). Venetian printers and publishers in the fifteenth century. *La Bibliofilia, 100*(2–3), 157.

Nicol, D. M. (1992). *Byzantium and Venice; A Study in Diplomatic and Cultural Relations*. Cambridge: Cambridge Univerity Press.

Ninfo, A., Fontana, A., Mozzi, P., & Ferrarese, F. (2009). The map of Altinum, Ancestor of Venice. *Science, 325*, 577.

No Grandi Navi. (n.d.). Retrieved from https://www.facebook.com/comitatonograndinavi.

Norwich, J. J. (1977). *Venice; The Rise to Empire*. London: Allen Lane.

Norwich, J. J. (2001, September 30). The religion of empire; Review of Venice Lion City by Gary Will. *Los Angeles Times Book Review*, pp. 1–2.

Nuovo, A. (1987). *Il caorano arabo ritrovato* (Venezia, P. e A. Paganini, tra l'agosto 1537 e l'agosto 1538. *La Bibliofilia, 89*(3), 237–271.

Nuovo, A. (2103). *The Book Trade in the Italian Reanissance*. Leiden: Brill.

Nurminen, M. T. (2015). *The Mapmaker's World; A Cultural History of the European World Map*. London: Pool.

O'Doherty, M. (2011). Fra Mauro's world map (c. 1448-1449). *Wasafini, 26*(2), 30–36.

Oldest Evidence of stone tool use and meat-eating among humn ancestors discovered: Lucy's species butchered meat. (2010). Retrieved from https://www.sciencedaily.com/releases/2010/08/100811135039.htm.

Olson, S. (2003). *Mapping Human History; Genes, Race and Our Common Origins*. New York: Mariner Books.

Ongania, F. (1895). *Early Venetian Printing Illustrated*. Venice, London, and New York: Ongania, Nimmo, and Charles Scribners's Sons.

Orbasli, A. (2000). *Tourists in Historical Towns, Urban Conservation, and the Heritage Management*. Abington: Taylor & Francis.

Osheim, D. J. (2011). Plague and foreign threats to public health in Early Modern Venice. *Mediterranean Historical Review, 26*(1), 67–80.

Ottoni, E., DeResende, B. D., & Izar, P. (2005). Watching the best nutcrackers: What capuchin monkeys (*Cebus apella*) know about others' tool-using skills. *Animal Cognition, 8*(4), 215–219.

Oxford Univeristy. (2018). 15cBOOKTRADE. Retrieved from http://15c booktrade.ox.ac.uk.

Parsons, M. (2014, October 31). Ireland's "oldest known separate map" expected to fetch 3 million. *The Irish Times*. Retrieved from https://www.irishtimes .com/culture/ireland-s-oldest-known-separate-map-expected-to-fet ch-3-million-1.1982656.

Pavanetto, L. (2019). *Mappe ed Esploratori sulle Rotte dalla Serenissima*. Vittorio Veneto: Dario de Bastiani.

Pendergrast, M. (2003). *Mirror Mirror; A History of the Human Love Affair with Reflection*. New York: Basic Books.

Penrose, B. (1952). *Travel and Discovery in the Renaissance 1420–1620*. Cambridge: Harvard University Press.

Penzo, G. (2002). *Barche Veneziane. Catalogo Illustrato dei Piani di Costruzione*. Venezia: Il Leggio, Sottomarina.

Pickford, J. (2018, January 6). Birth of the knowledge economy. *Financial Times*.

Pizzati, L. (2007). *Venetian-English, English-Venetian: When in Venice Do as the Venetians*. Bloomington, Indiana: AuthorHouse.

Poli, D. D. (2011). *Il Museo del Merletto*. Venezia: Marsilio.

Pollan, M. (2001). *The Botany of Desire*. New York: Random House.

Poore, E. M. (2015). Ruling the market: how Venice dominated the early music printing world. *Musical Offerings, 6*(1), 49–60.

Porter, D. (1999). *Health, Civilization and the State*. London: Routledge.

Povoledo, E. (2019). Venice flooding brings city to "its knees." Retrieved from https://www.nytimes.com/2019/11/13/world/europe/venice-flood. html?searchResultPosition=1.

Prager, F. (1944). A history of intellectual property from 1545 to 1787. *Journal of the Patent Office Society, 26*(11), 711–760.

Priani, E. (2017). "Shrouded in a dark fog"; Comparison of the disgnosis of pellagra in Venice and general paralysis of the insane in the United Kingdom, 840-1900." *History of Psychiatry, 8*(2), 166–181.

Puga, D., & Trefler, D. (2104). International trade and institutional change; Medieval Venice's response to globalization. *Quarterly Jounal of Economics, March,* 753–821.

Pullan, B. (1971). *Rich and Poor in Renaissance Venice: the Social Institution of a Catholic State, to 1620.* Cambridge: Harvard University Press.

Queller, D. E. (1986). *The Venetian Patriciate: Reality Versus Myth.* Urbana: University of Illinois Press.

Reader, J. (2004). *Cities.* New York: Atlantic Monthly Press.

Reese, G. (1934). The first printed collection of part-music. *The Musical Quarterly, 20*(1), 39–6.

Renaissance colour palette. (n.d.). Retrieved from http://www.visual-arts-cork .com/artist-paints/renaissance-colour-palette.htm.

Renier-Michiel, G. (1994). *Origine della Feste Veneziane.* Venezia: Filippi Editore.

Riello, G. (2013). *Cotton; The Fabric That Made the Modern World.* Cambridge: Cambridge University Press.

Rosand, E. (1991). *Opera in Seventeenth-Century Venice: The Creation of a Genre.* Berkeley: University of California Press.

Ross, W. (2015). The death of Venice: Corrupt officials, mass tourism, and soaring property prices have stifled life in the city. Retrieved May 14, 2020, from https://www.independent.co.uk/news/world/europe/the-death-of -venice-corrupt-officials-mass-tourism-and-soaring-property-prices-have -stifled-life-in-10251434.html.

Salt in Venice. (n.d.). Retrieved from http://www.venicethefuture.com/schede /uk/174?aliusid=174.

Sama, C. M. (2003). *Selected Writings of Eighteenth Century Venetian Woman of Letters.* Chicago: University of Chicago Press.

Sama, C. M. (2004). Liberty, equality, frivolity! An Italian critique of fashion periodicals. *Eighteenth Century Studies, 37*(3), 389–414.

Sani, B. (2007). *Rosalba Carriera: 1673–1757: Maestra dell Pastello nell'Europa Ancien Régime.* Torino: Umberto Allmande.

Save Venice. (n.d.). Retrieved from https://www.savevenice.org.

Schuessler, J. R. (February 26, 2015). A tribute to the printer Aldus Manutius and the roots of the paperback. *The New York Times.*

Schultz, J. (1978). Jacopo de' Barbari's view of Venice: Map making, city views and moralized geography before the year 1500. *Art Bulletin*, *60*, 425–474.

Sciama, L. D. (2003). *A Venetian Island; Environment, History and Change in Burano*. New York: Berghahn Books.

Seraphim, H., Sheeran, P., & Pilato, M. (2018). Over-tourism and the fall of Venice as a destination. *Journal of Destination Marketing & Management*, *9*, 374–376.

Servadio, G. (2018). *L'Italiano piu' Famoso del Mondo: Vita e Avventure di Giovanni Battista Belzoni*. Rome: Bompiani.

Settis, S. (2014). *If Venice Dies*. New York: New Vessel Press.

Seville, A. (1999). The Italian roots of the lottery. *History Today*, *49*(3), 17–19.

Shannon, P. M., Alemsegede, A., Marean, C. W., Ynn, J. G., Reed, D., Geraads, D., Bearat, H. A. (2010). Evidence for stone-tool-assisted consumption of animal tissue before 3.39 million years ago at Dikika, Ethopia. *Nature*, *466*(7308), 857–860.

Sievers, G. W. (2015). *111 Places in Venice That You Should Go*. Cologne: Emons Verlag.

Silver, M. G. (2019). The city that launched the publishing industry. Retrieved July 9, 2019, from http://www.bbc.com/travel/story/20190708-the-city-that-launched-the-publishing-industry.

Small, M. F. (1996). An Anthropologist's Attic. *Scientific American, July*, 82–85.

Somers Cocks, A. (2004). Flooding: Why does Venice flood? In J. DaMosto & C. Fletcher (Eds.), *The Science of Saving Venice* (pp. 31–43). Turino: Umberto Allemandi & Co.

Somers Cocks, A. (2007). Introduction. In J. DaMosto, T. Morel, R. Gibin, S. Tonin, F. Fracchia, R. Agnoletto, E. Tantucci (Eds.), *The Venice Report: Demography, Tourism, Financing and Change of Use of Buildings* (pp. 7–9). Cambridge: University of Cambridge Press.

Stahl, A. M. (2000). *Zecca; The Mint of Venice*. Baltimore: The Johns Hopkins University Press.

Steinberg, A., & Wylie, J. (1990). Counterfeiting nature: Artistic innovation and cultural crisis in Renaissance Venice. *Comparative Studies in Society and History*, *32*(1), 54–88.

Stephens, M. (1988). *A History of the News*. New York: Viking.

Suchak, M., & DeWaal, F. B. M. (2012). Monkeys benefit from reciprocity without the cognitive burden. *Proceedings of the National Academy of Sciences*, *109*(38), 15191–15196.

Swetz, F. J. (1987). *Capitalism and Arithmetic; The New Math of the 15th Century*. La Salle: Open Court.

Tabaczynska, E. (1968). Studies in Glass History and Design, ed. R. (Gresham Press, 1968) pp. 20–23. In J. Charleston, W. Evans, & A. E. Werner (Eds.), *VIIIth International Congress on Glass*. (pp. 20–23). New York: Gresham Press.

Tassini, G., & Sadoulet, L. (2017). Lessons from the Merchants of Venice. Retrieved from https://knowledge.insead.edu/blog/insead-blog/lessons -from-the-merchants-of-venice-6501.

Testa, S. (2011). *E le Chiamano Navi*. Venezia: Corte del Fontego.

Tiboni, F. (2012). The sewen boat from Cavanella d'Adige (Veneto, Italy); Excavation and first analysis. In J. Gawronski, A. van Holk, & J. Schokkenbroek (Eds.), *Ships and Maritime Landscapes: Proceedings of the Thirteenth International Symposium on Boat and Ship Archaeology* (pp. 290– 293). Amsterdam: Barkhius.

Tinti, P. (2018). Francesca Griffo da Bologna. Retrieved from http://www .griffoggl.com/en/biografia/.

Tognotti, E. (2013). Lessons from the history of quarantine, from plague to Influenza A. *Emerging Infectious Diseases*, *19*(2), 254–259.

Tosi, L., Rizzetto, F., Zecchin, M., Brancolini, G., & Baradello, I. (2009). Morphostratigraphic framework of the Venice Lagoon (Italy) by very shallow water CHRS surveys: Evidence of radical changes triggered by human-induced river diversions. *Geophysical Research Letters*, *36*(9), L09406.

Toso Fei, A. (2008). *Veneziaenigma*. Triviso: Elzivero.

Toso Fei, A. (2010). *The Grand Canal; Mysteries, Anecdotes, and Curiosities About the Most Beautiful Boulevard in Venice*. Venice: STUDIOLT2.

Towner, J. (1985). The Grand Tour: A key phase in the history of touirsm. *Annals of Tourism Research*, *12*, 297–333.

Tran, T., Signoli, M., Fozzati, L., Aboudharam, G., Raoult, D., & Drancourt, M. (2011). High throughput, multiplexed pathogen detection authenticates plague waves in Medieval Venice, Italy. *PLOSOne*, *6*(3), 1–5.

Truss, L. (2004). *Eats, Shoots & Leaves; The Zero Tollerance Approach to Punctuation*. New York: Avery.

Tuchscherer, M. (2003). Coffee in the Red Sea area from the sixteenth to the ninteeth century. In W. G. Clarence-Smith & S. Topik (Eds.), *The Global Coffee Economy in Africa, Asia and Latin America, 1500–1989* (pp. 50–66). Cambridge: Cambridge University Press.

Tucker, C. M. (2011). *Coffee Culture: Local Experinces, Global Connections*. New York: Routledge.

Turner, J. (2015). The spice that built Venice. Retrieved from https://www.smithsonianmag.com/travel/spice-trade-pepper-venice-180956856/.

Umgiesser, G. (n.d.). S.H.Y.F.E.M. (Shallow Water Hyrodynamic Finite Element Model). Retrieved January 1, 2017, from https://sites.google.com/site/shyfem/application-1/lagoons/venice.

Valsecchi, M. C. (2007). Mass plague graves found on Venice "quarantine" island. Retrieved from http://news.nationalgeographic.com/news/2007/08/070829-venice-plague_2.html.

VanSchaik, C. P., & Knott, C. D. (2001). Geographic variation in tool use on Neesia fruits in orangutans. *American Journal of Physical Anthropology, 114,* 331–342.

Vega, F. E. (2008). The rise of coffee. *American Scientist, March-Apri,* 138–145.

Venetian Cartography. (n.d.). Retrieved from http://veniceatlas.epfl.ch/atlas/gis-and-databases/urban/venetian-cartography/.

Venetian Lagoon. (n.d.). Retrieved from https://www.facebook.com/RevolvyEarth.

Venezia Autentica. (n.d.). Retrieved from https://veneziaautentica.com.

Venice Calls. (n.d.). Retrieved from https://www.venicecalls.com.

Venice in Peril. (n.d.). Retrieved from https://www.veniceinperil.org.

Vidal, M. (2017). Preface. In A. Messina (Ed.), *The History of Perfume in Venice.* Venice: Lineadacqua.

Visalberghi, E., Addessi, E., Truppa, V., Spagnoletti, N., Ottoni, E., Izar, P., & Fragaszy, D. (2009). Selection of Effective Stone Tool Use by Wild Bearded Capuchin Monkeys. *Current Biology, 19*(3), 213–217.

Vogel, K. A. (2010). Fra Mauro and the modern globe. *Globe Studies, 57/58,* 81–92.

Wake, C. H. (1979). The changing pattern of Europe's pepper and spice imports, ca 1400–1700. *The Journal of European Economic History, 8*(2), 361–403.

Walker, J. (1999). Gambling and Venetian noblemen c. 1500–1700. *Past & Present, 162*(28–69).

Wallert, A. (1998). *From Tempura to Oil Paint: Changes in Venetian Painitng, 1460–1560*. Amsterdam: Rijksmuseum.

Wheelis, M. (2002). Biological warfare at the 1346 siege of Caffa. *Historical Review, 8*(9), 971–975.

Whitcombe, C. (2004). *Copyright in the Renaissance: Prints and the Privilegio in Sixteenth-Century Venice and Rome*. Leiden: Brill.

Whiten, A., Goodall, J., McGrew, W., Nishida, T., Reynolds, V., Sugiyama, Y., Boesch, C. (1999). Culture in Chimpanzees. *Nature, 399*, 682–685.

Wild, A. (2004). *Coffee*. New York: Norton.

Willan, A., Cherniavsky, M., & Claflin, K. (2012). *The Cookbook Library*. Berkeley: University of California Press.

Wills, G. (2001). *Venice, Lion City, The Religion of Empire*. New York: Simon and Schuster.

Wilson, E. O. (2017). *The Origins of Creativity*. New York: Liveright.

Winkler, E. (2019). Was Shakespeare a Woman? Retrieved from https://www.theatlantic.com/magazine/archive/2019/06/who-is-shakespeare-emilia-bassano/588076/.

Woodward, D. (1987). Medieval mappamundi. In *The History of Cartography, Volume 1* (pp. 203–255). Chicago: University of Chicago Press.

Woodward, D. (1995). *Maps as Prints in the Italian Renaissance; Makers, Distributers and Conservers*. London: The British Library.

Worrall, S. (2015). The Ghetto's hidden beauty. *Smithsonian Magazine, Winter*, 72–75.

Wykes, A. (1964). *The Complete Illustrated Guide to Gambling*. New York: Doubleday.

Zampano, G. (2019). Mayor of Venice Askes UNESCO to Place City on World Heritage Blacklist. Retrieved from https://www.luxurytraveladvisor.com/destinations/mayor-venice-asks-unesco-to-place-city-world-heritage-blacklist.

Zampieri, F., Zanatta, A., Elmaghawry, M., Bonati, M. R., & Thiene, G. (2013). Origin and development of modern medicine at the Univerity of Padua and the role of the "Serenissima" Republic of Venice. *Global Cardiology Science and Practice, 21*, 150–162.

Zanertti, M. (n.d.). *Escursioni Laguna Nord Venezia*. Verona, Italia: Cierre Edizioni.

Zorzi, A. (1999). *Venice 697–1797; A City, A Republic, An Empire*. Milan: Arnoldo Mondadori.

Notes

Chapter One

1   Sixty years of potato washing by Japanese macaques on Koshima Island:
    https://www.youtube.com/watch?v=Fo7ar-ISpCo.
2   Potatoes do not grow on Koshima. Usually, the monkeys ate whatever
    greenery grew on the island.
3   (Dondi, 2018)
4   (Kennedy, 2016) p. 217
5   (Ottoni, DeResende, & Izar, 2005; Visalberghi et al., 2009)
6   (VanSchaik & Knott, 2001)
7   (Whiten et al., 1999)
8   (Whiten et al., 1999)
9   (Shannon et al., 2010)
10 ("Oldest Evidence of stone tool use and meat-eating among human
    ancestors discovered: Lucy's species butchered meat," 2010)
11 (Kennedy, 2016)
12 (Altshuller & Shapiro, 1956)
13 (Kennedy, 2016)
14 (S. Johnson, 2010)
15 (Kennedy, 2016)
16 (S. Johnson, 2010) p. 28–29
17 (Fuentes, 2017; Wilson, 2017)
18 (D. Cheney, Seyfarth, & Smuts, 1986)
19 (Lessing, 2001)
20 (Florida, 2005), p. 4
21 (Kennedy, 2016)
22 (Reader, 2004)
23 (Reader, 2004) p. 7
24 (Florida, 2005)
25 (Florida, 2005)
26 (S. Johnson, 2010)
27 (Reader, 2004) p. 4
28 (Norwich, 2001) p. 1–2

Chapter Two

1 United Nations figure from (Argardy & Alder, 2005)
2 (Kummu, de Moel, Ward, & Varis, 2011)
3 (Gibbons, 2001)
4 (H. F. Brown, 1893)
5 (Umgiesser, n.d.)
6 ("Venetian Lagoon," n.d.)
7 (Brambati, Carbognin, Quaia, Tiatini, & Tosi, 2003)
8 (Brambati et al., 2003) (Goy, 1989)
9 (Goy, 1989)
10 (Zorzi, 1999)
11 (Housley, Ammerman, & McClennen, 2004)
12 (Keahey, 2002)(Housley et al., 2004)
13 (Benazzi et al., 2011)
14 (Tosi, Rizzetto, Zecchin, Brancolini, & Baradello, 2009)
15 (Zanertti, n.d.)(Crouzet-Pavan, 2000)
16 (Benazzi et al., 2011)
17 (Muir, 1981) As Mary Beard points out in her book about Roman history, this invention of origin stories is common (Beard, 2015). It's always about trying to substantiate current politics with made-up versions of how things started. I recently said this to a friend, and she nodded and said, "Right, like the American Revolution." Talking about this in Philadelphia comes with deeply felt sarcasm.
18 (Muir, 1981)
19 Author Gary Wills says Venetians are "autochthonous," meaning "sprung from their own turf," which gives them a sense of unity and specialness (Wills, 2001).
20 (Molmenti, 1906)
21 (Miozzi, 1957)
22 (Molmenti, 1906)
23 The role of Byzantium in Venetian history explains why the Basilica San Marco is a Byzantine church with Byzantine mosaics. Eventually, Venice dominated the economics of Byzantium with their trading skills (Nicol, 1992).
24 (Irvine, 2008; Ninfo, Fontana, Mozzi, & Ferrarese, 2009)(Molmenti, 1906)
25 (Norwich, 1977)
26 (H. F. Brown, 1893)
27 (Ammerman, 2003)
28 (Muir, 1981)
29 (Crouzet-Pavan, 2000)
30 (P. Brown, 1991)
31 (Brown 1893, p.6)
32 ibid.

33   (Muir, 1981)

34   (Canal, Fersuoch, Spector, & Zambon, 1989)(Ammerman, Pearson, Kuniholm, Selleck, & Vio, 2017)

35   (Canal, 2013)

36   (Calliari, Canal, Cavazzoni, & Lazzarini, 2001)

37   (Tiboni, 2012)

38   (from footnote in Housley, Ammerman, & McClennen, 2004)

39   (Ammerman, 2001)(Housley et al., 2004)

40   (Morris, 1960)

41   (Canal, 2013; Crouzet-Pavan, 2000)

42   (Housley et al., 2004)

43   (Leciejewicz, Tabaczynska, & Tabaczynski, 1977)

44   (Ammerman, DeMin, Housley, & McClennen, 1995)

45   (Ammerman et al., 1995)

46   (Ammerman et al. 2017, p. 1621)

47   *Barena* in *veneziano* is actually ghebi, and there are five categories important to Venetians. ("Divulgazione; Banca Dati Ambientale sulla Laguna di Venezia," n.d.)

48   (Ammerman, DeMin, & Housley, 1992)(Housley et al., 2004)

49   (Ammerman et al. 2017, p. 1621)

50   (Ammerman et al., 1995)

51   (Ammerman et al., 1995)

52   (Norwich, 1977)

53   (Ammerman, McClennen, DeMin, & Housley, 1999)

54   (Housley et al., 2004) The problem with Venice today is that the land has sunk due to groundwater usage while global sea levels have risen, so what was once three feet above the highest water level is now often under it, hundreds of times a year. When the tides come in, they also rush through the drainage system, which no longer drains but is flooded and comes up through the drains into the city. A process to fill in parts of Venice with more landfill and raise the worst areas, such as Piazza San Marco, which is Venice's lowest point, has been attempted, but still the water rises.

55   (Ammerman et al., 1999)

56   (Zorzi, 1999)

57   (Crouzet-Pavan, 2000, p.45)

58   (Goy, 1989)

59   (Goy 1989, 1997)

Chapter Three

1   Mappamundi does not really mean "map of the world." It means "cloth of the world" or, more precisely, "napkin of the world," since mappa means "napkin" in Latin.

2   (Lane, 1973)

3   Harvey says maps were virtually unknown in the Middle Ages, and that there was not even a word for "map" in everyday language, even Latin. What we might call maps from that time are simply graphic representations of a landscape or region, an add-on as it were. (Harvey, 1991)

4   (Pavanetto, 2019).

5   (Harvey, 1991; Marino, 1992).

6   Artist and cartographer Cristoforo Sorte drew maps of the mainland that depicted irrigation and drainage.

7   (Cosgrove, 1992)

8   (Campbell, 1987; Edson, 2007)

9   (Campbell, 1987)

10 (Campbell, 1987)

11 (Harvey, 1991)

12 Marco Polo brought a magnetic compass back from China in the 13th century, but the idea of a lodestone pointing north was already in Europe. Polo claimed his compass was for divining things, not for navigation.

13 (Harvey, 1991)

14 The first printed portolan chart appeared in 1539 and was drawn by Venetian cartographer Giovanni Vavassore.

15 (Harvey, 1991)

16 See chapter 8 for the accomplishments of Sanudo's son, also named Marino Sanudo. (Edson, 2007; Kittler & Holdsworth, 2014)

17 (Edson, 2007)

18 (Fiorani, 2005; Harvey, 1991; Woodward, 1995)

19 (Harvey, 1991)

20 (Edson, 2007; Nurminen, 2015)

21 (Woodward, 1987)

22 (Carlton, 2012)

23 (Cattaneo, 2003; Falchetta, 2006; Vogel, 2010)

24 For a fictional account of Fra Mauro see (Cowan, 2007)

25 Rumor has it that the "trusted source" that Fra Mauro referred to was de' Conti, although it's odd that he was not explicit in this reference.

26 (Cattaneo, 2011)

27 (Boelhower, 2018)

28 (Edson, 2007)

29 Some historians believe that the "copy" in the Biblioteca Marciana in Venice is actually the first map and the lost one that was sent to Portugal is a copy. But "copy" in this case means a genuine hand-drawn item, not a Xerox, photo, or print of the original.

30 (Cattaneo, 2011; Edson, 2007; Falchetta, 2006)

31 (Boelhower, 2018)

32 (O'Doherty, 2011) p. 36

33 (Cattaneo, 2011)

34 (O'Doherty, 2011)

35 (O'Doherty, 2011)

36 (Parsons, 2014)

37 (Boelhower, 2018) p. 280

38 (Edson, 2007)

39 The pioneering and cosmological achievement of Fra Mauro's map has been recognized in the naming of a lunar crater on the near side of the moon.

40 (Kittler & Holdsworth, 2014)

41 (Nurminen, 2015)

42 For an analysis of changing (or not) Venetian city maps, see the website for the Venice Atlas project, where a series of maps of Venice are overlaid in a video ("Venetian Cartography," n.d.).

43 (Kittler & Holdsworth, 2014)

44 (Schultz, 1978)

45 (Kittler & Holdsworth, 2014)

46 ("Lessage Workshop at Homo faber," 2018)

47 (Armstrong, 1996)

48 (Armstrong, 1996). Bordone was also the first person to describe Pizarro's conquest of Peru in his island atlas.

49 (Kittler & Holdsworth, 2014)

50 (Carlton, 2012)

51 (Fiorani, 2005)

52 They were frescoed by Giacomo Gastaldi (Carlton, 2012).

53 ("Venetian Cartography," n.d.)

54 (Carlton, 2012; Nurminen, 2015)

55 (Carlton, 2012; Woodward, 1995)

56 (Woodward, 1995)

57 (Milanesi, 2016)

58 Coronelli's globes are about five feet in diameter. But he also made one thirty feet wide with a door. Viewers could crowd inside the globe and imagine what it was like to be at the center of the earth.

59 Years after the heyday of Venetian mapmaking, in 1829 Venetian geographer Adriano di Rodolfo Balbi and Andre-Michel Guerry constructed the first comparative choropleth map (a map depicting a particular variable using colors), which depicted the ratio of crime to education. Recent examples of choropleth maps are the basis for The New York Times column "The Upshot."

60 Although many people playing the pool game probably don't know there was a real person named Marco Polo. And even those who do rarely know that he was Venetian, even after landing at Venice's Marco Polo International Airport.

61 Missionaries and traders had already been in Western Asia by 1260, but none had gone as far east as the Polos (Larner, 1999).

62 It's possible to stand in the campo where the Polos reportedly lived. There is no sign about them, but the area is lovely.

63 (Bergreen, 2007; Larner, 1999)

64 (Larner, 1999) p. 1

65 (Edson, 2007)

66 (di Robilant, 1011)

67 (Edson, 2007)

68 (Penrose, 1952)

69 (Penrose, 1952)

70 Another Venetian expat—someone not just exploring or trading but actually living for a long time abroad, as the Polos and de' Conti had—was Niccolao Manucci, who spent his life at the Mughal court in India, wrote about it (in four volumes), and was the first European to describe a harem in 1698.

71 (Crone, 2016)

72 (Lane, 1973)

73 (Penrose, 1952)

74 The Zen brothers were also supposedly explorers but their story is highly disputed (Lane, 1973).

75 (Penrose, 1952)

76 (Mancall, 2006)

77 (Mancall, 2006)

78 Ramusio's book also contains the first mention of tea in European literature.

79 (Penrose, 1952)

80 (Penrose, 1952)

81 Venetians also managed some risky emigration early on. For example, Pietro Cesare Alberti, a citizen of Venice, was the very first Italian-American immigrant into the New World. He landed at the Dutch colony of Manhattan in 1635 and started a farm in Brooklyn.

82 (Servadio, 2018)

83 (Belzoni, 2001; Hume, 2011)

84 During a regatta near Venice, I was in a small spectator boat that tracked alongside the racing boats (with six rowers each). Most amazing to me was the fact that the Venetian spectators stood casually in their small motorized boats as they bounced along, holding an umbrella to fend off the rain, and also managing to take photographs and videos. My rowing teacher also said to me one day that Venetian kids are easy to teach rowing because they immediately "get it," as they have been traveling by boat since the day they were born, even now in the age of cars, trains, and planes.

85 (Penzo, 2002)

86 (Zorzi, 199).

87 (Cattaneo, 2011)

88 But then Galileo was also sometimes wrong. Attempting to solve the problem of increasing the speed of a galley, Galileo suggested that the oars be much longer outside the boat than inside. It took one of the Lords of the Arsenal and a Venetian, Giacomo Contarini, to explain to Galileo that an

extremely long oar would easily work because the force of the water on the blade would extend so far from the fulcrum of the stroke (Ferreiro, 2010).

89　(Bedini, 1985)

90　According to a display at the Museo Storico Navale in Venice, in 1757 Benedetto Civran invented the "camello," a system of pontoons that could leverage a boat out of mud, but I can find no confirmation of this invention.

91　I recently read the phrase "my arsenal of yarn," which shows how far this label spread.

92　(Lane, 1934)

93　(Lane, 1934)

94　(Lane, 1934, 1973; Wills, 2001)

95　For a detailed and excellent history of the types of boats built by the Venetian state, see (Lane, 1934).

96　(Lane, 1934)

97　There is extensive written evidence of how Venetians designed and built ships documented by designers (Hocker & McManamon, 2006).

98　The entire area of what is known today as Castello was full of shipyards, some private, as well as the Arsenal and manufacturers of items required by ships, such as rope and sails.

99　After the 16th century, slaves were sometimes used on galleys but they were well paid, could have their own small trading businesses, and could rise in rank to the admiralty (Lane, 1934).

100　(Davis, 1997)

101　(Lane, 1934)

102　(Wills, 2001)

Chapter Four

1　Presentation by Dr. Tim Ingold, "The Sustainability of Everything," Department of Anthropology, University of Pennsylvania, February 20, 2019.

2　See my article on loneliness on Medium.com.

3　(Benedict, 1934) p. 2.

4　Back when there were horses in Venice, they used them too, but today there are no horses or cars. Even bicycles are not allowed because they get in the way of walking.

5　(Barzini, 1964) p. 60-61

6　(Mackenny, 2000)

7　(Chojnacka, M., 2001)

8　(Chojnacka, M., 2001) p. 67

9　(Mackenny, 2000). That identifier is on the street signs painted high up on buildings. Words such as Parocia San Pantalon are the parish identified by the closest church.

10　(Ammerman, 2003)

11   (Lane, 1973)

12   (P. F. Brown, 2004; Wills, 2001)

13   (Crouzet-Pavan, 1999)

14   (Crouzet-Pavan, 1999)

15   For a charming and accurate description of how these wells worked, see
     Scientific American 7, no. 3 (July 19, 1862): 43.

16   In 1227 someone in Venice built the first brick hearth with a chimney breast
     and flu, which was copied all over the city and presumably protected against
     house fires.

17   (Toso Fei, 2008)

18   (Sievers, 2015)

19   Much later, in 1987, Venice extended their guardianship over citizens to all
     Venetian pets by being the first Italian city to adopt an Animal Rights Act.

20   (Grubb, 2000)

21   (Wills, 2001)

22   (Lane, 1934, 1973)

23   (Kostyolo, 2008, 2010)

24   (Morris, 1990)

25   (Fenlon, 2009)

26   (Crouzet-Pavan, 1999)

27   (Martin & Romano, 2000)

28   (Feldman, 2000; Muir, 1981)

29   Note the first Sensa symbolized a thanks from the pope to Venice for
     helping out in a war. The pope handed over one of his rings, and the doge
     threw it.

30   (Pizzati, 2007)(Feldman, 2000)

31   (Zorzi, 1999)

32   At a recent visit to the African-American Museum of Philadelphia, I was
     stunned to see a piece by artist Sonya Clark that contained two objects—a
     slab of marble with the word *sciavo* carved into it, and a block painted on
     the floor with the word *ciao*. Clark's explanation of this piece, called Lingua
     Franca, was right on point. She noted that ciao is from Venetian dialect and
     it comes from *sciavo vostro*, literally meaning "I am your servant." Clark's
     point was that this word we now use in a trendy way all over the world is
     based on a racial slur that comes from the days of African slavery, in which
     Venice was an active participant.

33   (Pizzati, 2007)

34   One odd, and controversial, change to Venice has been appearing in the
     street signs. These signs are black letters on a white background, high up on
     the sides of buildings (after all, there is no place for signs on poles on the
     thousands of crossroads in Venice). These signs used to all be in their original
     Venetian, but now the city has decided to make the words more readable
     for tourists. And so, a renaming and repainting campaign has been silently
     at work, transforming these signs by adding double letters—Venetian has

no double letters—and removing the telltale x's, which are the hallmark of Venetian. The result is a mishmash of names that are neither Venetian nor Italian, and therefore make no sense at all. A protest of sorts was launched in 2016 when someone with a ladder went all over the district of San Polo, presumably at night, and blacked out the extra letter with spray paint. Dubbed "The Bandit of San Polo," this person has reached a quiet and surely satisfying fame. And interestingly, these "defaced" signs remain.

35 (Pizzati, 2007)

36 (Wills, 2001) p. 150

37 Of course, there was also lots of of spying, jealousy, and backstabbing. There were once over one hundred lion heads with open mouths, called Bocce di Leone, inset in walls, where people could deposit denunciations of their fellow citizens. The remaining three mouths are in the courtyard of the Palazzo Ducale; on the broad street along the Giudecca Canal at the south end of Venice; and at the Church of San Martino in the Castello district. They are not a sign of rampant jealousy and paranoia, which the word denunciation feels like to a non-Italian speaker, but in fact a way for the populace to seek justice. If the accusation was anonymous, it went through a complicated process of scrutiny before anything was done. If signed, it also needed to be witnessed by two others to be taken seriously. The Bocce di Lione were a sort of Small Claims Court way before its time.

38 Humans seem to have a compulsion to make groups of "us" and "them," but nowhere is a clean division possible, given the endless migration of human groups and their penchant for sex and marriage across these arbitrarily drawn lines, which make no biological sense at all (Olson, 2003).

39 (Grubb, 2000)

40 (Wills, 2001)

41 (Wills, 2001) or any good history of Venice.

42 (Grubb, 2000)

43 (Chojnacki, 2000)

44 (Puga & Trefler, 2104)

45 (Queller, 1986)

46 (Chojnacki, S., 2000)

47 (Honour, 1990)

48 (Apellániz, 2013)

49 (Apellániz, 2013)

50 The Scuola Grande were also big on self-flagellation in public (Pullan, 1971).

51 (Black, 1989)

52 (Goy, 1997)

53 (P. F. Brown, 1977)

54 (Kostylo, 2008, 2010)

55 (Kostylo, 2010)

56 (Pullan, 1971)

57 The symbolic importance of Venetian craftsmen can be seen metaphorically

crawling up and over the arches of the 13th-century entryways of the Basilica San Marco; these reliefs are thought to be the oldest and most realistic portrayal of medieval laborers. There are even more on the tall columns in the Piazzetta next to the Palazzo Ducale. One column is carved with reliefs of shipbuilders, fishermen, and basket makers, while the other column sports fruit sellers, butchers, and cattle sellers. The Venetian workforce is also permanently carved across the columns and arches holding up the facade of the Palazzo Ducale—stone masons, jewelers, farmers, notaries and other such occupations (Fenlon, 2009).

58 (Friedman, 2019)

59 (Wills, 2001)

60 Veneziautentica.com

61 (Wills, 2001)

62 (Chojnacki, S., 2000; Labalme, 1981)

63 (Chojnacki, S., 2000)

64 See (Lavin, 2002) for an in-depth description of the many convents in Venice, including the corrupt ones.

65 (Ambrosini, 2000)

66 Her husband was apparently very supportive. He had the words "a very learned woman" carved into her headstone after she died in childbirth.

67 Modesta da Pozzo's pen name was Moderata Fonte, translated as "moderate fountain," contrasting with the translation of her given name, which means "modest well" (Labalme, 1981).

68 Marinella's book is reminiscent of the current *Women After All*, by Melvin Konner. (Konner, 2015)

69 (Labalme, 1981)

70 Caminèr Turra was the second female journalist in Italy. The honorific of "first" goes to Caterina Cracas, but there is little information about her work. See also Caterina Sama's detailed biography of Caminèr Turra's life and work (Sama, 2003)

71 (Sama, 2003, 2004)

72 (Sama, 2004) p. 391.

73 (Sama 2004)

74 (Calvani, 2011)

75 (Ambrosini, 2000)

76 (Clarke, 2015)

77 (Crouzet-Pavan, 1999)

78 (Ambrosini, 2000; Lavin, 2002)

79 That would be Frankfurt, Germany in 1462.

80 (Zorzi, 1999)

81 Ghèto means "spout" or a "throw" in Venetian, but not as the verb "to throw." The verb throw is *gettare* in Italian, but *sgiaventàr, slansàr, tiràr,* or *trar* in Venetian, so that makes no sense. Others have suggested the name comes from the Hebrew word with an added guttural sound, which would be

weird, since Venetian has no guttural sounds, and the word Ghetto is used in the 1516 state proclamation forming the official ghetto, so the area was not self-named by the Jews. The Venetian dictionary also says that ghéto (with the accent in the other direction) means "ghetto," which of course is the contemporary meaning (note the lack of two t's, which is typical of Venetian). The current iteration, then, incorporates the German Jews' so-called addition of h but does not add the additional t of Italian. And as a further aside, the Venetian word ghèto (the added h and the first kind of accent) means "Have you?"

82   (Calabi, 2017)

83   To find the Greek church, just leave Piazza San Marco to the east and quickly look up to see a very leaning church tower.

84   Twenty-two Italian cities followed Venice in sequestering Jews into ghettos (Bonfil, 1994).

85   Today Jews in Venice obviously live anywhere in the city including the ghetto. And in 2016, the ghetto community, the Jewish community, and the city of Venice celebrated the five hundredth anniversary of the ghetto with a performance of Shakespeare's Merchant of Venice performed outside in the campo of the Ghetto Nuovo. There was a retrial of Shylock, with Justice Ruth Bader Ginsburg presiding, and Shylock was found not guilty (Donadio, 2016).

86   (Bonfil, 1994)

87   (Wills, 2001)

88   (P. F. Brown, 2004) p.216.

89   (Bonfil, 1994; Worrall, 2015)

90   (Worrall, 2015)

91   (Calabi, 2017; Pullan, 1971)

92   (Adams, 1786). Adams eloquently summed up the history of Venice and didn't miss a detail of how the Venetian government worked. Venice had no constitutional document like the United States, but, like Great Britain, it had a body of legislation that formed its constitution (Lane, 1973).

93   (Lane, 1973) p.95

94   (Lane, 1973; Morris, 1960; Norwich, 1977; Wills, 2001)

95   (Zorzi, 1999)

96   (Madden, 2012)

97   (Wills, 2001)

98   For the clearest explanation of this convoluted process. see (Zorzi, 1999), p. 340.

99   And yet they tried. Venetian historian and writer Alberto Toso Fei explains that the word *broglio* is Italian for voter fraud. That word is actually *veneziano*, and it referred to the park of trees right outside the Palazzo Ducale, where people made all sorts of deals that exchanged votes for favors on all sorts of debated rule and regulations (Toso Fei, 2008).

100 Venetian historian and writer Alberto Toso Fei comments that the Venetian

voting system is why the electoral system of the United States has primaries and why the place where one puts their vote is called the ballot box. (Toso Fei, 2010)

101 In fact, the years before the 2020 US elections were marked by comparing money and the candidate with the biggest purse often predicted the winner, favoring those with rich friends or their own high bank accounts. The role of lobbies in the United States is presented as an intractible influence because politicians don't want to cut off the money streaming into their campaign coffers.

102 (Honour, 1990)

103 (Wills, 2001). In a sense, as Hough Honour puts it, Venice was the first constitutional monarchy (Honour, 1990). That is, Venice had a list of laws that formed a constitution and a head of government who was a symbolic ruler, but powerless.

104 (Zorzi, 1999)

105 (Zorzi, 1999)

106 As a long aside in the history of Venice, in the 1400s, Venice moved away from its own identity as a maritime republic with trading ports across the Mediterranean and began to take over land on the mainland. Gaining that land allowed some aristocrats to have vacation homes along the Brenta River, where they could escape during the hot and humid Venetian summers, and also enabled others to have and farm large tracts of land. Land control also meant Venice had access to all the lumber it needed to build ships and the lagoonal foundations of the city. That land buffer probably also protected Venice from assault from mainland Europe—a threat that came to life when Napoleon conquered the city in 1797. But during this land expansion period, Venice conquered territory and included it in the Venetian Republic. In doing so they claimed ownership to the important cities of Padua, Vicenza, Treviso, and Verona, among others, but they acted with the same kind of loose control that they had over ports in the Mediterranean. They put *providitori* in place, who acted as governors of a sort. More significantly, they did not rule by military might. Instead, the republic hired mercenary soldiers who then reported to non-military elite in the council system of the Venetian government. This was the first time that politicians and the military were combined, with the politicians in charge. Today, that system is the rule in Western countries, but it was unheard of back then and probably had more to do with the Venetian focus on making money than overtaking land and people. Instead, they hired their land army and put them under the long arm of the political state back home with general oversight by a *providitori* who probably didn't want to be away from Venice. In later centuries, this combination of politicians and the military was considered a military revolution, but for Venice, it was simply expedient. The state needed a loyal military and was willing to pay for it as long as someone in the central government was in charge.

107 (Khan, 1990; Mollin 2001)

108 (Labalme & Sanguineti White, 2008) and all fifty-eight volumes are available on Google Books.

109 (Toso Fei, 2010)

Chapter Five

1 (McNeill, 2010)

2 Of course, that negative and punishing belief system has been used to explain the AIDS epidemic, specifically among homosexual men and intravenous drug users, although that reasoning is revealed as false when there are places such as continental Africa, where AIDS appears mostly among heterosexual individuals.

3 (Frith, 2012)

4 The bacterium *Yersinia pestis* is named after French-Swiss bacteriologist Alexandre Yersin, who encountered the Black Death in Hong Kong in 1894.

5 There has been some speculation that the waves of mortality were from flu or anthrax or hemorrhagic fever, but archaeologists have confirmed using dental pulp from 173 skeletons recovered in the burial grounds of Lazzaretto Vecchio. According to PCR comparisons of the DNA of various possible other diseases, these individuals, buried during various waves of plague, were always struck down by the Black Death, although they might have also had another bacterial infection spread by body lice (Tran et al., 2011).

6 (Lane, 1973) p. 19; It has been widely assumed that ships coming from Kaffa in Crimea in 1347 brought the Black Plague when infected corpses were lofted into the city by Mongol attackers (making that event an early form of biological warfare). However, relatively recent research shows that although the city might have been infected by those plague-ridden bodies, there is no chance that that event alone was responsible for the Black Plague entering Europe (Wheelis, 2002).

7 (Lane, 1973)

8 (Konstantinidou, Mantadakis, Falagas, Sardi, & Samonis, 2009)

9 (Lane, 1973) p. 19

10 (Porter, 1999)

11 (Porter, 1999)

12 (Konstantinidou et al., 2009)

13 Quarantine in Italian is *quarantena*.

14 (Tognotti, 2013)

15 (Porter, 1999)

16 (Crawshaw, 2012)

17 (Valsecchi, 2007)

18 ("Lazaretto Nuovo," n.d.)

19   If you happen to be in Venice during the *carciofi* (artichoke) season, buy some of these beauties at the Rialto Market or any vegetable stand around town, along with a fennel bulb (*finocchio* in Italian). Without cooking anything, take off the outside leaves, and thinly slice both the artichokes and the fennel. Add olive oil, salt, and pepper, and taste these artichokes as they are meant to be eaten.

20   The Austrians used this island as a military post. They dismantled all the living quarters and administrative buildings and built a brick wall that now circles the island. A nice nature walk outside this wall gives a view of the marshy barene islands in the lagoon and the Dolomite mountains in the distance. No wall kept in the quarantined; there was just no way to get off the island.

21   (Ambrose, 2005)

22   (Ambrose, 2005)

23   The reclaiming of Lazzaretto Nuovo is owed to one archaeologist, Girolamo Fazzini, who has dedicated his life to clearing off the vegetation that consumed the island and various archaeological and restorative projects. Visitors can come to the island on Saturdays and Sundays in the summer for a tour and a nature walk of the periphery.

24   Reportedly bought by the current mayor of Venice in 2018 for $500,000.

25   (Osheim, 2011)

26   (Ginnaio, 2011)

27   (Priani, 2017)

28   Photographs can be seen here: https://www2.le.ac.uk/departments/history/research/grants/PreviousProjects/rough-skin.

29   (Marchini, 1995)

30   (Cattani & Fazzini, 2011)

31   (Crawshaw, 2011)

32   The Ospedale degli Innocenti in Florence was the first place to vaccinate against smallpox in 1756, but those inoculations were next available in Venice for free, making Venice the first city to offer vaccinations to its citizens at no cost because they recognized that contagious diseases were, and are, a public health hazard, and that for vaccination to kill off a disease, everyone has to be inoculated.

33   (Osheim, 2011)

34   (Tognotti, 2013)

35   (Tognotti, 2013)

36   The university claims that 30 percent of the population of Padua are university students, making Padua an intense example of a college town.

37   (Zampieri, Zanatta, Elmaghawry, Bonati, & Thiene, 2013)

38   The students of Padua, as a collective, also stand out. In 1945 the students were awarded the Gold Medal of Military Valor for their active and dangerous resistance against German occupation and fascism. They are the only university students to win this medal. Their proctor, a free-thinking

Communist, encouraged them to take a stand and do something, and they did, putting themselves in harm's way.

39 The university in Venice, called Ca'Foscari, was founded in 1868, seventy-one years after the fall of the republic.

40 (Zampieri et al., 2013)

41 (Zampieri et al., 2013)

42 Realdo Colombo was also an anatomist at Padua, but it is unclear if he discovered particular body parts or physiological mechanisms. Some claim Colombo first described the pulmonary system, but Muslim physicians had done so centuries before. Others have claimed he was the first to identify the clitoris as the site of female sexual pleasure, but apparently Fallopio made notes on this before Colombo. But he seems to be the first to use the word "placenta." Another Padua anatomical controversy focuses on the discovery of the pancreatic duct. It was first described either by German Georg Wirsung in the late 1500s when he was an anatomist there or by Giacomo Combier, who murdered Wirsung because he felt the discovery was his, or by Moritz Hoffman, a student of Wirsung, who claimed the pancreatic duct discovery for himself after Wirsung was dead.

43 Girolamo Fabrici d'Acquapendente has been christened the father of embryology because he described the embryo so perfectly. D'Acquapendente also invented the vertical tracheotomy and was the first person to suggest the use of placing a tube into that slit, although he never performed a tracheotomy himself. The procedure used today is based on his idea.

44 (Klestinec, 2011)

45 (Klestinec, 2004)

46 Fallopio also invented the aural speculum for looking into the ear canal.

47 (Mortazavi et al., 2013). Some have suggested that Realdo Colombo, not Fallopio, first described the clitoris as an organ of female sexual pleasure.

48 Galileo did much of his revolutionary thinking at Padua. While there he laid out the physics of levers and pulleys, wrote the law of inertia, proffered the idea that musical notes were wavelengths of air, and famously sided with Copernicus, which later gained him the title of heretic. While at Padua he perfected the telescope to the thirty-second power in 1609 and observed that the moon's surface is rough; the Milky Way is full of stars; Jupiter has many moons; and Venus has phases, among other astronomical highlights and firsts.

49 (Levett & Agarwal, 1979)

50 (Dacome, 2012)

51 (Zampieri et al., 2013)

52 It was also at the Orto Botanico di Padova that botanist Prospero Alpini first recognized the sexual division of plants and the need for fertilization, which became the basis for Linnaean taxonomy.

53 (Mandich, 1960)

54   (Mandich, 1960) p. 378

55   (Griffin, 2004)

56   (Toso Fei, 2008). Toso Fei also recounts that if you go to Campo San
      Stefano, stand in front of the pharmacy there, and look down, there are
      circular lines that were imprinted by the mortars where theriac was made
      outdoors centuries ago. A pharmacy is still selling potions at the same spot.

57   (Lovric, 2016)

58   The English word treacle is another name for theriac.

Chapter Six

1   (Small, 1996)

2   (McGrew, 1992; Whiten et al., 1999)

3   (Lennon, 2014)

4   (Lopez, 1971; Puga & Trefler, 2104)

5   (Gleeson-White, 2011; Lane, 1973)

6   (Morris, 1990)

7   (Zorzi, 1999)

8   (Freeland, 2012)

9   (Mandich, 1948)

10   (Zorzi, 1999)

11   (Crouzet-Pavan, 1999)

12   (Crouzet-Pavan, 1999) p. 156.

13   (Lane, 1973)

14   (Lane, 1973)

15   (Lane, 1973)

16   (Fenlon, 2009)

17   (Lane, 1973; Morris, 1990)

18   (Lane, 1973) p. 38.

19   Venetian citizens and others were also captured by pirates or rival cities
      such as Genoa and used as slaves from the 14th century on. This was not
      slave trade, but capture of enemies or opponents. Venice then put in place a
      system for buying back these Venetians (Davis, 2000).

20   (Galloway, 1977; Riello, 2013)

21   (Zorzi, 1999) p. 201.

22   (Lane, 1933a)

23   (Apellániz, 2013; Puga & Trefler, 2104; Turner, 2015)

24   (Tassini & Sadoulet, 2017)

25   (Apellániz, 2013)

26   (Toso Fei, 2010)s

27   The Zibaldone was in the hands of the Venetian da Canal family until the
      18th century (Dotson, 1994).

28   (Dotson, 1994)

29   (Dotson, 1994)

30 Another merchant manual, *La practica della mercatur*, compiled by Florentine Francescao Balducci Pegolotti, was published in 1340.
31 (Puga & Trefler, 2104)
32 (Crouzet-Pavan, 1999; Kurlansky, 2002)
33 (Crouzet-Pavan, 1999)
34 ("Salt in Venice," n.d.)
35 (Kurlansky, 2002)
36 ("Salt in Venice," n.d.)
37 (Crouzet-Pavan, 1999; "Salt in Venice," n.d.)
38 (Kurlansky, 2002) p. 85
39 (Lane, 1973)
40 (Toso Fei, 2010)
41 (Mozzato, 2006; Riello, 2013)
42 (Riello, 2013)
43 (Lane, 1973)
44 Interestingly, some historians believe that Shakespeare might have been a woman named Emilia Bassano, who was the daughter of a Venetian family that had immigrated to London before her birth. Even as a possible collaborator with Shakespeare, she would have been able to explain much about Venice (Winkler, 2019).
45 (Tassini & Sadoulet, 2017)
46 (Puga & Trefler, 2104)
47 (Tassini & Sadoulet, 2017) p. 1.
48 (Lopez & Raymond, 1955)
49 (Turner, 2015)
50 (Turner, 2015; Wake, 1979)
51 (Lane, 1933b)
52 The discussion of the silk industry in Venice and its mainland relies on Molà (Molà, 2000)
53 Today the Veneto is experiencing a similar vegetation specialty, or monocrop, of the grapes that produce prosecco. Old vineyards that once produced specialty and boutique wines have been dug up and replanted with prosecco grapes as prosecco becomes one of the most sold and exported wines of the region.
54 (Lanaro, 2006)
55 See Molà, chapter 6, for discussion of various patents on machinery for the silk industry (Molà, 2000).
56 (Molà, 2000)
57 (Demo, 2014)
58 (Deall Valentina, 2006)
59 (Caracausi, 2014)
60 (Kingdom, n.d.)
61 (Sciama, 2003)
62 (Morelli, n.d.-a)

63 It is a common story of handmade needle work to be both the provenience of women and poorly underpaid. For example, knitters on Shetland, Scotland, knit for money. They sold their work to passing sailors and eventually were exploited for their workmanship. Shetland today is experiencing a revival of their particular styles, yarns, and motif all done in yarn from native Shetland sheep. But some of the women at the center of this revival, now in their seventies, once earned money by hand-knitting colorful sweater yokes and selling them to knitwear companies that attached machine-made sweater bodies. See https://www.shetlandwoolweek.com.

64 (Morelli, n.d.-a)

65 (Poli, 2011)

66 Venetian female prisoners continue to produce handmade goods and sell them. See https://malefattevenezia.it and their shops in San Polo that sells goods made in their workshop near the prison on the Giudecca.

67 (Poli, 2011)

68 (Sciama, 2003)

69 It seems that the countess and her school revived the art of Burano lace, which had sort of disappeared, through one eighty-year-old woman named Vicenza Memo (aka Cencia Scarparoila), who explained the technique to a schoolteacher, who then passed it along to young girls on Burano, although it's hard to imagine that one person on the whole island still knew how to make lace in the late 1800s (Marina Vidal, n.d.).

70 (Sciama, 2003)

71 (Sciama, 2003)

72 (Mandich, 1960)

73 For the history of copyright per se, see chapter 8.

74 The original patent statue from 19 March, 1474 (translated): "There are men in this city, and also there come other persons every day from different places by reason of its greatness and goodness, who have most clever minds, capable of devising and inventing all kinds of ingenious contrivances. And should it be legislated that the works and contrivances invented by them could not be copied and made by others so that they are deprived of their honour, men of such kind would exert their minds, invent and make things that would be of no small utility and benefit to our State. Therefore, the decision has been made that, by authority of this Council, any person in this city who makes any new and ingenious contrivances not made heretofore in our Dominion, shall, as soon as it is perfected so that it can be used and exercised, give notice of the same to the office of our Provveditori di Comun, having been forbidden up to ten years to any other person in any territory and place of ours to make a contrivance in the form and resemblance of that one without the consent and license of the author. And if nevertheless someone should make it, the aforesaid author and inventor will have the liberty to cite him before

any office of this city, which office will force the aforesaid infringer to pay him the sum of one hundred ducats and immediately destroy the contrivance. But our Government will be free, at its complete discretion, to take and use for its needs any of the said contrivances and instruments, with this condition, however, that no one other than the authors shall operate them."

75   (Johns, 2009)
76   (Maitte, 2014)
77   (Mandich, 1948)
78   (Frumkin, 1945; Prager, 1944)
79   (Molà, 2000)
80   (Messinis, 2017)
81   (Messinis, 2017)
82   (Vidal, 2017)
83   (Messinis, 2017)
84   (Grandi, 2014)
85   (Grandi, 2014; Messinis, 2017)
86   (Grandi, 2014)
87   (Messinis, 2017)
88   (Messinis, 2017)
89   (Eldridge, 2015)
90   https://cleopatrasboudoir.blogspot.com/2013/02 /elizabeth-ardens-venetian-beauty.html.
91   (Eldridge, 2015)
92   As an aside, the first use of the word wig was in Shakespeare's *Two Gentlemen from Verona*, a city in the Venetian Republic.
93   (Toso Fei, 2008)
94   (Eldridge, 2015)
95   (Toso Fei, 2010)
96   (Messinis, 2017), see pages 91-96.
97   (Cortese, 2017)
98   (McCray, 1999)
99   (Mass & Hunt, 2002; Tabaczynska, 1968)
100   (Magno, 2013b)
101   (Magno, 2013b)
102   Rich Venetians had had a taste of elegant dining in 1072 when the wife of Doge Giovanni picked up a gold fork and ate with it. The fork was said to have arrived with the dogeressa from her home in Constantinople.
103   (McCray, 1999)
104   (McCray, 1998)
105   (McCray, 1999)
106   (Magno, 2013b)
107   (McCray, 1999)
108   (Magno, 2013b)

109   At first, Murano wouldn't allow foreigners to join the guild but after the plague, non-Venetians were welcome to replace those who had died. Murano was thereafter full for "foreign" laborer, that is glass workers from outside Venice, who were not allowed to join the guild but did much of the hard labor or making glass, such as keeping the furnaces going.

110   (Maitte, 2014)

111   (Pendergrast, 2003)

112   (Macfarlane & Martin, 2002). There is also some indication that Barovier was not the only one making cristallo at the time (McCray, 1998).

113   (Mass & Hunt, 2002)

114   (Magno, 2013b)

115   (McCray, 1999)

116   (V. Brown, 2015)

117   (V. Brown, 2015)

118   (Magno, 2013b)

119   Interestingly, Briati set up in Venice rather than Murano, thereby setting himself apart from the Muranese, which was probably a good idea.

120   The skill of glassblowers not only brings joy; it can also bring death. Turns out there is a weapon called the "Venetian stiletto," a glass blade that can be jammed into a chest and then broken with a twist of the wrist. It leaves only a whisper of a scratch, but plunged into the heart, it is deadly, and so the weapon of assassins (Toso Fei, 2010)

121   (Magno, 2013b)

122   It doesn't take much to discern the difference between Murano glass and a cheap imitation. It's the clarity, the colors, the finish of the piece and all its internal swirls and edges, as well as the price. If it says "Made in Italy," that means not on Murano but in some mainland factory employing cheap materials and substandard methods with no heritage of Murano at all. And if a shop owner says, "It's Murano" but doesn't display the MADE IN MURANO decal, stay away. Many shops with centuries of experience can build you a chandelier and ship it to your door. These are the people you can implicitly trust to wrap and ship your newly purchased glass.

123   (Morelli, n.d.-b)

124   (Matthew, 2010)

125   (Matthew, 2010)

126   (Steinberg & Wylie, 1990)

127   (Wallert, 1998)

128   (Berrie & Matthew, 2006)

129   (L. C. Matthew, 2002; L. C. Matthew & Berrie, 2010)

130   (Berrie & Matthew, 2006)

131   (Wallert, 1998)

132   (Berrie & Matthew, 2006)

133   ("Colorito," n.d.)

134   ("Renaissance colour palette," n.d.)

135　(L. C. Matthew & Berrie, 2010)

136　(Berrie & Matthew, 2006)

137　(Wills, 2001)

138　(Sani, 2007)

139　(Toso Fei, 2010)

140　(Farago, 2018)

141　("Letizia Battaglia. Photography as a life choice," 2019)

142　(Puga & Trefler, 2104)

143　(P. F. Brown, 2004)

144　(P. F. Brown, 2000)

145　They were so serious about these laws that everyone was encouraged to report transgressions; a slave could earn freedom by reporting on his or her masters (P. F. Brown, 2004)

146　(P. F. Brown, 2000)

147　At the 2019 Christmas market of the famous glassblowing furnace Ercole Moretti, I bought a strand of these peals that had been made in the 1960s. They are perfectly formed balls pearlized with the grey luster so perfect that they look like pop-beads. I also purchased many cards of pearlized glass buttons that were manufactured in the 1960s.

148　(P. F. Brown, 2000)

149　(Goldthwaite, 1987)

Chapter Seven

1　(Jaeggi & Gurven, 2013)

2　(D. L. Cheney & Seyfarth, 1990)

3　(Suchak & DeWaal, 2012)

4　(Brosnan & DeWaal, 2005)

5　(Ferguson, 2000)

6　(Ferguson, 2000)

7　(Ferguson, 2000)

8　(Ferguson, 2000)

9　(Ferguson, 2000) p. 30

10　(Frankopan, 2016)

11　(Swetz 1987)

12　(Swetz, 1987) p. 15.

13　The Encyclopedia of Money says that Venice was the capitalist forerunner of other cities set in water, such as modern Honk Kong. https://encyclopedia-of-money.blogspot.com/2015/09/venetian-ducat.html

14　(Gleeson-White, 2011)

15　(Ferguson, 2000; Gleeson-White, 2011)

16　Perhaps that history is why, in 1709, Venetian physicist and mathematician Giovanni Poleni constructed a sort of pinwheel counting machine that is the precursor to the modern calculator (Zorzi, 1999).

17 (Stahl, 2000)
18 (Stahl, 2000)
19 (Lane, 1985)
20 (Stahl, 2000)
21 (Lane, 1985)
22 (Stahl, 2000)
23 (Puga & Trefler, 2104)
24 (Gleeson-White, 2011)
25 (Gleeson-White, 2011)
26 (Lane, 1985)
27 (Haraniya, 2018)
28 (Honour, 1990)
29 (Toso Fei, 2010)
30 (Lane, 1985)
31 (Lane, 1937)
32 (Gleeson-White, 2011)
33 (Barone 1989)
34 (Lane, 1985)
35 (Lane, 1937)
36 (Dunbar, 1892)
37 (DeRoover, 1948)
38 (Toso Fei, 2010)
39 (Gleeson-White, 2011)
40 (Lane, 1937)
41 (Lane, 1937) p. 195
42 (Lane, 1937; Mueller, 1997)
43 (DeRoover, 1948; Lane, 1937)
44 (Lane, 1937; Mueller, 1997)
45 (Dunbar, 1892)
46 (Dunbar, 1892)
47 Venetians were also instrumental in establishing the first stock exchange, which was founded in 1309 in Bruges, Belgium. It was started by the Della Borsa family, who had long ago moved to Bruges from Venice, set up in the main square, and changed their name to the Dutch version. They eventually moved their business to Antwerp. The Italian word borsa means "purse" and also "stock exchange." It translates as *beurze* in Dutch (Barone, 1989). *Ter beurze* in Dutch means "on the stock exchange."
48 (Dunbar, 1892)
49 (Dunbar, 1892)
50 (Bryer, 1993)
51 (Gleeson-White, 2011)
52 (Lane, 1973; Toso Fei, 2010)
53 (Gleeson-White, 2011)
54 Reading Gleeson-White's (Gleeson-White, 2011) later chapters, where

she shows how double-entry bookkeeping has influenced not only global finance but also science and the humanities, I realized that my training in animal behavior, and the evolutionary theory that underpins that work, relies on an analysis of costs and benefits (to reproductive success) for any behavioral strategy. Costs and benefits are debits and credits.

55 For example (Lall Niham, 1986) or (Martinelli, 1977)

56 There is gossip that Da Vinci and Pacioli were lovers, but who knows.

57 Pacioli received a ten-year copyright by the Senate and was able to renew that for twenty years on reprints. He was one of the first authors to receive a copyright for a written work (Gleeson-White, 2011)

58 (R. G. Brown & Johnson, 1963; Geysbeek, 1914)

59 (Gleeson-White, 2011)

60 (R. G. Brown & Johnson, 1963)

61 (R. G. Brown & Johnson, 1963)

62 (Gleeson-White, 2011)

63 (R. G. Brown & Johnson, 1963)

64 (Gleeson-White, 2011)

65 From (Gleeson-White, 2011) p. 113

66 (Fischer, 2000)

67 (Gleeson-White, 2011)

68 (R. G. Brown & Johnson, 1963)

69 (Bryer, 1993)

70 (Swetz, 1987)

71 Cited in (Gleeson-White, 2011) p. 161 and again p. 162.

72 (Gleeson-White, 2011)

73 Cited in (Gleeson-White, 2011) p. 162.

74 (Gleeson-White, 2011). Gleeson-White also explains that the word capital comes from *capitale*, a Latin derivative of *caput*, which means "chief," "head," and "property."

75 (Chiapello, 2007)

76 (Braudel, 1982) is the latest English translation. The three-volume original, in French, was published in 1967 (volume 1) and 1979 (volumes 2 and 3). See also (Lanaro, 2006).

77 (Chiapello, 2007)

78 (Marx, 1867; Marx & Engels, 1848)

79 (Freeland, 2012) Freeland also suggests that the same is happening in the United States today. Our super-rich oligarchs don't care about the middle classes because instead of needing them as a market, they can now turn to middle classes around the world. She claims that oligarchs help themselves but can't maintain an economic system that requires flexibility and inclusion of markets (and workers) from all levels of society. Unrest and civil war are the result. She says that the Serrata (see chapter 4) turned Venice from a trading power into a museum. See also (Puga & Trefler, 2104)

80 (Chiapello, 2007)

81   (McCray, 1999) p. 16
82   (Freeland, 2012)
83   (Gleeson-White, 2011)
84   (Lanaro, 2006)

Chapter Eight

1   (Dondi & Harris, 2013b)
2   (H. F. Brown, 1891)
3   (Barolini, 1992)
4   (Ongania, 1895)
5   (Dondi, 2018)
6   (Lowry, 1991)
7   (H. F. Brown, 1891)
8   The first book printed by an Italian, a Venetian priest, came out in 1475 (Honour, 1990; Magno, 2013a)
9   Once the printing business was booming, the Council of Ten butted in and required printers to have a license (not a monopoly like a privilege, but a legal clearance to be a printing house) in 1526. That requirement meant, of course, that the powerful government committee could review all printed works. The need for a license was not enforced for twenty years, but eventually all printers were licensed.
10   (Magno, 2013a)
11   (Dondi & Harris, 2013b; Magno, 2013a)
12   (Magno, 2013a)
13   (Wills, 2001) p. 308.
14   (Dondi, 2018)
15   (di Robilant, 1011; Honour, 1990)
16   (Bernstein, 1998; Magno, 2013a)
17   There was no guild or other oversight body specifically for printers, but eventually they joined a *scuole*. Foreign-born printers became members of guilds that included other foreign businessmen. And when the Scuola San Rocco opened in 1480, the printers signed up and made it their favorite confraternity. These confraternities were, of course, important for any thriving business. Here, contacts were made and deals set, and the *scuole* were an arena for sharing techniques (Agee, 1983).
18   (Magno, 2013a; Swetz, 1987)
19   See examples of printed pages from de Spira, Jenson, Manutius in (Library, 2003).
20   (Whitcombe, 2004)
21   (Dondi, 2018)
22   Note the publication of the first pocket atlas by Giacomo Gastaldi in 1548 that was mentioned in chapter 3. Even with a smartphone GPS available to some, a small pocket atlas is still the best guide for navigating an unknown

city. No one but Venetians could possibly get about the city, or any city, these days without a small folded map.

23 (Reese, 1934)
24 (Bernstein, 1998)
25 (Poore, 2015)
26 (Bernstein, 1998)
27 (Agee, 1983; Poore, 2015)
28 (Bernstein, 1998)
29 (Bernstein, 1998; Poore, 2015)
30 (Bernstein, 2001)
31 (H. F. Brown, 1891; Whitcombe, 2004)
32 (Magno, 2013a)
33 (Grendler, 1975)
34 In 1573, well-known Venetian painter Paolo Veronese was commissioned to paint a Last Supper for a convent in Venice. When the Inquisition saw the finished work, they were horrified by his artistic liberties. Veronese had added dogs, jesters, cats, and Germans to the feast and left off the penitent Mary Magdalene, a traditional guest in other religiously symbolic compositions of the dinner. Confronted by the Inquisition, Veronese ignored their accusations and simply renamed the painting as a secular supper. The Inquisition left him alone after that.
35 (Magno, 2013a) p. 23. See also
36 (Magno, 2013a; Nuovo, 2103) Note that the Frankfurt book fair is, even today, the most important annual market for publishers and authors.
37 (Dondi, 2018; Needham, 1998)
38 (Clark, 2018; Pickford, 2018)
39 (Oxford Univeristy, 2018) http://15cbooktrade.ox.ac.uk.
40 (Dondi, 2018) Title of exhibition: Printing Evolution Exhibition 450-1500; Fifty Years That Changed Europe Correr Museum, fall 2018, Venice.
41 (Dondi, 2018) and (Oxford Univeristy, 2018)
42 (Dondi, 2018)
43 (Dondi, 2018; Dondi & Harris, 2013a; Lowry, 1979; Magno, 2013a)
44 (Eisenstein, 1979)
45 (May, 2002)
46 (Kostylo, 2010; May, 2002)
47 (Long, 1991)
48 (Kostylo, 2008)
49 In 2015, The Grolier Club in New York City, which is America's "oldest and largest society of bibliophiles," celebrated the five hundredth anniversary of Manutius's life with the exhibit Aldus Manutius: A Legacy More Lasting Than Bronze. The exhibit displayed a collection of 150 Manutius books borrowed from sources around the world.
50 (Barolini, 1992)
51 (Lowry, 1979)

52 Although this house may not be his actual press. It might just be a few doors down (Knoops, 2018). Also, after leaving Venice during the War of Cambrai and returning, he moved his press into Torresani's workshop in San Paternian on the other side of the Grand Canal (Barolini, 1992)

53 (Davies, 1995; Dazzi, 1969; Lowry, 1979)

54 (Lowry, 1979; Magno, 2013a)

55 The New York Times recently reviewed a new tiny book format printed by Dutton. These youth-oriented fiction books are about the size of a cellphone, are printed on onion skin paper, and the pages flip upwards, like a flipbook. https://www.nytimes.com/2018/10/29/business /mini-books-pocket-john-green.html?action=click&module=Top%20 Stories&pgtype=Homepage.

56 (Schuessler, 2105)

57 (Barolini, 1992; Magno, 2013a)

58 Private libraries were common and often extensive. For example, in Venice, Cardinal Grimiani had fifteen thousand books, historian of Venice Marino Sanudo, six thousand. See (Magno, 2013a) p. 32.

59 (Hudson, 2015; Magno, 2013a; Schuessler, 2105)

60 (Schuessler, 2105)

61 (Barolini, 1992)

62 (Schuessler, 2105)

63 (Magno, 2013a)

64 Colophon comes from the Greek word Kolophōn, which means "finishing touch."

65 This book still haunts us. *Hypnerotomachia Poliphili* is the basis for the modern popular novel *The Rule of Four* by Ian Caldwell (2004) and Robin Sloan's 2012 best seller *Mr. Penumbra's 24-hour Bookstore.*

66 An anchor colophon is used today by the current Anchor Books, a subsidiary of Doubleday. The Anchor imprint was established in 1953 as a paperback publisher, the oldest in the United States.

67 (Schuessler, 2105)

68 (Brainerd, 1992; Nadeau, 2015)

69 (H. F. Brown, 1891; Carpo & Benson, 2001)

70 (Wills, 2001)

71 (Schuessler, 2105)

72 (Tinti, 2018)

73 (Wills, 2001)

74 (Barolini, 1992)

75 ("Griffo Classico," n.d.)

76 ("Bembo," n.d.)

77 (Blumenthal, 1973)

78 (Carpo & Benson, 2001)

79 (Truss, 2004)

80 (Truss, 2004)

81 (Truss, 2004) p.79.

82 (Truss, 2004)

83 As a university professor who edits and grades thousands of term papers, I can say with authority that the semicolon is the most overused punctuation mark of our time.

84 (Magno, 2013a)

85 (Barolini, 1992)

86 (Barolini, 1992)

87 (Barolini, 1992)

88 (Barolini, 1992)

89 (Library, 2003; Schuessler, 2105)

90 (Hunt, 2015; Wills, 2001)

91 (Wills, 2001)

92 (Wills, 2001) p. 306

93 (Farrington, 2015)

94 The author, Colonna, was a priest of a different sort, one who got in trouble for defending paganism.

95 (Barolini, 1992) p. 6

96 The following "first" books are not mentioned here because they grace other chapters: 1561 Alchemist Isabella Cortese publishes in Venice *The Secrets of Lady Isabella Cortese*, the first book of cosmetology (ch. 4); 1614 Santorio Santorio publishes *De Statica Medicina* and invents evidence-based medicine (ch. 6); 1698 Venetian expatriate Niccolao Manucci publishes his four-volume work, *Storia do Mogor*, about life in the Mughal court of India and gives the first Western description of a harem (ch. 3); 1703 Physician and medical professor at Padua Bernardino Ramazzini publishes the second edition of *De Morbis Artificium Diatriba* (Diseases of workers, first edition published in 1700 in Modena) and establishes the discipline of occupational medicine (ch. 6); 1761 Anatomist and pathologist at Padua Giovanni Battista Morgagni publishes his five-volume *De Sedibus et causis morborum per anatomen indagatis* (Of the seats and causes of diseases investigated through anatomy) and establishes the field of pathology (ch. 6); 1494 Luca Pacioli publishes first book of algebra, accounting, and codifies double-entry bookkeeping (ch. 7); 1472 Paolo Bagellardo publishes *De Infantium Aegritudinibus et Remediis*, the first printed book on pediatrics, which is also the first printed medical book (ch. 6); 1503 The first printed Venetian merchant manual *Tariffa de pexi e mesurei*, written and published by Venetian Bartolomeo Paxi (ch. 7); 1528 Pietro Coppo publishes *Portolano*, one of the first rutters, or mariner's handbook, a geography of information for maritime navigation (ch. 2); 1543 Anatomist and physician at the University of Padua Andreas Versalius pubishes the first illustrated anatomy book *De Humini Corpus Fabrica* with accurate and finely drawn illustrations based on personal

observation (ch 6); 1558 Luigi Alvise Cornaro, born in Venice and living in Padua, at age eighty-three writes *Della Vita Sobria*, the first treatise on how to live a long life, thereby inventing the self-help book (ch. 6); 1591 Paduan botanist Prospero Alpini publishes in Venice the first description of the coffee plant and coffee as a drink in *De Medicina Egyptiorum* (ch. 9).

97 See (Magno, 2013a) chapter 4 for an entertaining and enlightening history of this discovery. Also (Nuovo, 1987).

98 (Minuzzi, 2016)

99 At first, this city clock boasted only a lovely clockface with Roman numerals. But it was revamped in 1858 by Luigi di Luci, who added Arabic numbers on drums, making it the first digital city clock.

100 (Willan, Cherniavsky, & Claflin, 2012)

101 (Magno, 2013a)

102 (Wills, 2001) p, 311

103 (Wills, 2001)

104 (Carpo & Benson, 2001; Wills, 2001)

105 (Carpo & Benson, 2001) p. 45

106 (Carpo & Benson, 2001)

107 (Carpo & Benson, 2001)

108 (Carpo & Benson, 2001)

109 (Bernstein, 1998)

110 Gary Wills calls the Palladian style "Thomas Jefferson's architectural Bible" (Wills, 2001) p. 313.

111 (Wills, 2001) p. 313.

112 On a modern note about proportions and how the human eye and brain need them, Da Vinci's Vitruvian Man drawing is now held by Venice's Academia Gallery, the art museum of Venice. The title is an honorific for the Roman architect Vitruvius mentioned above.

113 (Honour, 1990)

114 (Toso Fei, 2010)

115 (McIntyre, 1987)

116 (McIntyre, 1987)

117 (McIntyre, 1987)

118 (Stephens, 1988)s

119 Renowned Venetian publisher of maps and books during the 18th century, Giambattista also issued a weekly journal about business and intellectual life in Venice, and this, too, could be considered a kind of early newspaper that played an important role in the development of the newspaper business.

120 Il Gazzettino online: https://www.ilgazzettino.it.

121 (McIntyre, 1987)

122 As a repeated visitor to Venice, I never bother to look up "what's on" because I know as soon as I walk the city I'll see posters of current exhibits, concerts, lectures, and protests.

Chapter Nine

1   (Renier-Michiel, 1994)
2   (Anselme & Robinson, 2013)
3   (Kendall, 1987)
4   (Anselme, 2013)
5   How to stop gambling? Anselme and Robson have an answer based on the research: "At a societal level, one approach allowing to address pathological gambling might be that gamblers at casinos can win more often than they lose but only very small gains (similar to the wagered amounts) to render gambling persistence less attractive." (Anselme & Robinson, 2013).
6   (Toso Fei, 2008)
7   (Toso Fei, 2008)
8   A third column is said to have landed in the water, but no one has found it.
9   (Toso Fei, 2008)
10 (Walker, 1999)
11 (Walker, 1999)
12 The party starts with the very traditional and very charming Festa delle Marie. Even today, this parade begins at the Church of San Pietro in Castello, which is the official city church (not the basilica). Twelve young women in medieval dress are presented to the crowd in remembrance of a time in 973 when twelve poor girls were kidnapped by Dalmatian pirates. In an act of community charity, these girls had been attached to noble families so that they would have adequate dowries for marriage. The doge sent Venetian sailors after the pirates and recovered the girls and their jewels, and today's parade is a celebration of that rescue. There was also a brief period starting in 1271 when the girls were replaced by giant wooden dolls because the crowd had become lascivious about seeing the yearly virgins on display. Those giant dolls were soon copied by craftsmen and sold to onlookers with the name "marionettes" (little Marys), from which we have the name for puppets on strings.
13 Venice offered other socially amoral but greatly enjoyed pleasures, such as prostitution (see chapter 4).
14 (Walker, 1999)
15 (Labalme & Sanguineti White, 2008), although some have suggested the title goes to Florence (Seville, 1999).
16 (Labalme & Sanguineti White, 2008)
17 For illustrations of the original gambling devices and various card sets, see (Fiorin, 1989).
18 Walker points out that the veneziano word *zougar* (*giocare* in Italian, which means "to play") means both "play" and "gamble" (Walker, 1999)
19 (Dummett, 1980a, 1993)
20 (Hargrave, 1930)
21 (Hargrave, 1930)

22 Giorgio Ghidòil and his cards and other illustrations can be found on Salizada S. Antonin in the sestiere of Castello. https://www.veneziagallery.it

23 (Dummett, 1993)

24 (Hargrave, 1930)

25 (Hargrave, 1930)

26 (Wykes, 1964)

27 (Dummett, 1993; Madden, 2012)

28 Bassett evolved into faro, which was the game of choice for the early American West—the game that cowboys played in saloons (Dummett, 1993).

29 (Dummett, 1980b)

30 Another pack of cards called the trestine pack, which was found in the Veneto.

31 (Dummett, 1980b)

32 *Casino* in Venetian (and Italian) means "a mess or chaos." *Che casino* means "What a mess!" but the *veneziano* word *casinó* means "gambling house." Therefore, we have the word casino in the English language, indicating a gambling house.

33 (Lane, 1973)

34 These fallen nobles were called Barnabotti after the San Barnaba area.

35 (J. H. Johnson, 2011)

36 (Walker, 1999)

37 (Jonglez & Zoggoli, 2018)

38 (Fiorin, 1989)

39 (Toso Fei, 2010)

40 (E. J. Johnson, 2002)

41 (Glixon & Glixon, 2006)

42 (J. H. Johnson, 2011)

43 Commedia dell'arte did not spring up from a vacuum; the plays presumably have ancient Roman and Greek roots (Jordan, 2014)

44 The zani also included the servant and trickster Arlecchino, or Harlequin, that trickster dressed up in a costume of brightly colored diamond shapes, who still shows up at Halloween or masked balls.

45 (Barzini, 1964)

46 (Madden, 2012)

47 (Jordan, 2014)

48 (Kerr, 2015)

49 (Madden, 2012)

50 (Glixon & Glixon, 2006)

51 (E. J. Johnson, 2002; Jordan, 2014)(E. J. Johnson, 2002)

52 (Glixon & Glixon, 2006)

53 (E. J. Johnson, 2002)

54 (E. J. Johnson, 2002), p. 946.

55 Interestingly, young women trained as singers in the various convents were not allowed to be in operas (Glixon & Glixon, 2006).

56 (Freeman, 1996)

57 (Heller, 2003)(Glixon & Glixon, 2006)

58 (Rosand, 1991)

59 (Glixon & Glixon, 2006)s

60 (Madden, 2012)

61 See Berendt's detailed book on the rebuilding of La Fenice (Berendt, 2006).

62 (Tucker, 2011)

63 (Pollan, 2001; Wild, 2004)

64 (Tucker, 2011)

65 (Tuchscherer, 2003; Tucker, 2011)

66 (Vega, 2008)

67 (Toso Fei, 2010)

68 (Barone, 1989)

69 (Toso Fei, 2010)

70 In Italy there are many standard ways to have your coffee at a bar. Standing at the counter, one orders either an espresso, Americano, macchiato, or cafe latte, among many others. There is no latte, because that means "milk," so the correct order is "café latte." The biggest surprise beyond the flavor is the price, which is less than half of the same drink in America. Italians see coffee as a human right, which explains the low price, and also explains why baristas sometimes give a free cup to those who can't afford it.

71 (Toso Fei, 2010)

Afterword

1 (Silver, 2019)

2 (Towner, 1985)

3 (Ross, 2015)

4 (Somers Cocks, 2007)

5 (Bertocchi & Visentin, 2109)(Somers Cocks, 2007)

6 (Bertocchi & Visentin, 2109)

7 (Bertocchi & Visentin, 2109)

8 (Bertocchi & Visentin, 2109) p. 1.

9 (Bertocchi & Visentin, 2109) p.2

10 (Bertocchi & Visentin, 2109)

11 Settis writes that in 1950 the number of Venetian deaths matched the number of births, but by 2000 there were 1,058 deaths and only 404 births and surely that comparison is even more disparate today (Settis, 2014).

12 (Gmelch, 2010) p. 4.

13 (Gmelch, 2010)

14 ("Italy-Contribution of travel and tourism to GDP as share of GDP," 2018

15   (Alexis, 2017)

16   These figures are from 2010 and are presumably higher today (Gmelch, 2010).

17   Greenwood wrote that in 1989, before the explosion of cruise ship travel and mass tourism. (Greenwood, 1989)

18   (González, 2018)

19   (Alexis, 2017)

20   (Alexis, 2017)

21   (Bertocchi & Visentin, 2109)

22   (Alexis, 2017)

23   (Alexis, 2017) p. 291

24   During the Vogalonga regatta of hand-rowed boats in June 2019, another giant cruise ship motored right through the throng of tiny rowed boats at the start line. The contrast was startling, and why that ship didn't delay its entry into Venice, given the annual regatta that takes up the entire Giudecca Canal, is anybody's guess.

25   (González, 2018)

26   (González, 2018)

27   (González, 2018)

28   Elizabeth Becker writes, "The EPA estimates that a cruise ship carrying 2,000 passengers on the open sea will pollute the air with the same amount of sulfur dioxide as 31.1 million automobiles every day." p. 164.

29   (Mervosh, 2019)

30   (Becker, 2013)

31   (Testa, 2011)

32   (González, 2018; Settis, 2014)

33   (Somers Cocks, 2004)

34   (Somers Cocks, 2004)

35   (Povoledo, 2019)

36   (Gmelch, 2010)

37   (Greenwood, 1989)

38   (Fletcher & Da Mosto, 2004)

39   (Keahey, 2002)

40   (Orbasli, 2000)

41   The Italian constitution and provisions of the European Union state that anyone can travel freely through Italy and other countries of the EU.

42   (Casagrande, 2016)

43   (Dickinson, 2018)

44   For example, (Buckley, 2017; Coldwell, 2017; Giuffrida, 2017), and (Hardy, 2019).

45   (Seraphim, Sheeran, & Pilato, 2018)

46   (Zampano, 2019). To quote the mayor of Venice, who is at war with the Italian Transport Minister: Mr. Brugnaro added that the city doesn't feel represented anymore by Transport Minister Danilo Toninelli, who "has an

arrogance I've never seen in my life and pretends of having understood in half a day what I haven't understood in fifty-seven years."

47 ("Venice in Peril," n.d.)
48 ("Save Venice," n.d.)
49 ("No Grandi Navi," n.d.)
50 ("Venezia Autentica," n.d.)
51 ("Generazione 90," n.d.)
52 ("Venice Calls," n.d.)
53 (Casagrande, 2016)
54 (Casagrande, 2016; González, 2018)
55 (Heritage, n.d.)
56 (Bassi, 2019)

Index

Cartography, 44–59, 61, 65, 73
Casanova, Giacomo, xiii–xiv, 203
Casinos, 216–217. *See also* Gambling
Cassiodorus, Flavius, 36–37
Castelvetro, Giacomo, 121, 203
Castiglione, Baldassare, 194–195
Cattaneo, Alberto, 64
Cattaneo, Angelo, 50
Catullus, 193
Cellini, Pietro, 215
Cereta, Laura, 84
Chandeliers/lamps, 154
Chevron bead, 153
Chiapello, Eve, 182
Child labor laws, 11, 74
Chimpanzees, 4–7, 124, 164
Chojnacki, Monica, 73
Church and state, separation of, 8–9,
  90–94, 114
Cicero, 187, 193
Circulatory system, 111, 114
Circumnavigation, 61
Cities (book), 15
Cities/urban areas. *See also* Venice
  diversity in, 17, 70–89
  growth of, 15–16
  influential power of, 18
  inventions and, 14–19
  lifestyles in, xi–xii, 16–19, 28–30,
    209–224
  loyalty in, 18
  social connections in, 16–19
Clancy, Tom, 191
Classic books, 193–195. *See also* Books
Clement VII, Pope, 93
Clement VIII, Pope, 210
Climate change, 235–237, 242–243
Clinical trials, 115–117. *See also* Medicine
Clock, astronomical, 64, 77
Clock, digital, 77
Clock tower, 77, 203
Cloth, 133–143
Code breaking, 93
Codices, 185–186. *See also* Books

Coffee, 86, 210, 222–223
Coffeehouses, 86, 222–223
Collective group, 11–17
Colombo, Realdo, 114
Colonna, Francesco, 200
Columbus, Christopher, 58, 61
Comin, Jacopo, 156–157
Commedia dell'arte (acting troupes),
  218–220
Commerce
  art supplies, 155–159
  beauty products, 145–150
  cloth, 133–143
  consumerism and, 123–162
  cosmetics, 145–150
  diamonds, 160–161
  glassmaking, 9, 150–155
  hair dye, 145, 148–150
  lace, 139–143
  laws, 127, 133, 144, 159–161
  lumber, 133–136
  luxury goods, 136–143
  material goods, 124–125, 159, 167
  mercantile business, 73, 124–162, 167–
    182, 188, 214, 229–230
  merchant manuals, 132–133
  perfumes/fragrances, 145–147
  soaps, 145–147
  staples, 133–136
  trade and, 125–162
Commercial Revolution, 126–127, 143–
  144, 155–162
Community. *See also* Venice
  confraternities in, 16–17, 82–84,
    120–121
  definition of, 70–71
  festivals in, 78–79, 159, 208, 210–214,
    221–222
  genders in, 79–81, 83–86
  ghetto in, 79, 87–89
  government and, 73–75, 89–95
  guilds in, 82–83
  identity of, 13–33, 45–46, 68, 70–73,
    75–83, 90–95

Di Pietro della Gondola, Andrea
  (Palladio), 205
Di Robilant, Andrea, 58–59, 123, 137
Di Spira, Johannes, 187
Diacono, Giovanni, 35
Diamond trade, 160–161
Discovery, Age of, 56–63, 125–126
Disease, 97–110, 114–115, 118. *See also*
  Plagues
Diseases of Workers (book), 118
Disinfection methods, 105, 109, 237
Dissection techniques, 112–113, 118
Diva, stage, 219–221
Diversity, 17, 70–89
Divina Proporzione (book), 176
Diving suit, 64–65
Doges, 91–93
Dondolo, Marco, 216–217

E

Eats, Shoots & Leaves (book), 199
Eau de cologne, 147
Economy, 15–16, 163–180. *See also* Money
Edison, Thomas, 1
Edson, Evelyn, 50
Egyptologists, 62–63, 218
Electrotherapy, 108
Eleganze della Lingua Toscana e Latina
  (book), 202
Elements, 112
Elevator, 77
Elizabeth, Queen, 141–142, 148
Emo, Angelo, 64
Endocrinology, 118
Engel, Friedrich, 182
Entertainment, 218–221
Epidemics, 98. *See also* Plagues
Epistolae Familiares (book), 187
"Eureka moments," 2, 10
Evolution, 3–8, 14, 121
Exploration, Age of, 56–63, 125–126
Explorations, reasons for, 1–19
Eyeglasses, 11, 153–154
Eyes/iris, 114

F

Facial paste, 147–148
Facini, Bernardo, 64
Factory line, 63–68
Faggin, Federico, 206
Fallopian tubes, 114
Fallopio, Gabrielle, 97, 114
Farina, Giovanni Maria, 147
Fedele, Casandra, 84
Fei, Alberto Toso, 132, 212, 223
Feminism, 84–85
Ferguson, Niall, 165
Festivals, 78–79, 159, 208, 210–214, 221–
  222. *See also* Carnivals
Fibonacci, 168
Fights, 218–219
Film festivals, 221–222
Finances, 163–182. *See also* Banking;
  Money
Fiolarius, Domenicus, 151
Fitness apps, 117
Floating battery, 64
Flooding, xv–xvi, 226, 229, 235–241
Florida, Richard, 15, 17
Food allergies, 121
Ford, Richard, 13
Fortune-telling, 215
Fossils, 5–6, 28–29
Fourth Book (book), 204
Fracastoro, Girolamo, 8, 99
Friedman, Thomas, 83
Fuentes, Agustín, 12
Fumigation, 103, 105

G

Galen, 112, 113
Galilei, Galileo, 9, 10, 65, 96, 97, 111,
  112, 115, 122
Galleons, 8. *See also* Ships
Gambling, 81, 210–217
Games, 211, 213–217
Gardano, Antonio, 189
Gastaldi, Giacomo, 55
Gazettes, 207–208